# Photochemistry of Organic Molecules and of Matrix-Isolated Reactive Intermediates: Themed Issue Honoring Professor Rui Fausto for His Contributions to the Field

# Photochemistry of Organic Molecules and of Matrix-Isolated Reactive Intermediates: Themed Issue Honoring Professor Rui Fausto for His Contributions to the Field

Guest Editors

**Gulce Ogruc Ildiz**
**Licinia L. G. Justino**

Basel • Beijing • Wuhan • Barcelona • Belgrade • Novi Sad • Cluj • Manchester

*Guest Editors*

Gulce Ogruc Ildiz
Department of Physics
Istanbul Kultur University
Istanbul
Turkey

Licinia L. G. Justino
Department of Chemistry
University of Coimbra
Coimbra
Portugal

*Editorial Office*
MDPI AG
Grosspeteranlage 5
4052 Basel, Switzerland

This is a reprint of the Special Issue, published open access by the journal *Photochem* (ISSN 2673-7256), freely accessible at: www.mdpi.com/journal/photochem/special_issues/Rui_Fausto.

For citation purposes, cite each article independently as indicated on the article page online and using the guide below:

Lastname, A.A.; Lastname, B.B. Article Title. *Journal Name* **Year**, *Volume Number*, Page Range.

ISBN 978-3-7258-2604-9 (Hbk)
ISBN 978-3-7258-2603-2 (PDF)
https://doi.org/10.3390/books978-3-7258-2603-2

© 2024 by the authors. Articles in this book are Open Access and distributed under the Creative Commons Attribution (CC BY) license. The book as a whole is distributed by MDPI under the terms and conditions of the Creative Commons Attribution-NonCommercial-NoDerivs (CC BY-NC-ND) license (https://creativecommons.org/licenses/by-nc-nd/4.0/).

# Contents

About the Editors . . . . . . . . . . . . . . . . . . . . . . . . . . . . . . . . . . . . . . . . . . . . . . . . . . . . . . . . . . . . . vii

Preface . . . . . . . . . . . . . . . . . . . . . . . . . . . . . . . . . . . . . . . . . . . . . . . . . . . . . . . . . . . . . . . . . . . . ix

**Alexandra McKinnon, Brendan Moore, Pavle Djuricanin and Takamasa Momose**
UV Photolysis Study of *Para*-Aminobenzoic Acid Using Parahydrogen Matrix Isolated Spectroscopy
Reprinted from: *Photochem* 2022, 2, 88–101, https://doi.org/10.3390/photochem2010008 . . . . . 1

**Barbara Golec, Aleksander Gorski and Jacek Waluk**
Phosphorescence and Photophysical Parameters of Porphycene in Cryogenic Matrices
Reprinted from: *Photochem* 2022, 2, 217–224, https://doi.org/10.3390/photochem2010016 . . . . 15

**Sándor Góbi, Mirjam Balbisi and György Tarczay**
Local and Remote Conformational Switching in 2-Fluoro-4-Hydroxy Benzoic Acid
Reprinted from: *Photochem* 2022, 2, 102–121, https://doi.org/10.3390/photochem2010009 . . . . 23

**Michelle T. Custodio Castro, Carlos O. Della Védova, Helge Willner and Rosana M. Romano**
Ar-Matrix Studies of the Photochemical Reaction between $CS_2$ and ClF: Prereactive Complexes and Bond Isomerism of the Photoproducts
Reprinted from: *Photochem* 2022, 2, 765–778, https://doi.org/10.3390/photochem2030049 . . . . 43

**Jussi Ahokas, Timur Nikitin, Justyna Krupa, Iwona Kosendiak, Rui Fausto, Maria Wierzejewska and Jan Lundell**
Conformational-Dependent Photodissociation of Glycolic Acid in an Argon Matrix
Reprinted from: *Photochem* 2023, 3, 197–208, https://doi.org/10.3390/photochem3020013 . . . . 57

**António Jorge Lopes Jesus, Cláudio M. Nunes and Igor Reva**
Conformational Structure, Infrared Spectra and Light-Induced Transformations of Thymol Isolated in Noble Gas Cryomatrices
Reprinted from: *Photochem* 2022, 2, 405–422, https://doi.org/10.3390/photochem2020028 . . . . 69

**Hoai Pham, Madelyn Hunsley, Chou-Hsun Yang, Haobin Wang and Scott M. Reed**
Demonstration of a Stereospecific Photochemical Meta Effect
Reprinted from: *Photochem* 2022, 2, 69–76, https://doi.org/10.3390/photochem2010006 . . . . . 87

**Licínia L. G. Justino, Sofia Braz and M. Luísa Ramos**
Spectroscopic and DFT Study of Alizarin Red S Complexes of Ga(III) in Semi-Aqueous Solution
Reprinted from: *Photochem* 2023, 3, 61–81, https://doi.org/10.3390/photochem3010005 . . . . . 95

**Sonia Ilieva, Meglena Kandinska, Aleksey Vasilev and Diana Cheshmedzhieva**
Theoretical Modeling of Absorption and Fluorescent Characteristics of Cyanine Dyes
Reprinted from: *Photochem* 2022, 2, 202–216, https://doi.org/10.3390/photochem2010015 . . . . 116

**Luca Evangelisti, Sonia Melandri, Fabrizia Negri, Marcello Coreno, Kevin C. Prince and Assimo Maris**
UPS, XPS, NEXAFS and Computational Investigation of Acrylamide Monomer
Reprinted from: *Photochem* 2022, 2, 463–478, https://doi.org/10.3390/photochem2030032 . . . . 131

**Monika Kopec, Beata Romanowska-Pietrasiak and Halina Abramczyk**
Decoding Breast Cancer Metabolism: Hunting BRCA Mutations by Raman Spectroscopy
Reprinted from: *Photochem* 2022, 2, 752–764, https://doi.org/10.3390/photochem2030048 . . . . 147

**Mahesh Subburu, Ramesh Gade, Prabhakar Chetti and Someshwar Pola**
Photooxidation of 2,2′-(Ethyne-1,2-diyl)dianilines: An Enhanced Photocatalytic Properties of New Salophen-Based Zn(II) Complexes
Reprinted from: *Photochem* **2022**, *2*, 358–375, https://doi.org/10.3390/photochem2020025 . . . . **160**

# About the Editors

**Gulce Ogruc Ildiz**

Gulce Ogruc Ildiz was born in Istanbul in 1977. After graduating from Kültür College, she received her undergraduate degree from Istanbul Technical University in Physics Engineering (2001) and her master's degree from Istanbul Technical University Nuclear Energy Engineering (2004) and Istanbul Kültür University Business Administration (MBA) (2003) departments. She received her doctorate degrees from Istanbul University Department of Atomic and Molecular Physics (2008) and Istanbul Kültür University Department of Civil Engineering Project Management Program (2009). She was appointed as an Assistant Professor (2010) at the Istanbul Kültür University Department of Physics and as a Professor in 2020. Between the years 2015 and 2022, she was the Executive Secretary of Istanbul Kültür University, Technology and Project Support Unit (TPDB); Scientific Research Projects (BAP) Coordinator; Faculty Administrative Board; and Faculty Board Member at IKU Faculty of Engineering, Faculty of Education, and Faculty of Arts and Sciences. She is the editor of several international journals in the field of spectroscopy. Gulce Ildiz's main scientific interests focus on the investigation of biologically relevant phenomena by Raman spectroscopy and chemometrics, matrix isolation infrared spectroscopy, and computational chemistry. She has been the Vice Rector of Istanbul Kültür University since August 2022.

**Licinia L. G. Justino**

Licínia L. G. Justino Simões was born in Coimbra, Portugal, in 1974. She obtained her undergraduate degree in Industrial Chemistry in 1996 from the Faculty of Sciences and Technology of the University of Coimbra and obtained her doctorate degree in Chemistry, specialization in Molecular Spectroscopy, in 2007, from the same university. She was awarded the Prof. Dr. António Jorge Andrade de Gouveia Prize by the Chemistry Department of the Faculty of Sciences and Technology of the University of Coimbra in 1996. Licínia Justino currently holds a researcher position at the University of Coimbra, working at the Computational Chemistry, Spectroscopy, and Thermodynamics Research Group. Her studies and interests are grounded in the search for next-generation efficient and sustainable materials based on inexpensive resources, anchored both on pioneering design principles and advanced spectroscopic and computational studies. Targeted applications aim to address societal challenges and include the development of sustainable materials for use in catalysis, photovoltaic systems, lighting, sensing, and other electronic and optoelectronic applications. Underlying these efforts, there is also the goal of contributing to a more fundamental understanding and prediction of structure-property-function relationships to contribute to a rational design of molecules tailored for specific purposes. She has co-authored several book chapters and published extensively, covering topics ranging from fundamental chemistry to materials science. She has also participated in numerous research projects, including several in collaboration with industry partners.

# Preface

This Special Issue of *Photochem* is dedicated to Prof. Rui Fausto in recognition of his influential contributions to the photochemistry of organic molecules and matrix-isolated reactive intermediates. This Special Issue's 12 articles span key topics in these areas, underscoring the impact of Prof. Fausto's work.

Rui Fausto was born in Coimbra, Portugal, graduated from the University of Coimbra in 1984, and obtained his Ph.D. in chemistry in 1988. After a postdoctoral tenure at the National Research Council of Canada, he returned to Coimbra, establishing a pioneering research program in molecular spectroscopy and computational chemistry. In 1994, he founded the Laboratory for Molecular Cryospectroscopy and Biospectroscopy (LMCB), now globally recognized for low-temperature spectroscopy and photochemistry. LMCB's research areas include matrix-isolation infrared spectroscopy and the study of chemical intermediates.

Rui Fausto is a Professor and Coordinator of the Computational Chemistry, Spectroscopy, and Thermodynamics Research Group at the University of Coimbra and holds the ERA-Chair Spectroscopy@IKU at Istanbul Kültür University (Türkiye). He is President of the Steering Committee of the European Congress on Molecular Spectroscopy (EUCMOS) and serves on several scientific editorial boards. Throughout his career, he has held leading roles within the University of Coimbra, including President of the Academic Council and Institute for Interdisciplinary Research and Vice President of the Faculty of Science and Technology's Scientific and Directive Boards. He has also served as President of the Physical-Chemistry Division of the Portuguese Chemical Society. Recognized for his scientific excellence, he received the RSC Journal Grant for International Authors in 2002 and the Portuguese Science Foundation's Excellence Prize in 2004 and 2005. With over 50 scientific books/book chapters and over 500 articles, Rui Fausto is a leading figure in his field.

Prof. Fausto is especially praised for pioneering vibrational excitation as a method to induce chemical reactivity in matrix-isolated organic molecules, which appears as a powerful and elegant strategy for the manipulation of chemical structures. His work on infrared and ultraviolet-induced photochemistry of reactive intermediates has opened new avenues for understanding chemical reactivity in organic compounds. Recently, his research on quantum mechanical tunneling strongly contributed to the renaissance of this field of research.

Beyond his scientific achievements, Rui Fausto is a dedicated mentor and advocate for music, painting, and human rights, exemplifying his broad intellectual and humanitarian outlook.

We extend our gratitude to the colleagues and reviewers who contributed to this Special Issue, honoring Prof. Fausto's legacy and advancing the scientific excellence showcased here.

Gulce Ogruc Ildiz and Licinia L. G. Justino
*Guest Editors*

Article

# UV Photolysis Study of *Para*-Aminobenzoic Acid Using Parahydrogen Matrix Isolated Spectroscopy

Alexandra McKinnon, Brendan Moore, Pavle Djuricanin and Takamasa Momose *

Department of Chemistry, The University of British Columbia, Vancouver, BC V6T 1Z1, Canada; alxmck@chem.ubc.ca (A.M.); bmoore@chem.ubc.ca (B.M.); pavledju@chem.ubc.ca (P.D.)
* Correspondence: momose@chem.ubc.ca; Tel.: +1-604-822-5265; Fax: +1-604-822-2847

**Abstract:** Many sunscreen chemical agents are designed to absorb UVB radiation (and in some cases UVA) to protect the skin from sunlight, but UV absorption is often accompanied by photodissociation of the chemical agent, which may reduce its UV absorption capacity. Therefore, it is important to understand the photochemical processes of sunscreen agents. In this study, the photolysis of *para*-aminobenzoic acid (PABA), one of the original sunscreen chemical agents, at three different UV ranges (UVA: 355 nm, UVB: >280 nm, and UVC: 266 nm and 213 nm) was investigated using parahydrogen ($pH_2$) matrix isolation Fourier-Transform Infrared (FTIR) Spectroscopy. PABA was found to be stable under UVA (355 nm) irradiation, while it dissociated into 4-aminylbenzoic acid (the PABA radical) through the loss of an amino hydrogen atom under UVB (>280 nm) and UVC (266 nm and 213 nm) irradiation. The radical production supports a proposed mechanism of carcinogenic PABA-thymine adduct formation. The infrared spectrum of the PABA radical was analyzed by referring to quantum chemical calculations, and two conformers were found in solid $pH_2$. The PABA radicals were stable in solid $pH_2$ for hours after irradiation. The *trans*-hydrocarboxyl (HOCO) radical was also observed as a minor secondary photoproduct of PABA following 213 nm irradiation. This work shows that $pH_2$ matrix isolation spectroscopy is effective for photochemical studies of sunscreen agents.

**Keywords:** PABA; PABA radical; photolysis; matrix isolation; parahydrogen

Citation: McKinnon, A.; Moore, B.; Djuricanin, P.; Momose, T. UV Photolysis Study of *Para*-Aminobenzoic Acid Using Parahydrogen Matrix Isolated Spectroscopy. *Photochem* **2022**, *2*, 88–101. https://doi.org/10.3390/photochem2010008

Academic Editor: Gulce Ogruc Ildiz and Licinia L.G. Justino

Received: 1 January 2022
Accepted: 21 January 2022
Published: 26 January 2022

**Publisher's Note:** MDPI stays neutral with regard to jurisdictional claims in published maps and institutional affiliations.

**Copyright:** © 2022 by the authors. Licensee MDPI, Basel, Switzerland. This article is an open access article distributed under the terms and conditions of the Creative Commons Attribution (CC BY) license (https://creativecommons.org/licenses/by/4.0/).

## 1. Introduction

The ozone layer in the stratosphere absorbs all UVC light and some UVB light, preventing most of this harmful radiation from reaching the surface of the earth. However, as the ozone hole grows, more UVB light reaches the surface of the earth, increasing the rate of skin cancer in humans [1]. Introduced in 1943, *para*-aminobenzoic acid (PABA) was one of the first active ingredients in sunscreen and subsequently was a component for both vitamin synthesis and for combating premature hair greying [2,3]. PABA's absorption capabilities of UV light make it a useful agent for preventing sun burns and cancerous skin tumors, which has been proven by experiments on mice [4,5]. However, electronic excitations of molecules via UV radiation often cause photodissociation of molecules, resulting in the degradation of UV absorption capabilities. Thus, understanding both the degradation process and the UV resistance of any sunscreen materials, such as PABA, are important. As PABA was once one of the most broadly used sunscreens, it is an ideal starting point for photolysis studies of sunscreen molecules [6,7]. Indeed, Chan et al. emphasized that PABA becomes reactive under UV irradiation. In the presence of thymine, a nucleic acid, photolyzed PABA (supposedly through the undetected intermediate radical) forms a carcinogenic adduct, breaking DNA strands [8,9].

In the gas phase, the $S_0 \leftarrow S_1$ transition of PABA occurs at 292.6 nm [10]. PABA's UV absorption has been measured in 10 different organic solvents, with maximum absorptions ranging from 280.9 nm in *n*-butyl acetate to 289.0 nm in methanol and ethanol [11]. A previous study reported that PABA's UV absorption in water has an absorption maximum of 260 nm and that there is a second absorption band at a wavelength less than 240 nm [12]. UV photolysis studies of PABA in solution show that PABA undergoes self-reactions following photolysis [13]. In anaerobic conditions, Shaw et al. reported two photoproducts following 254 nm and >290nm irradiation: 4-(4′-aminophenyl)aminobenzoic acid and 4-(2′-amino-5′-carboxyl)aminobenzoic acid [13]. Figure 1 shows the two photoproducts of PABA photolysis in solution, reporting that 4-aminylbenzoic acid (referred to as the PABA radical hereafter) was likely the predecessor of the two photoproducts, but due to rapid self-reactions, they were unable to isolate the radical in the solution [13]. Shaw et al. also reported that the PABA-thymine adduct was produced by photolyzing PABA in the presence of thymine, showing that the PABA radical was also the likely predecessor of this adduct, but were still unable to detect the PABA radical [9]. Other previous PABA photolysis studies in solution reported hydroxyl radical production, though PABA's elementary photolytic processes were unknown [14].

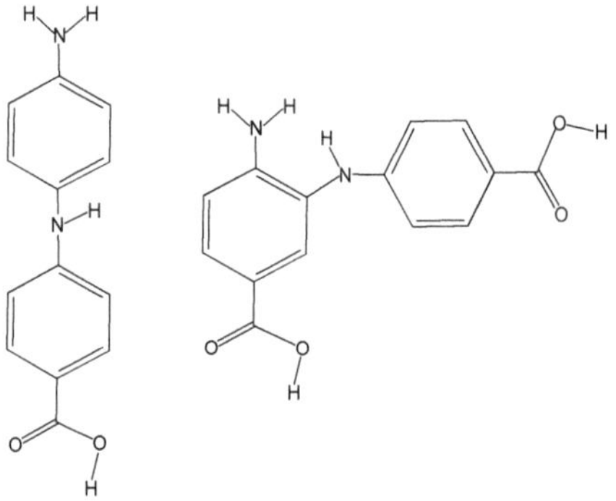

**Figure 1.** The two photoproducts that Shaw et al. observed following both 254 nm and >290 nm PABA photolysis in anaerobic solution [13]. (**Left**): 4-(4′-aminophenyl)aminobenzoic acid. (**Right**): 4-(2′-amino-5′-carboxyl)aminobenzoic acid.

Matrix isolation allows the observation of elementary pathways by stabilizing reactive species and preventing self-reactions. Parahydrogen ($pH_2$) was introduced as a matrix host at the end of the twentieth century [15]. A $pH_2$ matrix has properties that are advantageous over other matrix hosts. Due to its weak cage effects, molecules produced through in-situ photolysis in $pH_2$ can escape the lattice site and are isolated in a cryogenic environment, thus avoiding any further radical recombination reactions [16]. This reduces the number of possible photoproducts that can form following photolysis, simplifying photolysis mechanisms and improving accuracy in assigning photoproducts. Furthermore, a $pH_2$ matrix provides sharper, narrower spectral peaks, which can be studied in more detail by infrared spectroscopy [15–17]. In this study, the $pH_2$ matrix isolation spectroscopy was used to investigate the photodegradation of PABA. A $pH_2$ matrix sample containing PABA was irradiated with three different UV ranges (UVA: 355 nm, UVB: >280 nm, and UVC: 266 nm and 213 nm), and the resulting photoproducts were detected by Fourier-Transform Infrared (FTIR) Spectroscopy.

## 2. Materials and Methods

Normal hydrogen gas (99.99%, Praxair Canada Inc., Mississauga, Canada) was converted to >99.9% $pH_2$ with (FeOH)O as a magnetic converter at 14.1 K using a closed cycle He fridge [18]. The $pH_2$ was then deposited onto a BaF$_2$ window inside a cryostat at 3.8 K. PABA (>99%, Sigma–Aldrich, Oakville, Canada) was sublimed and deposited from inside a Knudsen cell heated to 110 °C at the same time as the $pH_2$ deposition for approximately 1 h, trapping PABA inside the $pH_2$ matrix [19]. Approximately 10–60 ppm PABA was deposited. The concentration of PABA and its photoproducts were estimated from the peak area, the calculated intensity of the peak, and the crystal thickness [20]. Table S1 in the ESI shows the peaks chosen for PABA and the photoproducts, as well as the modes' intensities. The thicknesses of the samples ranged from 1.2 mm to 1.8 mm, which were determined by previously developed protocols, which include the peak area of the 4495–4520 cm$^{-1}$ $pH_2$ peak [20]. After deposition, PABA was irradiated with 213, 266, or 355 nm UV pulses generated by a Nd:YAG laser (Litron, Nano-SG 150-10 10 Hz repetition rate, Rugby, England) or with broadband radiation at >280 nm using a 450 W xenon lamp (Ushio, UXL-451, Tokyo, Japan) equipped with a 290 nm long-pass filter (Toshiba UV 29, Tokyo, Japan). As the photon density employed in this work was different at different wavelengths, quantitative comparison of the photodissociation rates is not discussed in this work. The wavelengths were chosen based on the previous PABA UV absorption studies [10,12]. Furthermore, 355 nm is past the top absorption band, >280 nm is near the peak of the first absorption band, 266 nm is in between the two absorption bands, and 213 nm is near the second absorption band. Figure 2 depicts the experimental setup. In the cryostat, both the $pH_2$ gas line and the Knudsen cell tip point directly towards the BaF$_2$ window and both are equidistant from the window. The UV light travels at a 45° angle towards the window. The irradiation time of each wavelength varied, ranging from 30 to 60 min. Infrared spectra were collected by a Fourier Transform infrared spectrometer (Bruker, IFS 125 HR, Ettlingen, Germany) with 0.1 cm$^{-1}$ resolution averaged over 100 scans. Vibrational frequency calculations of PABA and the PABA radical were carried out using Gaussian 09 [21]. B3LYP/cc-pVDZ was used because of its reliability to calculate vibrational frequencies of small organic molecules, as well as its low computational cost.

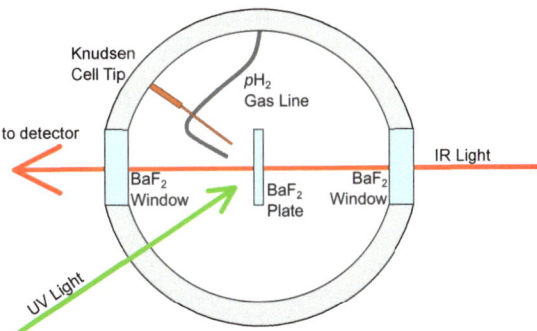

**Figure 2.** The experimental setup of the cryostat used for irradiation experiments.

## 3. Results and Discussion

### 3.1. Matrix-Isolation Infrared Spectroscopy of PABA

There are two PABA conformers, one of which is found to be substantially more stable. The geometries of the two PABA conformers are shown in Figure 3. Their optimized geometries obtained using B3LYP/cc-pVDZ are listed in Table S2 of the ESI. Conformer (I), which has a *cis* carboxyl group, was found to be more stable than Conformer (II), in which the carboxyl group is *trans*. In both conformers, the carboxyl groups reside in the plane of the phenyl ring. In contrast, the two amino hydrogen atoms are out of the plane of the phenyl ring with a H-N-C-C dihedral angle of 23.5° for both conformers. The zero-point energy corrected barrier height between the 23.5° and the −23.5° H-N-C-C dihedral angle is 7.51 kJ mol$^{-1}$ for Conformer (I). The less stable conformer of PABA (Conformer (II)) is 28.13 kJ mol$^{-1}$ higher in energy than the most stable conformer of PABA (Conformer (I)). The zero-point energy corrected barrier to conversion from Conformer (I) to Conformer (II) is 49.77 kJ mol$^{-1}$, and from Conformer (II) to Conformer (I) is 21.64 kJ mol$^{-1}$. Conformer (I) has a calculated conformation population distribution of 99.98% at the sublimation temperature.

**Figure 3.** The most stable conformer (Conformer (I)) and the less stable conformer (Conformer (II)) of PABA. Red: oxygen, blue: nitrogen, black: carbon, and white: hydrogen.

Figure 4 compares the $pH_2$-isolated IR spectrum of PABA and theoretically calculates vibrational transitions for both conformers of PABA. The entire PABA IR spectrum in the spectral range between 3675 cm$^{-1}$ and 675 cm$^{-1}$ is shown in Figure S1 in the ESI. The observed peaks shown in Figure 4 match well with those of Conformer (I). Table 1 displays the theoretical vibrational frequencies of Conformer (I) and observed frequencies in $pH_2$, as well as those in the gas and solid states previously reported [22,23]. The 30 vibrational modes of PABA can be assigned in this spectral region. According to the calculations, most of the remaining modes reside at frequencies lower than 650 cm$^{-1}$, as listed in Table S3 in the ESI. The $NH_2$ asymmetric stretching mode at 3437 cm$^{-1}$ shows a doublet with peaks separated by 1.1 cm$^{-1}$, and the $NH_2$ symmetric stretching mode at 3532 cm$^{-1}$ shows a doublet with peaks separated by 1.3 cm$^{-1}$. These doublets can be attributed to the $NH_2$ inversion tunneling splitting previously discussed for aniline [24]. The vibrational modes of the HOCO unit also show doublet peaks with slightly larger separations. The H-C-O bending at 1364–1368 cm$^{-1}$ has a separation of 3.6 cm$^{-1}$, the C=O stretching at 1744–1746 cm$^{-1}$ has a separation of 2.4 cm$^{-1}$, and the O-H stretching at 3580–3583 cm$^{-1}$, has a separation of 2.6 cm$^{-1}$. These splittings may also be due to tunneling motion. Although the HOCO unit is in the plane of the phenyl ring, the phenyl ring itself is slightly out of plane due to the twisted $NH_2$, which could induce tunneling splitting in the vibrational modes of the HOCO unit.

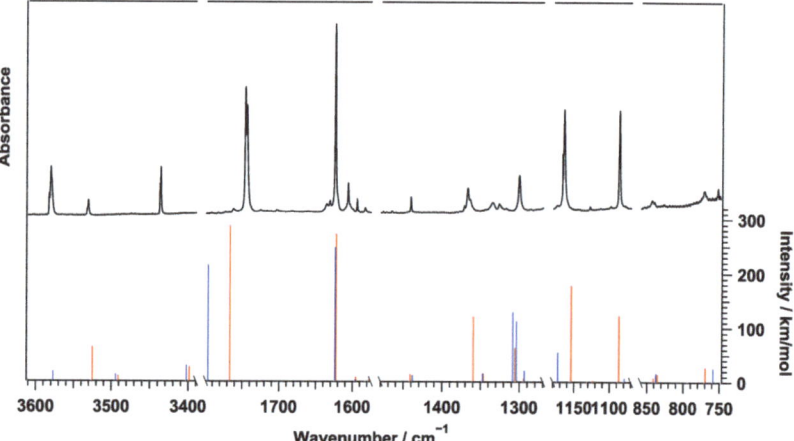

**Figure 4.** The IR Spectrum of PABA in $pH_2$ compared to the anharmonic B3LYP/cc-pVDZ intensities of PABA Conformer (I) (red) and Conformer (II) (blue).

Table 1. Observed and calculated wavenumbers (cm$^{-1}$) of PABA and their intensities (km mol$^{-1}$). [#]

| Vibrational Mode | Theoretical [†] Conformer (I) | Solid State [22] | Gas Phase [23] | pH$_2$ Matrix [‡] |
|---|---|---|---|---|
| C=C-C Wag | 697.92 (23.09) | | 694 | 698.0 (20.62) |
| O-C-O Scissor | 725.94 (7.59) | | 734 | 732.3 (4.13) |
| C-C-C Bend [§] | 770.40 (27.35) | 773.43 | 770 | 771.6 (10.92) |
| C-C-C Bend [§] | 836.13 (14.97) | 832.14 | 840 | 840 * |
| C=C-C Scissor | 841.92 (8.41) | | | 843 * |
| C-H Bend | 947.20 (0.74) | | | 954.6 (0.18) |
| H-C-C Scissor | 964.63 (0.21) | | | 967.4 (0.12) |
| H-C-C Rock | 1006.49 (26.28) | | | 1012.6 (0.68) |
| NH$_2$ Rock | 1046.07 (17.85) | | | 1048.5 (6.52) |
| NH$_2$ Rock [§] | 1087.64 (123.18) | 1065.26 | 1080 | 1086.6 (54.61) |
| C-O Stretch | 1124.00 (3.82) | | | 1127.9 (1.74) |
| H-C-C Bend [§] | 1154.98 (178.25) | 1169.17 | 1166 | 1164.8 (81.01) |
| C-O-H Scissor | 1187.34 (38.55) | | 1198 | 1194.7 (19.26) |
| C-H Bend | 1295.27 (7.09) | | 1260 | 1274.6 (1.23) |
| C-O Stretch | 1305.85 (63.39) | 1288.92 | 1286 | 1300.6 (32.48) |
| C-O Stretch | 1347.04 (16.10) | | 1325 | 1335 * |
| H-C-O Bend | 1360.55 (121.57) | 1367.66 | 1360 | 1364.2, 1367.8 (41.33) |
| C=C Stretch | 1441.77 (14.14) | 1421.41 | 1434 | 1440.9 (6.18) |
| C=C Asym. Str. | 1513.15 (17.55) | | 1518 | 1523.8 (13.82) |
| C=C Asym. Str. | 1578.09 (5.01) | | | 1583.3 (2.96) |
| NH$_2$ Scissor | 1596.06 (8.61) | | | 1606.7 (12.26) |
| NH$_2$ Scissor [§] | 1623.35 (271.97) | 1620.18 | 1622 | 1624.2 (100.00) |
| C=O Stretch [§] | 1767.10 (287.04) | 1690.90 | 1754 | 1743.9, 1746.3 (109.75) |
| C-H Asym. Str. | 3010.33 (17.31) | | 3010 | 3016.2 (0.70) |
| C-H Asym. Str. | 3016.28 (14.37) | | | 3028.9 (2.57) |
| C-H Sym. Str. | 3061.17 (1.94) | | 3044 | 3052.3 (3.25) |
| C-H Sym. Str. | 3070.83 (11.15) | | 3074 | 3079.3 (3.01) |
| NH$_2$ Asym. Stetch [§] | 3399.32 (26.90) | 3460.70 | 3422 | 3436.2, 3437.3 (19.59) |
| NH$_2$ Sym. Stretch [§] | 3491.38 (10.41) | | 3510 | 3531.0, 3532.3 (8.59) |
| O-H Stretch [§] | 3525.95 (64.46) | 3607.25 | 3586 | 3580.4, 3583.0 (19.32) |

[#] Intensities are written in parentheses. [†] Calculated with B3LYP/cc-pVDZ with anharmonic approximation [21].
[‡] This work. Intensities in the parenthesis are normalised to the 1624.2 cm$^{-1}$ transition. Doublets have both peaks listed and their combined intensity in parentheses. * Two or more PABA peaks overlap. [§] Reference [22].

As shown in Figure 4, most of the observed peaks are assigned to Conformer (I). According to the calculation (Table S3 in the ESI), Conformer (II) has a strong peak at 1624.8 cm$^{-1}$ for the NH$_2$ scissoring mode and at 1796.5 cm$^{-1}$ for the C=O stretching mode. These peaks are predicted to be shifted to slightly higher frequencies than the corresponding Conformer (I) peaks. As shown in Figure S1 of the ESI, weak peaks are observed at 1631 cm$^{-1}$ and 1762 cm$^{-1}$, both of which are slightly higher frequencies than the corresponding Conformer (I) peaks. These weak peaks may be assigned to Conformer (II), whose predicted wavenumbers are 1624.8 cm$^{-1}$ and 1796.5 cm$^{-1}$, respectively. Unassigned peaks at 1317 cm$^{-1}$ and 1326 cm$^{-1}$ may also be attributed to the C-N and C-O stretching modes of Conformer (II), whose predicted wavenumbers are 1304.5 cm$^{-1}$ and 1309.1 cm$^{-1}$, respectively. Although these assignments are still tentative, the intensity of Conformer (II) is significantly weaker than that of Conformer (I), which indicates that the majority of PABA isolated in pH$_2$ is of Conformer (I) geometry. Previous gas phase studies also concluded that PABA has only one stable rotamer [11,25,26].

### 3.2. UVA: 355 nm Irradiation

Following 10 min of 20 mJ of 355 nm irradiation, negligible photolysis of PABA was observed. The concentration of PABA decreased by 0.3% and no photoproducts were observed, thus confirming that there is no photodissociation of PABA at 355 nm.

*3.3. UVB : >280 nm Irradiation*

Irradiation of PABA using the xenon lamp with a 290 nm long-pass filter reduced the PABA peaks and generated new peaks. The top trace in Figure 5 shows a difference spectrum after irradiating PABA with >280 nm radiation minus the spectrum before irradiation. Several new peaks were clearly seen after the >280 nm irradiation. The entire difference IR spectrum is shown in Figure S2 in the ESI. Figure 6 shows the temporal behaviour of these newly formed peaks at 774 cm$^{-1}$ and 1078 cm$^{-1}$ as a function of the >280 nm irradiation time. Both peaks showed a doublet structure, and the temporal behaviour of each component of the doublets are shown. All peaks showed the same irradiation behaviour with a single exponential increase. The fitted rate constants agree within 20% between the four peaks: $0.026 \pm 0.001$ min$^{-1}$ and $0.023 \pm 0.001$ min$^{-1}$ for the peaks at 1079.5 cm$^{-1}$ and 1077.2 cm$^{-1}$, respectively, and $0.021 \pm 0.003$ min$^{-1}$ and $0.023 \pm 0.003$ min$^{-1}$ for the peaks at 774.3 cm$^{-1}$ and 774.6 cm$^{-1}$, respectively. The same irradiation behaviour indicates that these peaks are due to the same photoproduct or species produced from the same photoprocess (see Section 3.6).

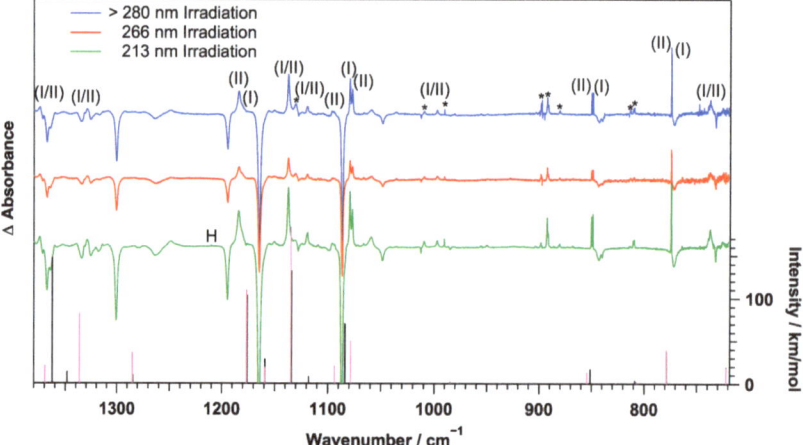

**Figure 5.** Difference spectra of PABA irradiated by 60 min >280 nm (blue), 30 min 266 nm (red), and 40 min 213 nm (green) UV radiation. Sticks are the calculated vibrational peaks of PABA radical Conformer (I) (black) and Conformer (II) (pink) using anharmonic B3LYP/cc-pVDZ. (I) = Conformer (I), (II) = Conformer (II), H = HOCO Radical, and * = Unassigned Peak.

The single exponential behaviour also indicates that these are primary photoproducts. Indeed, the decay of the PABA concentration roughly matches to the increase of the concentration of these photoproducts. Figure 7a compares the change in the concentration of PABA and that of the photoproduct under >280 nm photolysis. It can be seen that there is a close anti-correlation between the increase in photoproduct transitions and the decrease in PABA. PABA's decay rate constant, as shown in Figure 7a, is $0.022 \pm 0.002$ min$^{-1}$, which agrees with the photoproduct's production rate constant.

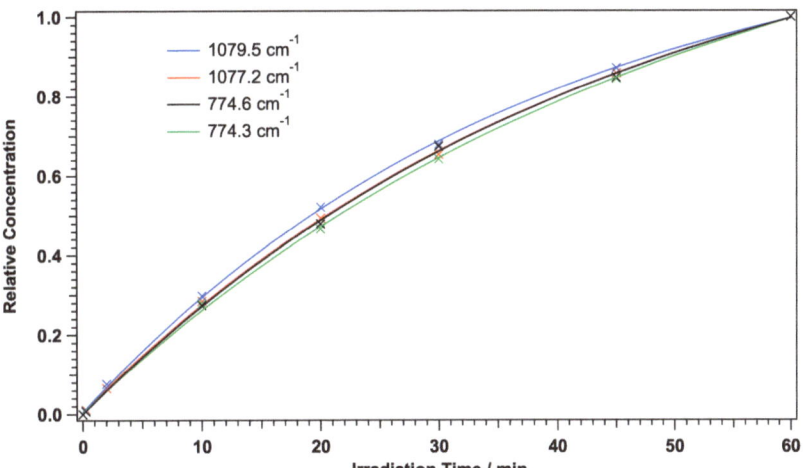

**Figure 6.** Temporal behaviour of four different photoproduct peaks following >280 nm irradiation. The y-axis is scaled to such that the value at 60 min becomes 1 for each peak. The scale factors are 8.74, 7.03, 18.79, and 18.47 for the peaks at 1079.5 cm$^{-1}$, 1077.2 cm$^{-1}$, 774.6 cm$^{-1}$, and 774.3 cm$^{-1}$, respectively.

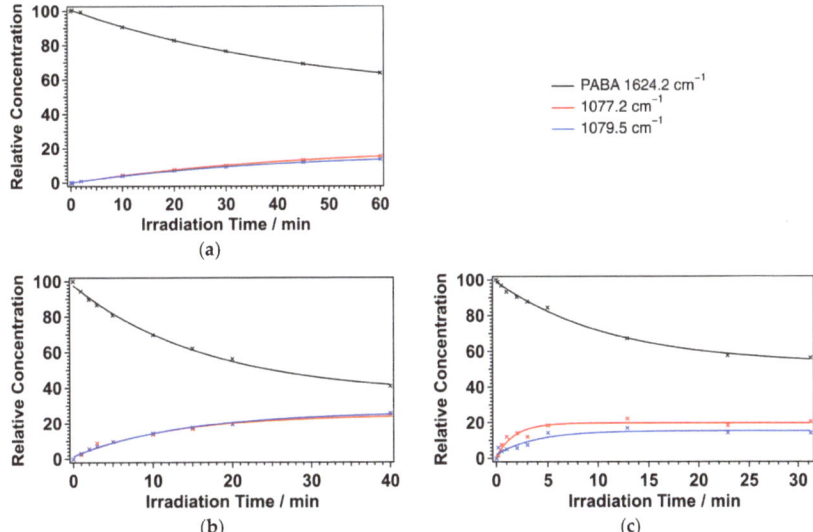

**Figure 7.** Temporal behaviour of the photoproduct peaks at 1077.2 cm$^{-1}$ (red) and 1079.5 cm$^{-1}$ (blue): (**a**) >280 nm; (**b**) 266 nm; (**c**) 213 nm. All data are fitted with a single exponential function. The duration of irradiation times at different wavelengths were different, since the photon density at different wavelengths was different.

Most of the observed photoproduct transition wavenumbers were similar to those of the PABA peaks. Therefore, it can be assumed that the photoproduct is similar in structure to PABA and that the photoproduct likely did not result from cleavage of the aromatic ring. The three most likely photoproducts are aniline (or the anilino radical) through the dissociation of the carboxyl group, benzoic acid (or the benzoic acid radical) through the dissociation of the amino group, and 4-aminylbenzoic acid (the PABA radical) through the loss of one amino hydrogen atom. Table S4 in the ESI shows the calculated vibrational transitions of the aromatic-centered anilino radical, and Table S5 in the ESI shows the calculated aromatic-centered benzoic acid radical transitions. Neither of these peaks were consistent with the photoproduct peaks observed here. In addition, neither the HOCO radical at 1845.1 cm$^{-1}$ in $pH_2$ [27] nor $NH_2$ at 1527.85 cm$^{-1}$ or $NH_3$ at 968.1 cm$^{-1}$ [21,28,29] was observed via the >280 nm irradiation. It is also noted that protonated aniline [30,31] and benzoic acid [32], which could be by-products of the formation of the anilino radical or the benzoic acid radical, were also not detected.

Therefore, the remaining assignment possibility is 4-aminylbenzoic acid (the nitrogen-centered PABA radical). Table 2 compares the experimental frequencies and intensities of the theoretical calculations of the PABA radical of the two possible conformers (see Section 3.6), and Table S6 in the ESI lists all of the PABA radical fundamental modes of these two possible conformers. The observed frequencies and their intensities agree well with the calculated frequencies, indicating that the primary photoproduct of PABA is the PABA radical.

**Table 2.** The PABA radical vibrational mode assignments and their wavenumbers (cm$^{-1}$). Intensities are written in parentheses.

| Mode | Sym | Assignment | Theoretical [†] Conformer (I) | Theoretical [†] Conformer (II) | $pH_2$ Matrix [‡] | Conformer Assignment |
|---|---|---|---|---|---|---|
| $\nu_9$ | A' | C-C-C Bend | 778.48 (38.0) | 778.74 (39.9) | 774.3 (16.5) | (I) |
|  |  |  |  |  | 774.6 (19.0) | (II) |
| $\nu_{11}$ |  | C-C-C Bend | 851.22 (17.0) | 854.31 (14.0) | 849.1 (13.0) | (I) |
|  |  |  |  |  | 850.5 (11.7) | (II) |
| $2\nu_6$ |  | O-H Bend | 1117.83 (8.3) | 1124.85 (6.9) | 1119.5 [§] (22.5) | (I)/(II) |
| $\nu_{20}$ | A'' | O-C-O Scissor | 721.73 (19.7) | 721.77 (38.0) | 737.6 (36.4) | (I)/(II) |
| $\nu_{22}$ |  | H-C-C Wag | 984.84 (2.0) | 984.84 (1.4) | 997.3 (8.3) | (I)/(II) |
| $\nu_{23}$ |  | H-C-C Rock | 1083.84 (70.4) | 1078.48 (49.8) | 1077.2 (35.3) | (II) |
|  |  |  |  |  | 1079.5 (49.0) | (I) |
| $\nu_{24}$ |  | O-C-C Stretch | 1091.25 (0.8) | 1093.57 (21.0) | 1095.0 (25.6) | (II) |
| $\nu_{25}$ |  | H-C-C Bend | 1134.02 (132.2) | 1134.69 (183.4) | 1137.4 (92.3) | (I)/(II) |
| $\nu_{27}$ |  | H-O-C Scissor | 1175.83 (103.5) | 1176.46 (109.9) | 1180.8 (24.0) | (I) |
|  |  |  |  |  | 1184.2 (100.0) | (II) |
| $\nu_{30}$ |  | H-N-C Stretch | 1348.41 (13.9) | 1336.48 (82.3) | 1329.3 * | (I)/(II) |
|  |  |  |  |  | 1332.0 * | (I)/(II) |
| $\nu_{31}$ |  | C-C-O Stretch | 1362.56 (147.4) | 1369.37 (20.8) | 1374.6 (13.5) | (I)/(II) |
| $\nu_{32}$ |  | C=C Asym. Str. | 1420.91 (15.3) | 1420.76 (14.7) | 1423.3 (5.8) | (I)/(II) |
|  |  |  |  |  | 1429.5 (5.9) | (I)/(II) |
| $\nu_{33}$ |  | C=C Sym. Str. | 1455.42 (19.6) | 1450.59 (6.3) | 1460.0 (1.1) | (II) |
|  |  |  |  |  | 1461.1 (3.3) | (I) |
| $\nu_{34}$ |  | C=C-C Asym. Str. | 1514.64 (0.7) | 1515.47 (20.5) | 1518.5 (4.5) | (II) |
| $\nu_{36}$ |  | C=O Stretch | 1764.16 (197.0) | 1765.90 (93.0) | 1744–1746 * | (I)/(II) |
| $\nu_{42}$ |  | O-H Stretch | 3531.75 (81.5) | 3535.61 (81.7) | 3570.9 (2.3) | (I)/(II) |
|  |  |  |  |  | 3572.7 (15.5) | (I)/(II) |
|  |  |  |  |  | 3574.3 (29.0) | (I)/(II) |
|  |  |  |  |  | 3575.5 (41.3) | (I)/(II) |

[†] Calculated with B3LYP/cc-pVDZ Anharmonic [21]. [§] Overtone peak. [‡] This work. Observed intensities are normalized to the 1184.2 cm$^{-1}$ transition. * Intensities not listed when PABA radical peaks overlap with PABA peaks. Modes with two peaks listed are doublets.

### 3.4. UVC: 266 nm Irradiation

Following 1.0 mJ of 266 nm irradiation with the Nd:YAG laser, the same photoproduct peak as the >280 nm irradiation was observed, as shown in the middle trace of Figure 5. Figure 7b shows the temporal behaviour of PABA and the PABA radical following 266 nm photolysis. The PABA decay rate constant matches the PABA radical's production rate constant, following 266 nm photolysis. The PABA radical is also the main primary photoproduct of 266 nm irradiation. This supports Shaw et al.'s discussion on the photolysis of PABA in solution, which determined that identical photoproducts were produced from both 254 nm and >290 nm irradiation [13].

During 266 nm irradiation, a small yet broad peak was noticed at 2124.1 cm$^{-1}$. Based on its position, this peak could be assigned to a ketene. The temporal behaviour of the peak at 2124.1 cm$^{-1}$ following the 266 nm photolysis is shown in Figure S3 of the ESI. There is not enough information to confidently determine which ketene it is, but it is confirmed that it is not methylketene[33]. From known intensities of other ketene-containing molecules, approximately <0.5 ppm of this ketene was isolated, and therefore, it is a minor primary photoproduct following the 266 nm photolysis [23].

*3.5. UVC: 213 nm Irradiation*

After 1.3 mJ of 213 nm irradiation with the Nd:YAG laser, the same photoproduct peaks to the 266 nm irradiation were observed, as shown in the bottom trace of Figure 5. Figure 7c shows the temporal behaviour of PABA and the PABA radical following 213 nm photolysis. The rate constant of PABA decay is almost the same as that of PABA radical production following 213 nm photolysis. The PABA radical is still the main photoproduct of 213 nm photolysis. Similar photolytic behaviour between the >280 nm, 266 nm, and 213 nm experiments suggests that the first and second excited state dissociations are the same.

In addition to the peaks of the PABA radical, a few weak peaks due to other photoproducts were detected after 213 nm photolysis. Table S7 in the ESI lists the peaks of other photoproducts. The unknown ketene peak at 2124.6 cm$^{-1}$ was also observed following 213 nm photolysis. The temporal behaviour of the peak at 2124.1 cm$^{-1}$ following the 213 nm photolysis is also shown in Figure S3 of the ESI. Unique to 213 nm photolysis, the *trans*-HOCO radical [27] was observed as a minor photoproduct. Figure 8a shows the most intense *trans*-HOCO radical peak and the temporal behaviour of the *trans*-HOCO radical following 213 nm photolysis. The *cis*-HOCO radical was not observed [34]. The HOCO unit dissociation may result in the formation of the *cis*-HOCO radical, but previous work noted that the *cis*-HOCO radical immediately undergoes a conformational change from the *cis* conformer to the much more stable *trans* conformer when isolated [27].

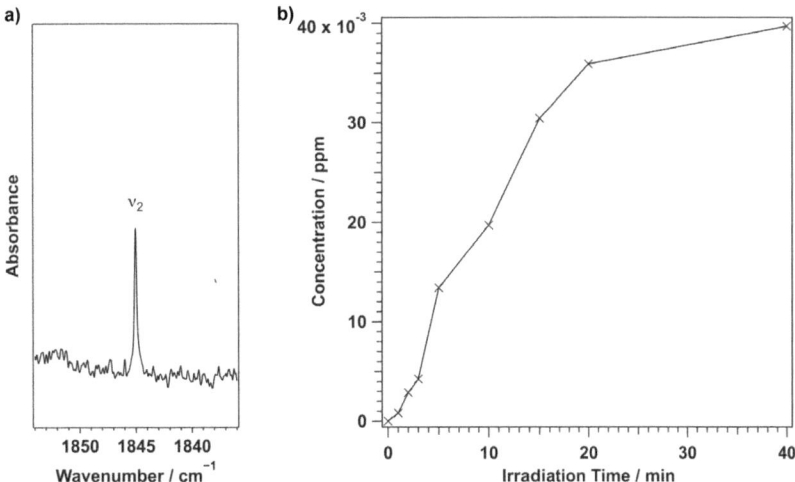

**Figure 8.** The most intense *trans*-HOCO radical peak (**a**) and its temporal behaviour (**b**) following 213 nm photolysis.

No clear co-product of the HOCO radical was observed. As shown in Figure 8b, the HOCO radical's production rate increased at early irradiation times. The curvature of the increase in the HOCO radical concentration in the early stages of the irradiation (<10 min) may be similar to the typical kinetic trend of secondary photoproduct, though it cannot be clearly concluded. The HOCO radical concentration plateaued around 20 minutes, and the CO and $CO_2$ concentrations increased following 213 nm photolysis [35,36]. This behaviour is the same as alanine photolysis [27]. Further work is needed to understand how HOCO radicals are produced by the photodissociation of PABA.

*3.6. The PABA Radical Conformational Analysis*

The PABA radical results from the loss of one amino hydrogen from PABA [13]. There are four possible PABA radical conformers. Two conformers have a *trans* carboxyl group and two have a *cis* carboxyl group. Similar to PABA, the two conformers with a *cis* carboxyl group are more stable than the *trans* carboxyl group conformers. The two conformers with a *trans* carboxyl group are both approximately 30 kJ mol$^{-1}$ higher in energy than the two conformers with a *cis* carboxyl group. Table S8 in the ESI shows the vibrational frequencies and intensities of Conformers (III) and (IV). The most and second most stable PABA radical conformers (the two conformers with the *cis* carboxyl group) are differentiated due to amino hydrogen lost from PABA and the position of the remaining amino hydrogen. Calculations determining the optimized geometry of the two most stable conformers of the PABA radical were carried out in Gaussian 09 using anharmonic B3LYP/cc-pVDZ. One conformer had the amino hydrogen *cis*, with respect to the PABA radical's carbonyl (Conformer (I)), and the other conformer had the amino hydrogen *trans*, with respect to the carbonyl (Conformer (II)). Figure 9 shows the structures of the four possible PABA radical conformers, two of which were observed in the $pH_2$ matrix. All of the atoms of the PABA radical were in the plane of the phenyl ring, including the remaining hydrogen atom on the amino group. Conformer (II) was 0.30 kJ mol$^{-1}$ less stable than Conformer (I). The zero-point energy corrected barrier to conversion for the H-N-C-C dihedral angle rotation from Conformer (I) to Conformer (II) was 54.89 kJ mol$^{-1}$, and Conformer (II) to Conformer (I) was 54.59 kJ mol$^{-1}$. Because the PABA N-H bond lengths shown in Table S1 are identical, and the conformer energies are nearly identical, it is likely that PABA does not exclusively dissociate to one radical conformer.

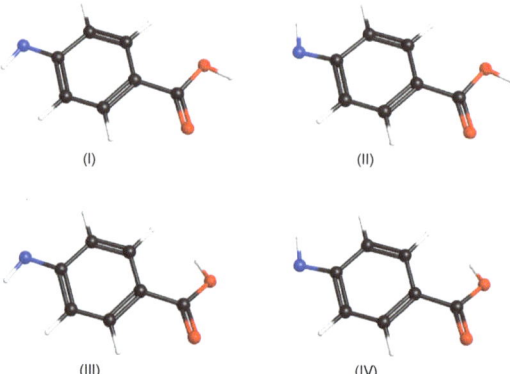

**Figure 9.** The four possible conformers of the PABA radical. Red: oxygen, blue: nitrogen, black: carbon, and white: hydrogen.

The experimental transitions assigned to the PABA radical produced irradiation of PABA at all three wavelengths, as shown in Table 2. Many PABA radical transitions observed in parahydrogen are doublets, while the corresponding PABA transitions were singlets. The doublets observed at $v_9$, $v_{11}$, and $v_{23}$ observed intensities matching the calculated relative intensities for Conformers (I) and (II). Figure 10 shows the $v_{34}$ mode, $v_{23}$ mode, $v_{11}$ mode, and $v_9$ mode of the PABA radical conformers. The separation of the doublets are 0.3 cm$^{-1}$, 1.4 cm$^{-1}$, and 2.3 cm$^{-1}$ for $v_9$, $v_{11}$, and $v_{23}$, respectively. The calculated frequency differences of 0.26 cm$^{-1}$, 3.09 cm$^{-1}$, and 5.36 cm$^{-1}$ for $v_9$, $v_{11}$, and $v_{23}$ are in agreement with the experimental frequency differences. Because the observed intensities and frequency differences of the experimental doublets closely match the expected differences between Conformers (I) and (II), these doublets likely arise from the production of both (I) and (II) of the PABA radical from photolysis of PABA. The assignment of Conformer (II) can be confirmed from $\nu_{34}$, where the calculated intensity is 1 km mol$^{-1}$ for Conformer (I) and 20 km mol$^{-1}$ for Conformer (II). This vibrational mode should only be observed for Conformer (II), and a singlet is indeed observed at 1518.5 cm$^{-1}$ for $\nu_{34}$. This, along with the observed doublets for $\nu_9$, $\nu_{11}$, and $\nu_{23}$ indicate that Conformers (I) and (II) are produced from photolysis of PABA. The doublet relative intensities are consistent with an approximately equal distribution between the two most stable conformers. The photodissociation of PABA yields a mixture of the most stable PABA radical conformers, as expected from the ground state geometry of PABA.

**Figure 10.** The $\nu_{34}$ mode, $\nu_{23}$ mode, $\nu_{11}$ mode, and $\nu_9$ mode of the PABA radical conformers.

*3.7. Overall Reaction*

From the observed photoproduct and its temporal behaviour, it is concluded that the following reaction scheme shown in Figure 11 is the only photodissociation process of PABA under UVB and UVC irradiation. PABA loses one of its amino hydrogen atoms, forming the PABA radical.

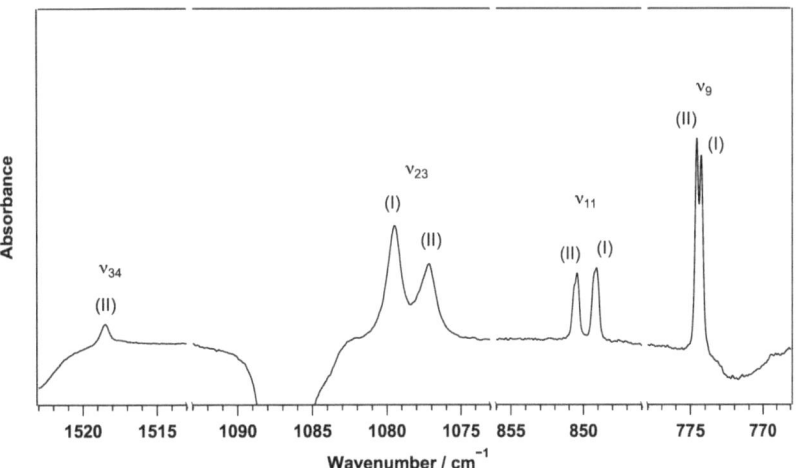

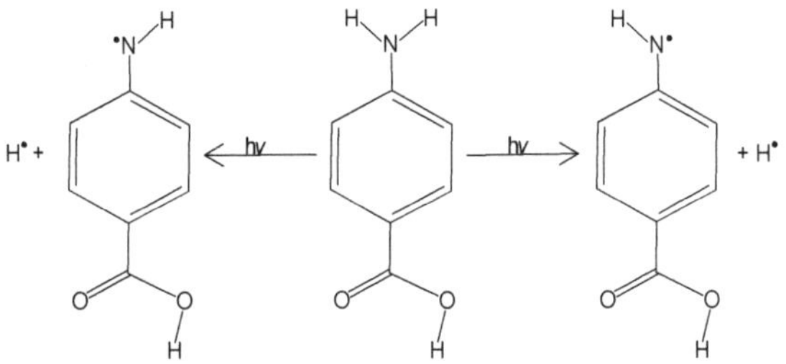

**Figure 11.** The UVB and UVC photodissociation process of PABA.

The production of the PABA radical following UV photolysis of PABA is consistent with the previous study by Mitchell et al. [25], who reported that PABA's first electronically excited state is at approximately 292 nm in the gas phase, where the electron density on the amino group decreases and the electron density on the carboxyl group increases [25].

They refer to PABA as a "push-pull" molecule, since it has an electron donating group (an amino group) on one end of the phenyl ring and an electron withdrawing group (a carboxyl group) on the other end at the *para* position of the phenyl ring. Due to the decrease of the electron density on the amino group, it is expected that the dissociation occurs on the amino group side. PABA, like other molecules of a similar structure, including 4-methyl-benzenamine, 2,6-dimethyl-benzenamine, 2,5-di-*tert*-butyl-benzenamine, and 2,4,6-tri-*tert*-butyl-benzenamine, forms a radical species resulting from the loss of an amino hydrogen upon irradiation [37].

Following photolysis, the system was left exposed to IR for 18 h and no spectral changes were detected. The PABA radical was very stable in solid $pH_2$. The reverse reaction of the PABA radical gaining a hydrogen atom via a tunneling reaction with a $H_2$ molecule or a H atom to reform PABA seems to be a very slow process in solid $pH_2$ at 3.8 K.

## 4. Conclusions

It was confirmed that PABA is resistant to UVA radiation, while it degrades upon UVB and UVC irradiation. The major primary photoproduct of PABA photolysis is the PABA radical whose infrared spectrum was observed for the first time. The confirmed presence of this radical reinforces Shaw et al.'s mechanism forming the PABA-thymine adduct, further confirming PABA's carcinogenic properties and DNA destroying capabilities [9]. Although, at present, ZnO and $TiO_2$ are most commonly used as sunscreen rather than PABA, humans still use PABA-containing products for other purposes [2,3]. Therefore, it is important to further investigate the chemical properties of the PABA radical.

In this work, we demonstrate that $pH_2$ matrix isolation spectroscopy is a powerful technique for the study of UV photochemical processes of sunscreen molecules. UV photolysis of other sunscreen agents, such as 2-ethylhexyl 4-(dimethylamino)benzoate (Padimate O) [38–40], would also be worth investigating using $pH_2$ matrix isolation spectroscopy.

**Supplementary Materials:** The following supporting information can be downloaded at: https://www.mdpi.com/article/10.3390/photochem2010008/s1, Electronic Supplementary Information (ESI) available: The observed wavenumbers of peaks used for integration (Table S1), the geometry of the two PABA Conformers (I) and (II) (Table S2), the entire spectrum of PABA in a solid $pH_2$ matrix (Figure S1), the frequencies and intensities of PABA Conformers (I) and (II) (Table S3), the aromatic-centred anilino radical (Table S4), the aromatic-centred benzoic acid radical (Table S5), the entire difference spectrum following >280 nm, 266 nm, and 213 nm irradiation (Figure S2), the frequencies and intensities of the PABA radical Conformer (I) and (II) (Table S6), other photoproducts' wavenumbers (Table S7), the temporal behaviour of the unknown ketene following 266 nm and 213 nm photolysis (Figure S3), and the vibrational frequencies and intensities of the PABA radical Conformers (III) and (IV) (Table S8).

**Author Contributions:** Methodology, T.M. and P.D.; investigation, A.M. and B.M.; formal analysis, A.M.; writing—original draft preparation, A.M.; writing—review and editing, A.M.; B.M., T.M. and P.D.; supervision, T.M.; funding acquisition, T.M. All authors have read and agreed to the published version of the manuscript.

**Funding:** This research was supported by a Natural Sciences and Engineering Research Council (NSERC) Discovery Grant in Canada and funds from the Canada Foundation for Innovation for the Centre for Research on Ultracold Systems (CRUCS).

**Institutional Review Board Statement:** Not Applicable.

**Informed Consent Statement:** Not Applicable.

**Acknowledgments:** A. M. thanks E. R. Miller for his technical assistance with the xenon lamp and E. R. Grant (UBC) for his thoughtful discussion on the thermodynamic stability of the conformers of PABA.

**Conflicts of Interest:** The authors declare no conflict of interest.

## References

1. Abarca, J.F.; Casiccia, C.C. Skin Cancer and Ultraviolet-B Radiation Under the Antartic Ozone Hole: Southern Chile, 1987–2000. *J. Am. Chem. Soc.* **2002**, *18*, 294–302.
2. Scott, H.W.; Dehority, B.A. Vitamin Requirements of Several Cellulolytic Rumen Bacteria. *J. Bacteriol.* **1964**, *89*, 1169–1175. [CrossRef] [PubMed]
3. Mahendiratta, S.; Sarma, P.; Kaur, H.; Kaur, S.; Kaur H.; Bansal, S.; Prasad, D.; Prajapat, M.; Upadhay, S.; Kumar, S.; et al. Premature Graying of Hair: Risk Factors, Co-Morbid Conditions, Pharmacotherapy and Reversal—A Systematic Review and Meta-Analysis. *Dermatol. Ther.* **2020**, *33*, 1–14. [CrossRef] [PubMed]
4. Snyder, D.S.; May, M. Ability of PABA to Protect Mammalian Skin from Ultraviolet Light-Induced Skin Tumors and Actinic Damage. *J. Investig. Dermatol.* **1975**, *65*, 543–546. [CrossRef]
5. Flindt-Hansen, H.; Thune, P.; Larsen, T.E. The Inhibiting Effect of PABA on Photocarcinogenesis. *Arch. Dermatol. Res.* **1990**, *282*, 38–41. [CrossRef]
6. Zhou, L.; Ji, Y.; Zeng, C.; Zhang, Y.; Wang, Z.; Yang, X. Aquatic Photodegradation of Sunscreen Agent p-Aminobenzoic Acid in the Presence of Dissolved Organic Matter. *Water Res.* **2013**, *47*, 153–162. [CrossRef]
7. Wong, T. Sunscreen Allergy and its Investigation. *Clin. Dermatol.* **2011**, *29*, 306–310. [CrossRef]
8. Chan, C.T.L.; Ma, C.; Chan, R.C.T.; Ou, H.M.; Xie, H.X.; Wong, A.K.W.; Wang, M.L.; Kwok, W.M. A Long Lasting Sunscreen Controversy of 4- Aminobenzoic Acid and 4-Dimethylaminobenzaldehyde Derivatives Resolved by Ultra-fast Spectroscopy Combined with Density Functional Theoretical Study. *Phys. Chem. Chem. Phys.* **2020**, *22*, 8006–8020. [CrossRef]
9. Shaw, A.A.; Wainschel, L.A.; Shetlar, M.D. Photoaddition of p-Aminobenzoic acid to Thymine and Thymidine. *Photochem. Photobiol.* **1992**, *55*, 657–663. [CrossRef]
10. Meijer, G.; de Vries, M.S.; Hunziker, H.E.; Wendt, H.R. Laser Desorption Jet-Cooling Spectroscopy of Para-Amino Benzoic Acid Monomer, Dimer, and Clusters. *J. Chem. Phys.* **1990**, *92*, 7625–7635. [CrossRef]
11. Magata, N. Solvent Effects on the Absorption and Fluorescence Spectra of Napthylamines and Isomeric Aminobenzoic Acids. *Bull. Chem. Soc. Jpn.* **1963**, *36*, 654–662. [CrossRef]
12. Lynch, K.; Pergolizzi, R.G. In vitro Method to Quantify UV Mediated DNA Damage. *J. Young Investig.* **2010**, *20*, 1–16.
13. Shaw, A.A.; Wainschel, L.A.; Shetlar, M.D. The Photochemistry of p-Aminobenzoic Acid. *Photochem. Photobiol.* **1992**, *55*, 647–656. [CrossRef] [PubMed]
14. Cismesia, A.P.; Nicholls, G.R.; Polfer, N.C. Amine vs Carboxylic Acid Protonation in *Ortho-*, *Meta-*, and *Para*-Aminobenzoic Acid: An IRMPD Spectroscopy Study. *J. Mol. Spectrosc.* **2017**, *332*, 79–85. [CrossRef] [PubMed]
15. Momose, T.; Fushitani, M.; Hoshina, H. Chemical Reactions in Quantum Crystals. *Int. Rev. Phys. Chem.* **2005**, *24*, 533–552. [CrossRef]
16. Momose, T.; Shida, T. Matrix-Isolation Spectroscopy Using Solid Parahydrogen as the Matrix: Application to High-Resolution Spectroscopy, Photochemistry, and Cryochemistry. *J. Chem. Phys.* **1998**, *108*, 4237–4241. [CrossRef]
17. Momose, T.; Hoshina, H.; Fushitani, M.; Katsuki, H. High-Resolution Spectroscopy and its Analysis of Ro-Vibrational Transitions of Molecules in Solid Parahydrogen. *Vib. Spectrosc.* **2004**, *34*, 95–108. [CrossRef]
18. Tom, B.A.; Bhasker, S.; Miyamoto, Y.; Momose, T.; McCall, B.J. Producing and Quantifying Enriched para-$H_2$. *Rev. Sci. Inst.* **2009**, *80*, 16108–16111. [CrossRef]
19. Wong, Y.T.A.; Toh, S.Y.; Djuricanin, P.; Momose, T. Conformational Composition and Population Analysis of β-alanine Isolated in Solid Parahydrogen and Argon Matrices. *J. Mol. Spectrosc.* **2015**, *310*, 23–31. [CrossRef]
20. Wakabayashi, T.; Momose, T.; Fajardo, M.E. Matrix Isolation Spectroscopy and Spectral Simulations of Isotopically Substituted $C_{60}$ Molecules. *J. Chem. Phys.* **2019**, *151*, 234301. [CrossRef]
21. Schmidt, W.; Polik, J. R. *WebMO Enterprise*; Version 17.0.012e; WebMO LLC: Holland, MI, USA, 2016. Available online: https://www.webmo.net (accessed on 13 July 2021).
22. Borah, B.; Gomti Devi, T. The Vibrational Study on the Molecular interaction of L- Proline and Para-Aminobenzoic Acid. *J. Mol. Struct.* **2020**, *1203*, 1–15. [CrossRef]
23. NIST Chemistry WebBook S.R.D. 69 Online. Available online: https://webbook.nist.gov/cgi/cbook.cgi?ID=C150130&Units=SI&Type=IR-SPEC&Index=0#IR-SPEC (accessed on 20 June 2021).
24. Fehrensen, B.; Luckhaus, D.; Quack, M. Inversion Tunneling in Aniline from High Resolution Infrared Spectroscopy and an Adiabatic Reaction Path Hamiltonian Approach. *Z. Phys. Chem.* **1999**, *209*, 1–19. [CrossRef]
25. Mitchell, D.M.; Morgan, P.J.; Pratt, D.W. Push-Pull Molecules in the Gas Phase: Stark Effect Measurements of the Permanent Dipole Moments of p-Aminobenzoic Acid in Its Ground and Electronically Excited States. *J. Phys. Chem. A* **2008**, *112*, 12597–12601. [CrossRef] [PubMed]
26. Fleisher, A.J.; Morgan, P.J.; Pratt, D.W. High-Resolution Electronic Spectroscopy Studies of *meta*-Aminobenzoic Acid in the Gas Phase Reveal the Origins of its Solvatochromic Behaviour. *J. Mol. Struct.* **2020**, *55*, 657–663.
27. Moore, B.; Toh, C.S.Y.; Wong, Y.T.A.; Bashiri, T.; McKinnon, A.; Wai, Y.; Lee, A.; Ovchinikov, P.; Chiang, C.; Djuricanin, P.; et al. Hydrocarboxyl Radical as a Product of α-Alanine Ultraviolet Photolysis. *J. Chem. Phys. Lett.* **2021**, *12*, 11992–11997. [CrossRef]
28. Ruzi, M.; Anderson, D.T. Matrix Isolation Spectroscopy and Nuclear Spin Conversion of $NH_3$ and $ND_3$ in Solid Parahydrogen. *J. Phys. Chem. A* **2013**, *117*, 9712–9724. [CrossRef]

29. Ruzi, M.; Anderson, D.T. Fourier Transform Infrared Studies of Ammonia Photochemistry in Solid Parahydrogen. *J. Phys. Chem. A* **2013**, *117*, 13832–13842. [CrossRef]
30. Gée, C.; Crépin, S.D.C.; Bréchignac, P. Infrared Spectroscopy of Aniline ($C_6H_5NH_2$) and its Cation in a Cryogenic Argon Matrix *Chem. Phys. Lett.* **2001**, *338*, 130–136.
31. Tsuge, M.; Chen, Y.H.; Lee, Y.P. Infrared Spectra of Isomers of Protonated Aniline in Solid *para*-Hydrogen. *J. Phys. Chem. A* **2020**, *124*, 2253–2263. [CrossRef]
32. Stephanian, S.; Reva, I.D.; Radchenko, E.D.; Sheina, G.G. Infrared Spectra of Benzoic Acid Monomers and Dimers in Argon Matrix. *Vib. Spectrosc.* **1996**, *11*, 123–133. [CrossRef]
33. Winther, F.; Meyer, S.; Nicolaisen, F.M. The Infrared Spectrum of Methylketene *J. Mol. Struct.* **2002**, *611*, 9–22. [CrossRef]
34. Ryazantsev, S.V.; Feldman, V.I. Matrix-Isolation Studies on the Radiation-Induced Chemistry in $H_2O/CO_2$ Systems: Reactions of Oxygen Atoms and Formation of HOCO Radical. *J. Phys. Chem.* **2014**, *119*, 2578–2586. [CrossRef] [PubMed]
35. Fajardo, M.E.; Lindsay, C.M.; Momose, T. Crystal Field Theory Analysis of Rovibrational Spectra of Carbon Monoxide Monomers Isolated in Solid Parahydrogen. *J. Chem. Phys.* **2009**, *130*, 244508–244517. [CrossRef] [PubMed]
36. Tam, S.; Fajardo, M.E. Observation of the High-Resolution Infrared Absorption Spectrum of $CO_2$ Molecules Isolated in Solid Parahydrogen. *Low. Temp. Phys.* **2000**, *26*, 889–898. [CrossRef]
37. Land, E.J.; Porter, G. Primary Photochemical Processes in Aromatic Molecules. *Trans. Faraday Soc.* **1963**, *59*, 2027–2037. [CrossRef]
38. Roscher, N.M.; Lindemann, M.K.; Kong, S.B.; G.Cho, C.; Jiang, P. Photodecomposition of Several Compounds Commonly Used as Sunscreen Agents. *J. Photochem. Photobiol. A Chem.* **1994**, *80*, 417–421. [CrossRef]
39. Fisher, M.S.; Menter, J.M.; Willis, I. Ultraviolet Radiation-Induced Suppression of Contact Hypersensitivity in Relation to Padimate O and Oxybenzone. *Soc. Investig. Dermatol.* **1989**, *92*, 337–341. [CrossRef]
40. Knowland, J.; McKenzie, E.A.; McHugh, P.J.; Cridland, N.A. Sunlight-Induced Mutagenicity of a Common Sunscreen Ingredient. *FEBS Lett.* **1993**, *324*, 309–313. [CrossRef]

Article

# Phosphorescence and Photophysical Parameters of Porphycene in Cryogenic Matrices

Barbara Golec [1], Aleksander Gorski [1] and Jacek Waluk [1,2,*]

[1] Institute of Physical Chemistry, Polish Academy of Sciences, Kasprzaka 44/52, 01-224 Warsaw, Poland; bgolec@ichf.edu.pl (B.G.); agorski@ichf.edu.pl (A.G.)
[2] Faculty of Mathematics and Science, Cardinal Stefan Wyszyński University, Dewajtis 5, 01-815 Warsaw, Poland
* Correspondence: jwaluk@ichf.edu.pl; Tel.: +48-22-343-3332

**Abstract:** Matrix isolation studies were carried out for porphycene, an isomer of porphyrin, embedded in solid nitrogen and xenon. The external heavy atom effect resulted in nearly a 100% population of the triplet state and in the appearance of phosphorescence, with the origin located at 10163 cm$^{-1}$. This energy is much lower than that corresponding to the $T_1$ position in porphyrin. This difference could be explained by postulating that the orbital origin corresponds in both isomers to the second excited singlet state, which lies much closer to $S_1$ in porphycene. Most of the vibrational frequencies observed in the phosphorescence spectrum correspond to totally symmetric modes, but several ones were assigned to the out-of-plane Bg vibrations. These bands are not observed in fluorescence, which suggests their possible role in vibronic-spin-orbit coupling.

**Keywords:** phosphorescence; triplet state origin; singlet-triplet energy gap; vibrational frequencies; porphyrin isomers

## 1. Introduction

The detailed characterization of the photophysical characteristics of a chromophore is a prerequisite for successful applications. In particular, the properties of the triplet state, such as its energy, yield of formation via intersystem crossing from the singlet state, lifetime, and the ability to generate singlet oxygen, are crucial when designing new materials, e.g., photosensitizers, photovoltaic cells, or light emitting diodes. Moreover, the triplet state parameters often determine the photostability of a molecule, since photodegradation usually involves the triplet state. It is therefore not surprising that large databases containing triplet state data are available for many popular chromophores [1]. One of them is porphyrin, justly called "pigment of life" [2]. In most porphyrins, the yield of $S_1$-$T_1$ intersystem crossing is high, exceeding 70–80%. The $S_1$-$T_1$ energy separation is about 3500 cm$^{-1}$. Investigations of parent, unsubstituted porphyrin in xenon matrices yielded values of 3683 and 3687 cm$^{-1}$, determined from the difference between the origins of absorption [3] and phosphorescence [4] of the two main sites observed in this environment. The triplet lifetimes in argon and xenon, 2.6 and 0.63 ms, respectively [4], are not much different, but the radiative constant of the triplet depopulation dramatically increases in the heavy atom matrix, which allows for the registration of vibronically resolved phosphorescence in the not so readily accessible spectral region of 8000–13,000 cm$^{-1}$.

Porphycene (Pc)—a structural isomer of porphyrin (Pr) (Scheme 1)—has gained much attention as a model for understanding single and double hydrogen transfer in the ground and electronically excited states [5] and as a promising agent for photodynamic therapy of cancer [6] and photoinactivation of bacteria [7]. Spectral and photophysical data for parent, unsubstituted porphycene and its derivatives are quite rich regarding the singlet state properties, but much less is known about the triplet characteristics. The triplet formation efficiency is lower than in porphyrin [5,8]. Moreover, it can be reduced practically to zero by multiple substitutions at the meso positions [9]. The triplet lifetimes, measured in solution [10,11] or lipid vesicles [12], are of the order of tens of microseconds. The triplet energies are considerably lower than in porphyrin (ca. 10,000 vs. 12,500 cm$^{-1}$). They were determined for several porphycenes from the emission studies in the near IR region, carried out at room temperature in degassed iodopropane solutions [13]; no spectra have been presented. For the parent porphycene, Nonell et al. reported the phosphorescence by comparing the emission in degassed and aerated bromobenzene solutions at room temperature [14]. The spectrum consisted of two broad bands, centered around 1000 and 1150 nm.

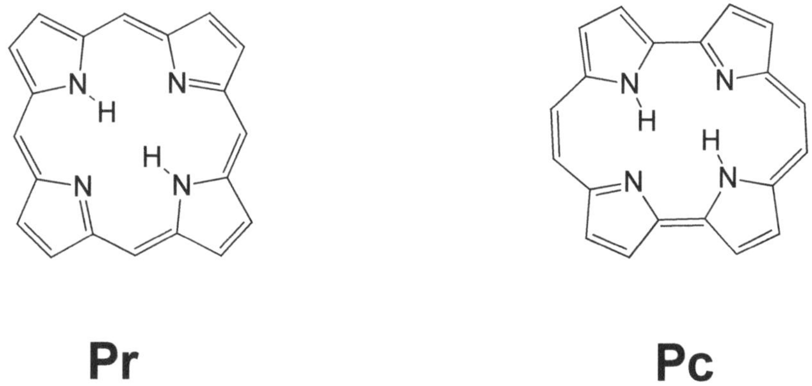

**Scheme 1.** (**Left**), porphyrin (porphine); (**right**), porphycene.

To the best of our knowledge, no vibrationally resolved phosphorescence spectra of porphycenes have been published so far. The goal of our work was to obtain such a spectrum by placing the chromophore in a rare gas matrix. Based on previous observations, xenon was used, in order to enhance the triplet formation yield and, possibly, to increase the radiative constant of $T_1$ depopulation. Next, in order to quantitatively study the external heavy atom effect, we measured the spectra in solid nitrogen. Triplet and singlet lifetimes as well as triplet formation efficiencies were compared for both matrices. Finally, we also performed quantum chemical calculations for porphyrin and porphycene, in order to understand the differences in singlet-triplet energy gaps and to assign the orbital origin of the triplet state.

## 2. Materials and Methods

Porphycene was synthesized and purified as described previously [15].

Nitrogen and xenon matrices were prepared on a sapphire deposition window attached to the cold finger of a closed-cycle helium cryostat (Displex 202, Advance Research Systems). Porphycene was sublimed into the rare gas by heating up to 390 K. The deposition window was kept at 10 K and 57 K during the deposition of nitrogen and xenon matrices, respectively.

Phosphorescence was measured with a home-made spectrometer based on a BEN-THAM DTMc300 double monochromator equipped with a TE cooled photomultiplier (Hamamatsu H10330C-75, 950–1700 nm registration range). Two different laser sources have been used for excitation: (i) a Powerchip 355 nm Laser (Teem Photonics) and (ii) a CW ring laser (Coherent 899), pumped by an argon laser (Coherent Innova 300).

Steady state fluorescence spectra and fluorescence decays were obtained using an Edinburgh FS 900 CDT / FL 900 CDT fluorometer (Edinburgh Analytical Instruments). A NanoLED diode (297 nm, 1 MHz repetition rate) was used for time-resolved measurements.

Absorption spectra were recorded using a Shimadzu UV2700 spectrophotometer.

Calculations were performed using the density functional theory (DFT) and its time-dependent variant (TD DFT), as implemented in the Amsterdam Modeling Suite (version 2021.10-4) [16]. BP86-D functional [17,18] and TZP basis set were used in these simulations.

## 3. Results and Discussion

Figure 1 shows the absorption spectra of Pc obtained for nitrogen and xenon matrices. The absorption of matrix-isolated Pc has been studied in detail before [19–22]. The presently obtained spectra are similar to the reported ones. The linewidths are somewhat narrower than in the previously published works, which enables a better resolution of a characteristic multiple-site structure observed in nitrogen. In contrast, only one dominant site, exhibiting a larger bandwidth, is observed for the xenon environment. Transitions to both $S_1$ and $S_2$ are red-shifted in xenon, by 180 and 230 $cm^{-1}$, respectively. This leads to the $S_1$-$S_2$ energy separation of 887 $cm^{-1}$ in nitrogen and 837 $cm^{-1}$ in xenon. The $S_1$-$S_2$ energy gap is thus significantly—nearly four times—smaller than in porphyrin, for which the reported values are about 3270 $cm^{-1}$ (in argon) and 3100 $cm^{-1}$ (in xenon).

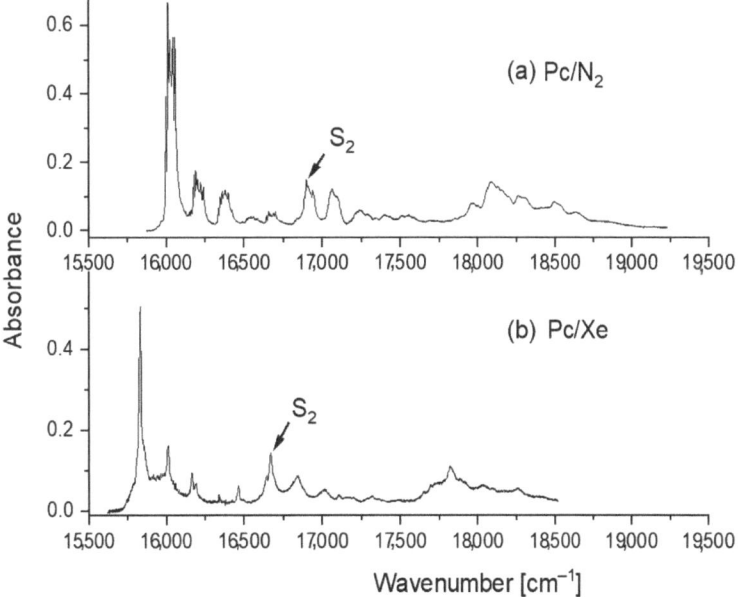

**Figure 1.** Absorption of Pc in (**a**) nitrogen and (**b**) xenon matrices. T = 10 K. The arrows mark the origin of the $S_0$-$S_2$ electronic transition.

Fluorescence spectra are shown in Figure 2. This experiment was performed with a broad excitation and low spectral resolution, so that the site structure is no longer observed. The purpose was to compare the intensities in both cases. The emission was much stronger in nitrogen. The relative quantum yields were estimated from the fluorescence decay times measured for both samples. Such a procedure assumes similar values of radiative decay constants in different environments, which has been demonstrated for porphycene [23]. The lifetime of fluorescence ($\tau_F$) in the nitrogen matrix, 21 ns, was 60 times longer than that measured for the xenon environment (Table 1). The fluorescence lifetimes of Pc measured in room temperature solutions are about 10 ns, whereas the quantum yields vary, depending on the solvent, between 0.39 and 0.50 [23]. This would suggest a rather low triplet formation efficiency in nitrogen. The experimentally obtained value is 0.36. On the other hand, the value obtained for the sample in xenon is practically 1, as could have been expected based on the dramatic shortening of the $S_1$ lifetime caused by the external heavy atom effect.

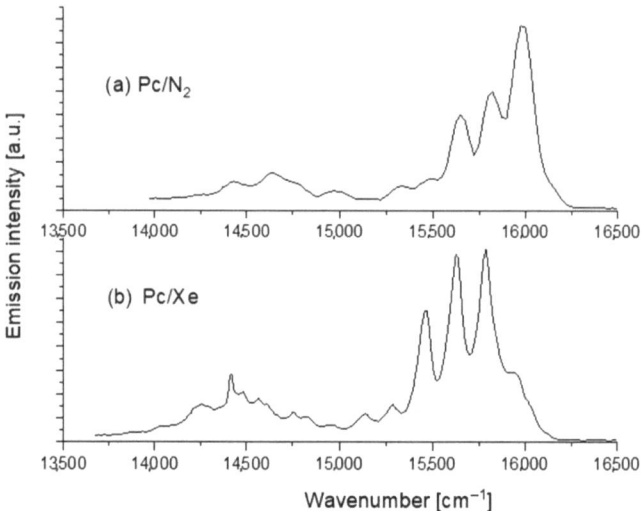

**Figure 2.** Fluorescence of Pc in (**a**) nitrogen and (**b**) xenon matrices. T = 10 K. Excitation wavelengths: (**a**) 370 nm and (**b**) 560 nm.

**Table 1.** Photophysical parameters of Pc in $N_2$ and Xe matrices.

|  | $\tau_F$, ns | $\tau_T$, μs | $\Phi_T$ |
| --- | --- | --- | --- |
| $N_2$ | 21 ± 2 | 2000 ± 40 | 0.36 ± 0.04 |
| Xe | 0.35 ± 0.1 | 45 ± 0.5 | 1 ± 0.05 |

The triplet decays reveal the opposite behavior (Table 1). The ratio of $T_1$ lifetimes ($\tau_T$), (2000 μs ($N_2$))/(45 μs (Xe)) is not much different from what is observed for the changes in the $T_1 \leftarrow S_1$ intersystem crossing rates. Interestingly, while the $T_1$ lifetimes are very similar for Pr in argon and Pc in nitrogen, in xenon it is the Pc triplet that decays much faster. A possible explanation is the larger value of the radiative constant of triplet depopulation in Pc. One should remember that the transition to the $S_1$ state is much weaker in Pr than in Pc.

While the phosphorescence was extremely weak in nitrogen-isolated Pc, it could be readily obtained in xenon matrices (Figure 3). The electronic ground state vibrational frequencies extracted from the phosphorescence spectrum are presented in Table 2. The same table also contains the frequencies previously obtained from the fluorescence of Pc in xenon. Because of the better spectral resolution in the present case, we could observe and assign additional lines, such as combinations of low frequency modes with the strong transition observed at 1554 cm$^{-1}$ (29 Ag). The assignments in Table 2 are based on our previous work that included frequencies obtained from IR, Raman, fluorescence, and inelastic neutron scattering data combined with quantum chemical calculations and isotopic substitution [24].

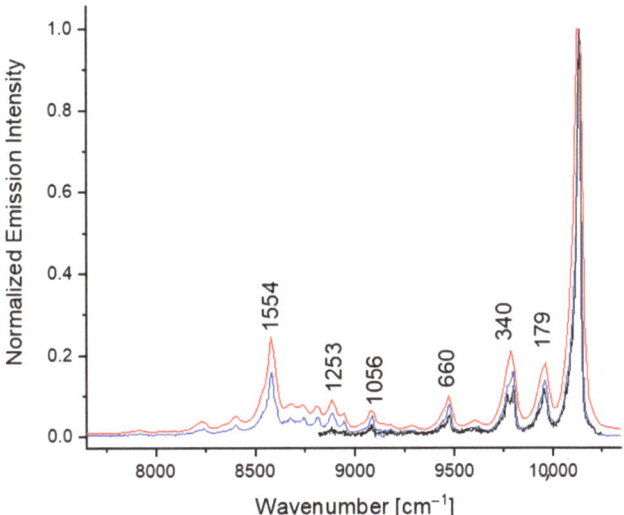

**Figure 3.** Phosphorescence of Pc in xenon matrices measured at three different spectral resolutions. T = 10 K. Excitation wavelength: 631.6 nm, corresponding to the location of the main site (see Figure 1b).

Most of the features present in the phosphorescence have their counterparts in the fluorescence spectrum. However, in contrast to fluorescence, which reveals only totally symmetric (Ag) modes, several bands in phosphorescence are assigned to out-of-plane Bg vibrations. This indicates a possible contribution of the vibronic-spin-orbit coupling mechanism [25] to the phosphorescence intensity.

Using the luminescence data available for both chromophores, we are now ready to compare the pattern of low-lying electronic states in Pc and Pr. The most conspicuous difference is the $S_1$-$T_1$ energy gap. While the origins of the $S_0$-$S_1$ transitions differ by less than 500 cm$^{-1}$ (16,270 cm$^{-1}$ in Pr vs. 15,821 cm$^{-1}$ in Pc), the $T_1$ state in Pc is located at an energy that is lower by nearly 2500 cm$^{-1}$ with respect to Pr (10,134 vs. 12,588 cm$^{-1}$). An opposite behavior is found for the relative positions of $S_1$ and $S_2$ in the two chromophores. The energy difference between the transitions to $S_1$ ($Q_x$) and $S_2$ ($Q_y$) in a xenon matrix exceeds 3100 cm$^{-1}$ in porphyrin [4], whereas in porphycene it amounts to only 838 cm$^{-1}$ [19].

We postulate that the lowering of the $S_2$ energy upon going from Pr to Pc is responsible for the large stabilization of the lowest triplet state in the latter. Calculations suggest that, in both molecules, the lowest triplet state is well described by the same electronic configuration, involving the HOMO–LUMO electron jump (Table 3). Such a configuration is dominant, for both Pr and Pc, in the $S_0$-$S_2$ transition, but not in $S_0$-$S_1$. This implies a larger singlet-triplet splitting for $S_2$. Assuming a similar value of the singlet-triplet splitting in both chromophores, one would expect a difference in the $T_1$ location corresponding to the difference in $S_2$ energies. This is indeed observed experimentally: the $T_1$ energies differ by ca. 2500 cm$^{-1}$, the energies of $S_2$ by ca. 2700 cm$^{-1}$.

**Table 2.** Vibrational frequencies (in cm$^{-1}$) observed in phosphorescence in xenon, compared with the previously reported data [19] obtained from fluorescence in the same matrix.

| Phosphorescence | Fluorescence | Assignment [a] |
|---|---|---|
| 179 | 184 | 2 Ag |
| 340 | 336 | 3 Ag |
| 368 | 358 | 4 Ag |
| 511 | 510 | 2 Ag + 3 Ag |
| 541 | 538 | 2 Ag + 4 Ag |
| 621 |  | 7 Bg |
| 660 | 665 | 7 Ag |
|  | 686 | 2 × 3 Ag |
| 696 |  | 9 Bg |
|  | 703 | 3 Ag + 4 Ag |
|  | 723 | 2 × 4 Ag |
| 847 |  | 1 Ag + 7 Bg |
|  | 863 | 9 Ag |
| 881 |  | 14 Bg |
| 959 | 961 | 10 Ag |
|  | 1010 | 3 × 3 Ag |
| 1056 |  | 14 Ag |
|  | 1079 | 3 × 4 Ag |
| 1192 | 1203 | 17 Ag |
| 1253 | 1265 | 19 Ag |
| 1322 |  | 20 Ag |
|  | 1354 | 21 Ag |
| 1392 | 1395 | 23 Ag |
| 1426 |  | 25 Ag |
| 1458 |  | 26 Ag |
| 1486 | 1497 | 27 Ag |
| 1554 | 1562 | 29 Ag |
|  | 1585 |  |
|  | 1614 | 30 Ag |
| 1730 |  | 2 Ag + 29 Ag |
| 1891 |  | 3 Ag + 29 Ag |
| 1918 |  | 4 Ag + 29 Ag |
| 2215 |  | 7 Ag + 29 Ag |

[a] based on ref. [24].

**Table 3.** Calculated transition energies.

|  | $S_1$ | $S_2$ | $T_1$ | $T_2$ |
|---|---|---|---|---|
| Pc | 17,191 [a]<br>(0.0927) [b]<br>54.4.% sH-L [c]<br>26% H-L | 18,226<br>(0.1549)<br>56.3% H-L<br>25.0% sH-L | 11,898<br>H-L | 12,449<br>sH-L |
| Pr | 17,505 (0.0007)<br>61.5% H-sL<br>37.5% sH-L | 18,563 (0.0004)<br>57.3% H-L<br>41.4% sH-sL | 14,087<br>H-L | 14,870<br>H-sL |

[a] cm$^{-1}$; [b] oscillator strength in parentheses; [c] dominant configurations; H, sH: highest and second highest occupied molecular orbital; L, sL: lowest and second lowest unoccupied molecular orbital.

The calculations accurately predict the difference in the triplet energies of Pr and Pc, (2200 cm$^{-1}$ vs. 2500 cm$^{-1}$ obtained in the experiment). They also correctly reproduce the lower energy of $S_1$ in Pc (300 cm$^{-1}$ vs. the observed value of 450 cm$^{-1}$). The $S_1$-$S_2$ separation is well reproduced for Pc, but for the $S_2$ state both the energy and the oscillator strength (relative to those of $S_1$) are predicted rather poorly.

## 4. Summary

Electronic spectroscopy studies of porphycene embedded in xenon matrices enabled an accurate determination of the location of the lowest triplet state. The vibronically resolved phosphorescence was recorded, exhibiting, in contrast to fluorescence, not only totally symmetric modes. The triplet location in porphycene is significantly red-shifted compared to porphyrin. This was explained by assigning the $T_1$ orbital origin to that of $S_2$, a state which is strongly stabilized with respect to $S_1$ upon passing from Pr to Pc.

In view of the crucial role the triplet state plays in various photophysical, photochemical, and photochemical processes, more detailed studies seem worthwhile, focusing on such issues as, e.g., substituent effects that could lead to the inversion of the two lowest lying triplet states in porphyrin and its isomers. Several relatively old papers suggest for Pr that the $T_1$ and $T_2$ states lie very close to each other and that their ordering may depend on the environment [4,26–29]. The same may be true for porphycene, as suggested by the calculated $T_1$-$T_2$ energy gap, which is even smaller than in porphyrin (Table 3).

**Author Contributions:** Conceptualization, J.W.; software, J.W.; formal analysis, B.G., A.G. and J.W.; investigation, B.G. and A.G.; writing—original draft preparation, J.W. All authors have read and agreed to the published version of the manuscript.

**Funding:** This research was funded by the Polish National Science Centre, grant numbers 2016/22/A/ST4/00029 and 2020/39/B/ST4/01956.

**Data Availability Statement:** Data is contained within the article.

**Conflicts of Interest:** The authors declare no conflict of interest.

## References

1. Montalti, M.; Credi, A.; Prodi, L.; Gandolfi, M.T. *Handbook of Photochemistry*, 3rd ed.; Montalti, M., Credi, A., Prodi, L., Gandolfi, M.T., Eds.; Taylor & Francis Group: Abingdon, UK, 2006.
2. Battersby, A.R. Tetrapyrroles: The Pigments of Life. *Nat. Prod. Rep.* **2000**, *17*, 507–526. [CrossRef] [PubMed]
3. Radziszewski, J.; Waluk, J.; Michl, J. FT Visible Absorption Spectroscopy of Porphine in Noble Gas Matrices. *J. Mol. Spectrosc.* **1990**, *140*, 373–389. [CrossRef]
4. Radziszewski, J.G.; Waluk, J.; Nepraš, M.; Michl, J. Fourier Transform Fluorescence and Phosphorescence of Porphine in Rare Gas Matrices. *J. Phys. Chem.* **1991**, *95*, 1963–1969. [CrossRef]
5. Waluk, J. Spectroscopy and Tautomerization Studies of Porphycenes. *Chem. Rev.* **2017**, *117*, 2447–2480. [CrossRef] [PubMed]
6. Stockert, J.C.; Cañete, M.; Juarranz, A.; Villanueva, A.; Horobin, R.W.; Borrell, J.; Teixidó, J.; Nonell, S. Porphycenes: Facts and Prospects in Photodynamic Therapy of Cancer. *Curr. Med. Chem.* **2007**, *14*, 997–1026. [CrossRef]
7. Polo, L.; Segalla, A.; Bertoloni, G.; Jori, G.; Schaffner, K.; Reddi, E. Polylysine-Porphycene Conjugates as Efficient Photosensitizers for the Inactivation of Microbial Pathogens. *J. Photochem. Photobiol. B Biol.* **2000**, *59*, 152–158. [CrossRef]
8. Planas, O.; Gallavardin, T.; Nonell, S. Unusual Properties of Asymmetric Porphycenes. In *Handbook of Porphyrin Science*; Kadish, K.M., Smith, K.M., Guilard, R., Eds.; World Scientific: Singapore, 2016; Volume 41, pp. 299–349.
9. Gil, M.; Dobkowski, J.; Wiosna-Sałyga, G.; Urbańska, N.; Fita, P.; Radzewicz, C.; Pietraszkiewicz, M.; Borowicz, P.; Marks, D.; Glasbeek, M.; et al. Unusual, Solvent Viscosity-Controlled Tautomerism and Photophysics: Meso-Alkylated Porphycenes. *J. Am. Chem. Soc.* **2010**, *132*, 13472–13485. [CrossRef]
10. Aramendia, P.F.; Redmond, R.W.; Nonell, S.; Schuster, W.; Braslavsky, S.E.; Schaffner, K.; Vogel, E. The Photophysical Properties of Porphycenes: Potential Photodynamic Therapy Agents. *Photochem. Photobiol.* **1986**, *44*, 555–559. [CrossRef]
11. Levanon, H.; Toporowicz, M.; Ofir, H.; Fessenden, R.W.; Das, P.K.; Vogel, E.; Köcher, M.; Pramod, K. Triplet-State Formation of Porphycenes—Intersystem Crossing Versus Sensitization Mechanisms. *J. Phys. Chem.* **1988**, *92*, 2429–2433. [CrossRef]
12. Redmond, R.W.; Valduga, G.; Nonell, S.; Braslavsky, S.E.; Schaffner, K.; Vogel, E.; Pramod, K.; Köcher, M. The Photophysical Properties of Porphycene Incorporated in Small Unilamellar Lipid Vesicles. *J. Photochem. Photobiol. B Biol.* **1989**, *3*, 193–207. [CrossRef]
13. Braslavsky, S.E.; Müller, M.; Mártire, D.O.; Pörting, S.; Bertolotti, S.G.; Chakravorti, S.; Koç-Weier, G.; Knipp, B.; Schaffner, K. Photophysical Properties of Porphycene Derivatives (18 π Porphyrinoids). *J. Photochem. Photobiol. B Biol.* **1997**, *40*, 191–198. [CrossRef]
14. Nonell, S.; Aramendía, P.F.; Heihoff, K.; Negri, R.M.; Braslavsky, S.E. Laser-Induced Optoacoustics Combined with near-Infrared Emission. An Alternative Approach for the Determination of Intersystem Crossing Quantum Yields Applied to Porphycenes. *J. Phys. Chem.* **1990**, *94*, 5879–5883. [CrossRef]
15. Urbańska, N.; Pietraszkiewicz, M.; Waluk, J. Efficient Synthesis of Porphycene. *J. Porphyr. Phthalocyanines* **2007**, *11*, 596–600. [CrossRef]

16. te Velde, G.; Bickelhaupt, F.M.; Baerends, E.J.; Fonseca Guerra, C.; van Gisbergen, S.J.A.; Snijders, J.G.; Ziegler, T. Chemistry with ADF. *J. Comput. Chem.* **2001**, *22*, 931–967. [CrossRef]
17. Becke, A.D. Density-Functional Exchange-Energy Approximation with Correct Asymptotic Behavior. *Phys. Rev. A* **1988**, *38*, 3098–3100. [CrossRef]
18. Perdew, J.P. Density-Functional Approximation for the Correlation Energy of the Inhomogeneous Electron Gas. *Phys. Rev. B* **1986**, *33*, 8822–8824, Erratum: Density-Functional Approximation for the Correlation Energy of the Inhomogeneous Electron Gas. *Phys. Rev. B* **1986**, *34*, 7406–7406. [CrossRef]
19. Starukhin, A.; Vogel, E.; Waluk, J. Electronic Spectra in Porphycenes in Rare Gas and Nitrogen Matrices. *J. Phys. Chem. A* **1998**, *102*, 9999–10006. [CrossRef]
20. Kyrychenko, A.; Waluk, J. Molecular Dynamics Simulations of Matrix Deposition. Iii. Site Structure Analysis for Porphycene in Argon and Xenon. *J. Chem. Phys.* **2005**, *123*, 064706. [CrossRef]
21. Kyrychenko, A.; Gawinkowski, S.; Urbańska, N.; Pietraszkiewicz, M.; Waluk, J. Matrix Isolation Spectroscopy and Molecular Dynamics Simulations for 2,7,12,17-Tetra-Tert-Butylporphycene in Argon and Xenon. *J. Chem. Phys.* **2007**, *127*, 134501. [CrossRef]
22. Gil, M.; Gorski, A.; Starukhin, A.; Waluk, J. Fluorescence Studies of Porphycene in Various Cryogenic Environments. *Low Temp. Phys.* **2019**, *45*, 656–662. [CrossRef]
23. Kijak, M.; Nawara, K.; Listkowski, A.; Masiera, N.; Buczyńska, J.; Urbańska, N.; Orzanowska, G.; Pietraszkiewicz, M.; Waluk, J. 2+2 Can Make Nearly a Thousand! Comparison of Di- and Tetra-Meso-Alkyl-Substituted Porphycenes. *J. Phys. Chem. A* **2020**, *124*, 4594–4604. [CrossRef] [PubMed]
24. Gawinkowski, S.; Walewski, Ł.; Vdovin, A.; Slenczka, A.; Rols, S.; Johnson, M.R.; Lesyng, B.; Waluk, J. Vibrations and Hydrogen Bonding in Porphycene. *Phys. Chem. Chem. Phys.* **2012**, *14*, 5489–5503. [CrossRef] [PubMed]
25. Penfold, T.J.; Gindensperger, E.; Daniel, C.; Marian, C.M. Spin-Vibronic Mechanism for Intersystem Crossing. *Chem. Rev.* **2018**, *118*, 6975–7025. [CrossRef] [PubMed]
26. van Dorp, W.G.; Soma, M.; Kooter, J.A.; van der Waals, J.H. Electron Spin Resonance in the Photo-Excited Triplet State of Free Base Porphin in a Single Crystal of *n*-Octane. *Mol. Phys.* **1974**, *28*, 1551–1568. [CrossRef]
27. Weiss, C.; Kobayashi, H.; Gouterman, M. Spectra of Porphyrins. Part III. Self-Consistent Molecular Orbital Calculations of Porphyrin and Related Ring Systems. *J. Mol. Spectrosc.* **1965**, *16*, 415–450. [CrossRef]
28. Knop, J.V.; Knop, A. Quantenchemische Und Spektroskopische Untersuchungen an Porphyrinen. I. Freie Base Porphin Und Metallo-Porphin. *Z. Für Naturforsch.* **1970**, *25*, 1720–1725. [CrossRef]
29. Sundbom, M. Semi-Empirical Molecular Orbital Studies of Neutral Porphin, $PH_2$, the Dianion $P^{2-}$ and the Dication $PH_4^{2+}$. *Acta Chem. Scand.* **1968**, *22*, 1317–1326. [CrossRef]

Article

# Local and Remote Conformational Switching in 2-Fluoro-4-Hydroxy Benzoic Acid

Sándor Góbi [1,*], Mirjam Balbisi [2] and György Tarczay [1,2,*]

[1] MTA-ELTE Lendület Laboratory Astrochemistry Research Group, Institute of Chemistry, ELTE Eötvös Loránd University, H-1518 Budapest, Hungary
[2] Laboratory of Molecular Spectroscopy, Institute of Chemistry, ELTE Eötvös Loránd University, H-1518 Budapest, Hungary; balbisimirjam@gmail.com
* Correspondence: sandor.gobi@ttk.elte.hu (S.G.); gyorgy.tarczay@ttk.elte.hu (G.T.)

**Abstract:** In this work, 2-F-4-OH benzoic acid was isolated in Ar matrices and conformational changes were induced by near-IR irradiating the sample. Upon deposition, three conformers could be observed in the matrix, denoted as **A1**, **A2**, and **D1**, respectively. **A1** and **A2** are *trans* carboxylic acids, i.e., there is an intramolecular H bond between the H and the carbonyl O atoms in the COOH group, whereas **D1** is a *cis* carboxylic acid with an intramolecular H bond between the F atom and the H atom in the COOH group, which otherwise has the same structure as **A1**. The difference between **A1** and **A2** is in the orientation of the carbonyl O atom with regard to the F atom, i.e., whether they are on the opposite or on the same side of the molecule, respectively. All three conformers have their H atom in their 4-OH group, facing the opposite direction with regard to the F atom. The stretching overtones of the 4-OH and the carboxylic OH groups were selectively excited in the case of each conformer. Unlike **A2**, which did not show any response to irradiation, **A1** could be converted to the higher energy form **D1**. The **D1** conformer spontaneously converts back to **A1** via tunneling; however, the conversion rate could be significantly increased by selectively exciting the OH vibrational overtones of **D1**. Quantum efficiencies have been determined for the 'local' or 'remote' excitations, i.e., when the carboxylic OH or the 4-OH group is excited in order to induce the rotamerization of the carboxylic OH group. Both 'local' and 'remote' conformational switching are induced by the same type of vibration, which allows for a direct comparison of how much energy is lost by energy dissipation during the two processes. The experimental findings indicate that the 'local' excitation is only marginally more efficient than the 'remote' one.

**Keywords:** conformational switching; near-IR laser irradiation; intramolecular vibrational energy relaxation (IVR); quantum efficiency; hydrogen atom tunneling; 2-fluoro-4-hydroxy benzoic acid; matrix isolation; IR spectroscopy

Citation: Góbi, S.; Balbisi, M.; Tarczay, G. Local and Remote Conformational Switching in 2-Fluoro-4-Hydroxy Benzoic Acid. *Photochem* **2022**, *2*, 102–121. https://doi.org/10.3390/photochem2010009

Academic Editors: Gulce Ogruc Ildiz, Licinia L.G. Justino and Stefanie Tschierlei

Received: 23 December 2021
Accepted: 23 January 2022
Published: 28 January 2022

**Publisher's Note:** MDPI stays neutral with regard to jurisdictional claims in published maps and institutional affiliations.

**Copyright:** © 2022 by the authors. Licensee MDPI, Basel, Switzerland. This article is an open access article distributed under the terms and conditions of the Creative Commons Attribution (CC BY) license (https://creativecommons.org/licenses/by/4.0/).

## 1. Introduction

To fully understand chemical reactions occurring in the environment, an investigation of the excited states of molecules and their relaxation is of the utmost importance. A vibrationally excited molecule may emit its excess energy via intra- or intermolecular processes. One example of the latter is when an excited molecule collides with a nearby one, transferring its excess energy in forms of vibrational, rotational, or translational energy. In contrast to this, during intramolecular vibrational energy transfer or intramolecular vibrational energy relaxation (IVR), the energy flows from one excited vibrational mode to another while the energy difference dissipates into the surroundings. This also means that in certain cases the excitation of a vibrational mode on one side of the molecule may result in a change in the geometry of a functional group on another, remote side of the molecule, called remote conformational switching.

The matrix isolation (MI) technique has been successfully used to study the conformational switching of molecules isolated in inert, nearly interaction-free matrices using mostly solidified noble or other inert gases (such as Ne, Ar, Kr, Xe, N$_2$, etc.) as matrix material. The intermolecular relaxation of matrix-isolated molecules is greatly limited, thereby allowing for the examination of the IVR processes. Furthermore, another primary advantage of the technique is that the resulting IR spectrum contains sharp absorption peaks, meaning that they do not overlap; thus, the different conformers can be distinguished from each other, and the induced changes in their population can be monitored. It is important to note, however, that in order to achieve this vibrationally induced conformational change, selective excitation of the vibrational modes is required, which necessitates the use of monochromatic laser irradiation sources.

The MI technique combined with near-IR laser is routinely performed to excite the vibrational overtones of hydroxyl (–OH) functional groups, thus efficiently inducing their rotation or even the conformational change of the whole carboxylic (–C(O)OH) group. This has been successfully done in the case of many different carboxylic acid monomers in Ar matrices, such as formic, [1–3] acetic, [4–6] trifluoroacetic, [7] tribromoacetic, [8] propionic, [9] 2-chloropropionic, [10] glycolic, [11] pyruvic, [12] oxamic, [13] and various dicarboxylic acids [14–16] as well as amino acids [17–22]. Furthermore, the OH rotamerization induced by selectively exciting the OH vibrational stretching overtone of cyclic or heterocyclic organic acid conformers has also been studied in Ar matrices for squaric, [23] 2-furoic, [24] and 2-fluorobenzoic acid [25]. It should be noted that the works above studied short-range IVR, where the rotamerization of a functional group is achieved by the excitation of the vibrational overtone of that particular group.

Of greater interest are those cases when the excitation and the rotamerization occurs in different parts of the molecule. Compared to its short-range counterpart, far fewer matrix-isolation studies have been devoted to these long-range IVR processes. The first one was cysteine, which showed the rotamerization of the thiol (–SH) group upon exciting the νOH stretching overtone (νOH), although it occurred along with other conformational changes in the molecule, thus rendering it not a selective method [21]. Nevertheless, this work pointed to the feasibility of remote switching using molecular antennas. The next step was to find a molecular system representing the first example for selective conformational control. Accordingly, the rotamerization of the thiol group in 2-thiocytosine was found to be reversibly carried out by pumping the amino (–NH$_2$) group [26]. The first remote switching using an NH group as a molecular antenna was done in the case of 6-methoxyindole, where the methoxy group could be selectively induced to rotate around the C–O bond [27]. Another molecule that greatly exemplifies this phenomenon is kojic acid, whose hydroxymethyl (–CH$_2$OH) moiety goes through rotamerization upon exciting the –OH group on the other side of the ring of the molecule [28]. The OH or NH$_2$ groups of 3-hydroxy-2-formyl-2H-azirine and 3-amino-2-formyl-2H-azirine, respectively, also serve as molecular antennas that facilitate conformational control over the heavy aldehyde moiety three bonds away [29]. The current 'record holder' system, investigated by matrix-isolation IR spectroscopy, is *E*-glutaconic acid, in which long-range IVR could act over eight covalent bonds to successfully induce the conformational change of an OH group [30].

In this work, we aimed to investigate the selective conformational switching of a matrix-isolated 2-fluoro-4-hydroxy derivative of benzoic acid (2-F-4-OH benzoic acid in short) achieved by narrowband near-IR laser irradiation. This molecule serves as a great example, in which both 'local' and 'remote' switching can be done by exciting the same type of vibration (i.e., OH overtone) selectively, and their efficiencies can be directly compared with each other. Preliminary experiments showed that upon the excitation of the $2\nu$(OH) stretching vibrational overtone in a hydroxy carboxylic acid, change in the geometry of the COOH group in another part of the molecule can be efficiently induced [31]. In other words, higher energy *cis* carboxylic acid can be generated from the more stable *trans* form by remote switching (using the notation of Pettersson et al. for the two types of carboxylic acids) [1]. Moreover, the *cis* isomer is assumed to slowly convert back to *trans* over time via tunneling, which can also be monitored by IR spectrometry. The experiments were supplemented by quantum-chemical computations, which made the spectral analysis easier by allowing for simple comparison between the simulated spectra and the experimental ones. It is important to emphasize that effective remote switching that acts over multiple bonds has been observed in only a handful of molecular systems. As such, in order to fully understand its mechanism, it is necessary to find more examples of this phenomenon and study its overall efficiency.

## 2. Experimental Methods

*2.1. MI-IR Experiments*

A commercial 2-F-4-OH benzoic acid ($\leq$100%) sample obtained from Sigma-Aldrich was used without further purification. The compound was inserted into a Knudsen cell with a heatable cartridge in a small quartz sample holder directly connected to the vacuum chamber. The adsorbed water and other volatiles had been removed the day before the experiments by slightly heating the sample compartment to 355 K and keeping it at that temperature for a couple of hours. A closed-cycle helium refrigeration system (Janis CCS-350R cold head cooled by a CTI Cryogenics 22 refrigerator) was used to cool down the CsI optical window to 13 K, which serves as the deposition temperature. The temperature of the cold window was measured by a silicon diode sensor connected to a LakeShore 321 digital temperature controller; the base pressure in the cooled chamber is usually found in the high $10^{-9}$ mbar region. A sample sublimation temperature of 375 $\pm$ 2 K was used during the experiment whereupon the vapors were mixed and co-deposited in a 1:1000 ratio with Ar (Messer, 99.9999%) on the window. The orientation of the window (at a relative angle of 45° with regard to the sample oven and the spectrometer beam) allows for a simultaneous deposition and spectral collection. For the latter, a Bruker IFS 55 FT-IR spectrometer equipped with an MCT detector cooled with liquid nitrogen was used. The transmission mid-IR spectra were taken during and after the deposition by averaging 16 and 128 scans, respectively, in the 2000–600 cm$^{-1}$ region with 1 cm$^{-1}$ resolution using KBr as a beam splitter and a low-pass filter with a cutoff wavenumber of 1830 cm$^{-1}$. In order to define the 2$\nu$(OH) stretching overtones of each conformer necessary for the laser irradiation experiments, preliminary near-IR spectra were collected with the same instrument, without the cutoff filter, utilizing a W lamp and a CaF$_2$ beam splitter in the spectral region of the 8000–2500 cm$^{-1}$ region at a resolution of 1 cm$^{-1}$.

*2.2. Near-IR Laser Irradiation*

The deposited matrices were in situ irradiated through the outer KBr window of the cryostat, applying a tunable narrowband laser light provided by the idler beam of an Optical Parametric Oscillator (OPO; GWU/Spectra-Physics VersaScan MB 240, fwhm $\approx$ 5 cm$^{-1}$). The OPO was pumped by a pulsed Nd:YAG laser (Spectra-Physics Quanta Ray Lab 150, $p \approx 2.1$–2.2 W, $\lambda$ = 355 nm, $f$ = 10 Hz, duration = 2–3 ns). The laser coming from the OPO device is perpendicular to the beam of the IR spectrometer, whereas the CsI cold window has a relative angle of 45° to both of them, allowing for an online spectral collection during laser irradiation. The wavelength of the laser beam was selected so that the 2$\nu$(OH) modes of the conformers of the matrix-isolated molecule were selectively excited. Accordingly, the sample was irradiated at 6952.0 cm$^{-1}$, 6994.9 cm$^{-1}$, 7077.1 cm$^{-1}$, and at 7093.2 cm$^{-1}$ for approximately 60 min in each case; the photon fluxes were measured to be (3.1 $\pm$ 0.6) $\times$ 10$^{17}$ cm$^{-2}$ s$^{-1}$ (6952.0 cm$^{-1}$), (3.2 $\pm$ 0.1) $\times$ 10$^{17}$ cm$^{-2}$ s$^{-1}$ (6994.9 cm$^{-1}$), (3.3 $\pm$ 0.1) $\times$ 10$^{17}$ cm$^{-2}$ s$^{-1}$ (7077.1 cm$^{-1}$), and (3.6 $\pm$ 0.1) $\times$ 10$^{17}$ cm$^{-2}$ s$^{-1}$ (7093.2 cm$^{-1}$), respectively. Mid-IR spectra were collected online in the meantime, averaging 16 scans (during the 6952.0 cm$^{-1}$ irradiation) and 12 scans (during the rest of the irradiations), respectively; 128 scans were taken before and after irradiation. After the irradiation experiments, the sample was left in the dark overnight while continuously collecting mid-IR spectra, averaging 128 scans every 15 min.

## 2.3. Theoretical Computations

The quantum chemical geometry optimizations, as well as the harmonic and anharmonic frequency computations, were carried out in Gaussian 09 (Rev. D01) with Becke's three-parameter hybrid functional B3LYP, using the non-local and local exchange correlation functionals as described by Lee–Yang–Parr and Vosko–Wilk–Nusair III, respectively [32–35]. Dunning's correlation-consistent cc-pVTZ basis set was used [36]. The anharmonic frequencies were used to simulate the vibrational spectra by convoluting them with Lorentzian functions with an FWHM of 0.5 cm$^{-1}$ in the Synspec software [37]. The isomerization barriers were estimated from the optimized transition states (TSs) using the Berny algorithm, as implemented in Gaussian [38]. This was followed by the computation of intrinsic reaction paths (IRPs) in the Cartesian coordinates [39]. The tunneling rates were estimated using the Wentzel–Kramers–Brillouin (WKB) model described in detail elsewhere [40].

## 3. Results and Discussion

### 3.1. Structure and Energy of the Conformers

The conformational structure of the molecule is primarily defined by its two functional groups, the 4-hydroxyl (–OH) and the carboxylic group (–COOH). The H atom on both moieties may have two orientations (the respective torsional angles are denoted by $\varphi_1$ and $\varphi_3$), whereas –COOH has two forms based on the alignment of its carbonyl (C=O) group ($\varphi_2$). This means $2^3 = 8$ different conformers, which are depicted in Scheme 1, in which the torsional angles most important to the molecular structure are also marked. The conformers with $\varphi_3$ close to 180° are named **A** ($\varphi_1 \approx 180°$) or **B** ($\varphi_1 \approx 0°$), respectively. The same applies to structures **C** and **D**, which can have a $\varphi_3$ value of 0° with $\varphi_1 \approx 0°$ or 180°, respectively, whereas the labels **1** or **2** after them differentiate the ones with $\varphi_2 \approx 180°$ and 0°, respectively. This means that the following notations for the eight conformers are used throughout the manuscript: **A1, A2, B1, B2, C1, C2, D1** and **D2**; their optimized geometries are listed Tables S1–S20 in the Supplementary Material.

According to the orientation of the groups discussed above, the conformers have different stabilities. The most stable conformer is **A1**, which is closely followed by **B1**; the only difference between them is the alignment of the 4-OH group. Apart from this, they both have a *trans*-COOH carboxylic group (i.e., there is a weak intramolecular H bond between the H atom of the OH group with the C=O oxygen atom), whereas the O atom in the C=O group is on the opposite side with regard to the F atom. It should be noted that the energy difference is predicted to be 0.8 kJ mol$^{-1}$, which is within the accuracy of the computational level. The next pair of conformers have slightly higher relative energies (by ca. 3 kJ mol$^{-1}$) denoted with **A2** and **B2**. They, similarly to **A1** and **B1**, only differ from each other in the orientation of the 4-OH hydrogen atom as well, but, unlike **A1** and **B1**, their C=O oxygen atoms are on the same side as the F atom. The next four conformers have *cis*-COOH groups, meaning inherently higher relative energies due to breakage of the abovementioned intramolecular H bond, which is counterbalanced to some extent in the case of the **C1** and **D1** by the formation of a less energetic H bond with the F atom. Their relative energy (with regard to the **A1** form) is approximately 8 kJ mol$^{-1}$; this value should more or less reflect on the difference of the H atom bond strength between the cases when the H atom creates the bond with the C=O oxygen and when it bonds with the F atom. Nevertheless, **C1** and **D1** only differ in the orientation of the 4-OH hydrogen atom, whereas their C=O oxygen atoms (similarly to **A1** and **B1**) are on the opposite side of the molecule with regard to the F atom, thus allowing for H bonding between the carboxylic H atom and the F atom. Lastly, **C2** and **D2** are unique in the sense that, due to steric effects, their –COOH group is not in the same plane as the aromatic ring, which means that the $\varphi_2$ torsional angle is closer to 30° instead of 0° and can be both positive and negative (i.e., the C=O oxygen atom and the carboxylic OH group are slightly above or below the plane of the aromatic ring). This results in the fact that both **C2** and **D2** are actually chiral and have two enantiomeric forms differentiated by an asterisk in Table 1 (i.e., **C2, C2\*, D2, D2\***), and the only difference between **C2** and **D2** is the orientation of the 4-OH group. Apart from this, their C=O oxygen atoms, similarly to **A2** and **B2**, is on the same side of the molecule with regard to the F atom and have *cis*-COOH groups (similarly to **C1** and **D1**). As a matter of fact, this causes distortion of the molecule as the COOH hydrogen atom becomes relatively close to the H atom on the C atom in position six of the ring. Due to the unfavorable effects listed above, there is no intramolecular H bond in them, and therefore, they are much less stable than the other conformers; their relative energy lies more than 30 kJ mol$^{-1}$ above the most stable **A1** form. This also implies that their Boltzmann population at the sample inlet temperature (375 ± 2 K) is expected to be significantly less than 1%, which prohibits their IR detection in the deposited matrix.

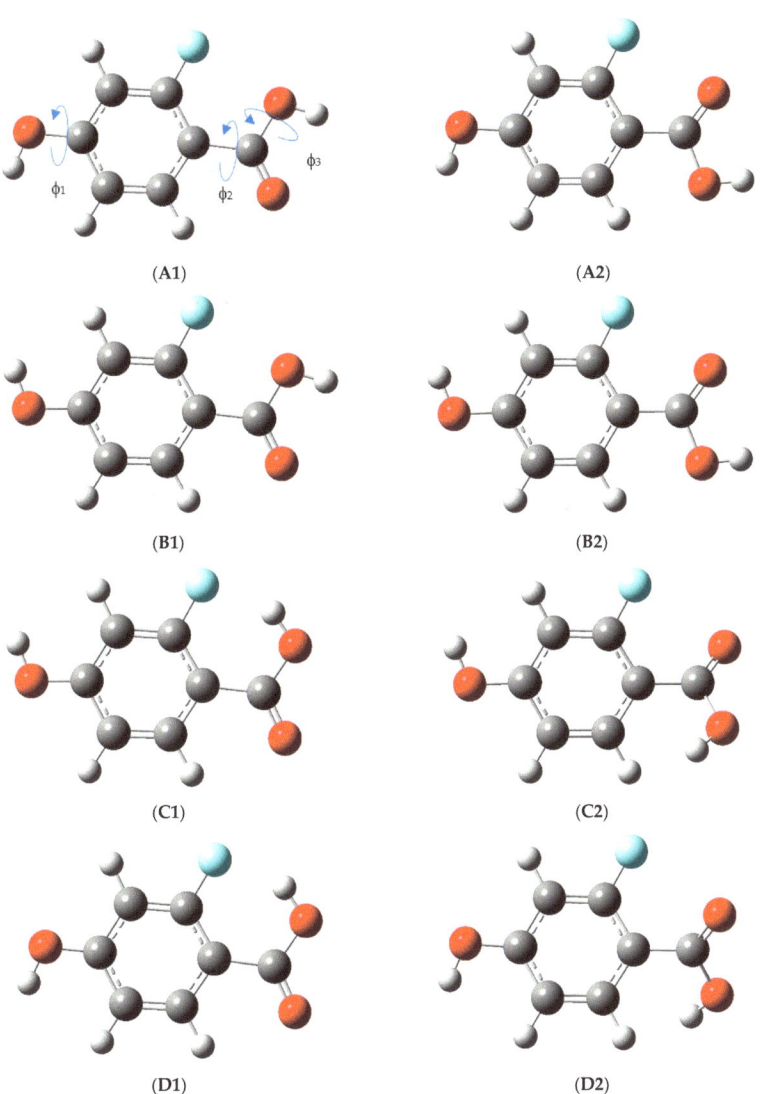

**Scheme 1.** Structures of the eight conformers of 2-F-4-OH benzoic acid, namely **A1, A2, B1, B2, C1, C2, D1,** and **D2** as denoted by the labels in bold and in parentheses below the structures. The three torsional angles $\varphi_1$, $\varphi_2$ and $\varphi_3$ are defined in structure **A1**. In case of **C2** and **D2**, only one of the two enantiomeric forms are depicted.

**Table 1.** Torsional angles (in degrees) and anharmonic zero-point energy-corrected relative energy values (kJ mol$^{-1}$) of the 2-F-4-OH-benzoic acid conformers.

|      | Torsional Angles [a]/° | | | $\Delta E_{ZPE,anharm}$ [b] |
|------|--------|--------|--------|--------|
|      | $\phi_1$ | $\phi_2$ | $\phi_3$ | |
| A1   | 180.0  | −180.0 | 180.0  | 0.0  |
| A2   | −180.0 | 0.0    | −180.0 | 3.4  |
| B1   | 0.0    | 180.0  | 180.0  | 0.8  |
| B2   | 0.0    | 0.0    | 180.0  | 3.5  |
| C1   | 0.0    | 180.0  | 0.0    | 9.1  |
| C2   | 0.8    | −33.6  | −11.2  | 30.0 |
| C2*  | −0.8   | 33.6   | 11.2   | 30.0 |
| D1   | 180.0  | 180.0  | 0.0    | 7.4  |
| D2   | 179.8  | −34.6  | −11.7  | 30.2 |
| D2*  | −179.8 | 34.6   | 11.7   | 30.4 |

[a] The torsional angles are defined in Scheme 1; [b] with regard to the minimum energy of −595.396627 hartree particle$^{-1}$.

The relative energies of the TSs, along with their optimized geometries, are summarized in Tables S21–S49 in the Supplementary Material. Based on the values listed there, the rotation of the 4-OH group along the $\varphi_1$ torsional angle has a barrier of approximately 20 kJ mol$^{-1}$ (A1↔B1, A2↔B2, C1↔D1, C2↔D2), and the same holds true for the rotation along $\varphi_2$ in the case of A1↔A2 and B1↔B2, but not for C1↔C2 and D1↔D2 where the barriers of the 'rightward' direction (C1→C2 and D1→D2, ≈30 kJ mol$^{-1}$) are much higher than those in the reverse one (C1←C2, D1←D2, <10 kJ mol$^{-1}$) due to the energy difference of the minima A1, B1 and A2, B2, respectively. Unsurprisingly, the barrier of the rotation around $\varphi_3$ is the highest one with values lying between 40–50 kJ mol$^{-1}$ (A1↔D1, B1↔C1), except if one of the conformers is much more stable than the other one. In these cases (A2↔D2, B2↔C2), the barrier heights are about 50 and 20 kJ mol$^{-1}$, respectively, depending on the direction of the transition. It should be noted that all the rotations around the three torsional angles may go either clockwise or counterclockwise, but the TSs produced by the two different rotational directions are mirror images of each other and, therefore, have identical energy and vibrational spectra. It is also worth mentioning that the TSs between the enantiomers C2↔C2* and D2↔D2* have a planar structure, and their energy is only slightly above the minima (≈5 kJ mol$^{-1}$, approximately 400 in cm$^{-1}$), which allows for their rapid interconversion upon exposure to the spectrometer beam source.

*3.2. Changes upon Near-IR Irradiation, Vibrational Analysis*

Some selected regions of the near- and mid-IR spectra of the sample molecule deposited in an Ar matrix are visualized in Figures 1a, 2a, 3a and 4a. Furthermore, by having a look at the mid-IR spectra in Figures 1b, 2c, 3c and 4c, one can see that irradiating some 2$\nu$(OH) stretching overtone bands results in a selective conformational switch. The spectral features can be divided into three different groups based on their general behavior upon irradiation: (a) the ones that increase, (b) the ones that decrease, and (c) the ones that do not change, which should belong to three different conformers or groups of conformers. One should also bear in mind that the enantiomers C2, C2* and D2, D2* cannot be differentiated by their IR spectrum. Furthermore, based on theoretical results, the conformers that differ only in the orientation of their 4-OH groups have almost identical vibrational frequencies; thus, they cannot be distinguished either (Tables S11–S20). This leads us to the conclusion that only four band groups should be detected in the matrix belonging to the conformer pairs (1) **A1/B1**, (2) **A2/B2**, (3) **C1/D1**, and (4) **C2/D2** (or **C2\*/D2\***). However, as mentioned above, only three groups of bands could be distinguished based on their response to laser irradiation. As such, the conformer pairs with the highest energy (i.e., group (4)) could be excluded based on the following considerations: First, they are 30 kJ mol$^{-1}$ above the most stable members of group (1), meaning that their Boltzmann population at the deposition temperature is expected to be 0.003%. For comparison, the values for the other groups are 72% for group (1), 23% for group (2), and 5% for group (3). It is important to note that the abundance of conformer groups in the freshly deposited matrix is very similar to these theoretically obtained ones and can be determined to be 61% for group (1), 33% for group (2), and 6% for group (3). Consequently, group (4) is not expected in the freshly deposited matrix. Secondly, no new bands arise during any of the near-IR excitation (only the relative intensities of the bands that are already present change), suggesting that members of this group are not generated upon irradiation. Furthermore, it can be assumed that only the most stable members of each of the three groups (i.e., **A1**, **A2**, and **D1**) are expected to be present in the matrix, since the higher energy members can be converted to them via tunneling. This hypothesis will be justified in Section 3.5 where an estimate on the tunneling rates are given.

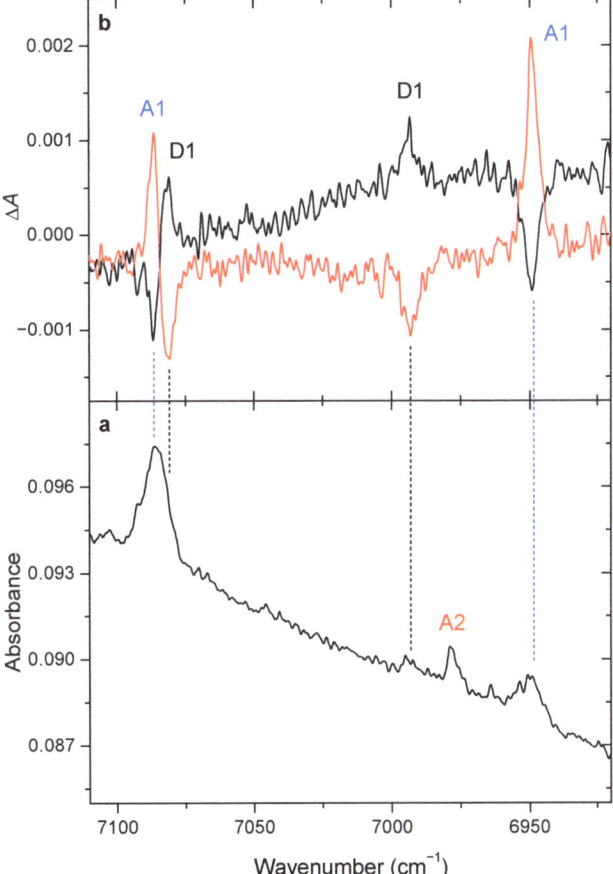

**Figure 1.** (a) Near-IR spectrum of the 2-F-4-OH benzoic acid in the 7110–6920 cm$^{-1}$ range after deposition; (b) difference spectrum after laser irradiating the matrix at 6952.0 cm$^{-1}$ (black trace) and at 6994.9 cm$^{-1}$ (red), respectively.

The bands belonging to **A1** all decrease when its 2$\nu$(OH) stretching overtone bands, found at 6952.0 cm$^{-1}$ and 7077.1 cm$^{-1}$, are excited. This occurs with the simultaneous increase in the bands belonging to **D1**, i.e., the **A1**→**D1** conversion is induced. Similarly, when the matrix is irradiated at 6994.9 cm$^{-1}$ and at 7093.2 cm$^{-1}$, the opposite can be observed, i.e., the **D1**→**A1** conversion; therefore these bands must belong to the 2$\nu$(OH) stretching overtones of the former conformer. It is important to note the findings described above show that no matter which OH group is actually excited (i.e., the 4-OH or the carboxylic one), all near-IR irradiation results in the same rotamerization process, which is the *cis–trans* rotamerization of the carboxylic moiety. As a consequence, the irradiation of the carboxylic OH group (6952.0 cm$^{-1}$ for **A1** and 6994.9 cm$^{-1}$ for **D1**) will be called 'local' excitation, whereas that of the 4-OH group (7077.1 cm$^{-1}$ for **A1** and 7093.2 cm$^{-1}$ for **D1**) will be called 'remote' excitation hereafter.

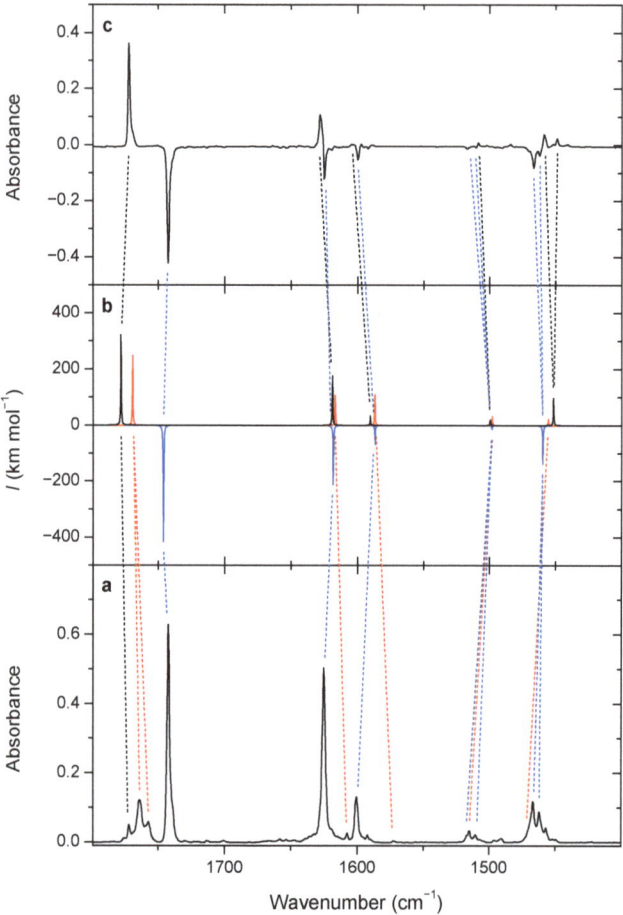

**Figure 2.** (a) Mid-IR spectrum of the 2-F-4-OH benzoic acid in the 1800–1400 cm$^{-1}$ range after deposition; (b) simulated spectrum of conformers **A1** (blue negative trace), **A2** (red), and **D1** (black) based on the anharmonic computations; (c) difference spectrum obtained by subtracting the spectrum taken after deposition from the one collected after 'locally' exciting **A1** at 6952.0 cm$^{-1}$.

In contrast to the results discussed above, the bands of **A2** are not affected by any of the irradiations, even if an excitation wavelength of 6978.8 cm$^{-1}$ is used, which is the frequency of its local carboxylic 2$\nu$(OH) stretching overtone. Accordingly, Table 2 sums up the general response of these three conformers to the excitation of various O–H stretching overtone bands. The different response to near-IR laser irradiation, accompanied by a comparison of the experimental and theoretical spectra, allows for assignment of the vibrational bands; Tables 3–5 lists their spectral assignment for part of the mid-IR spectral range.

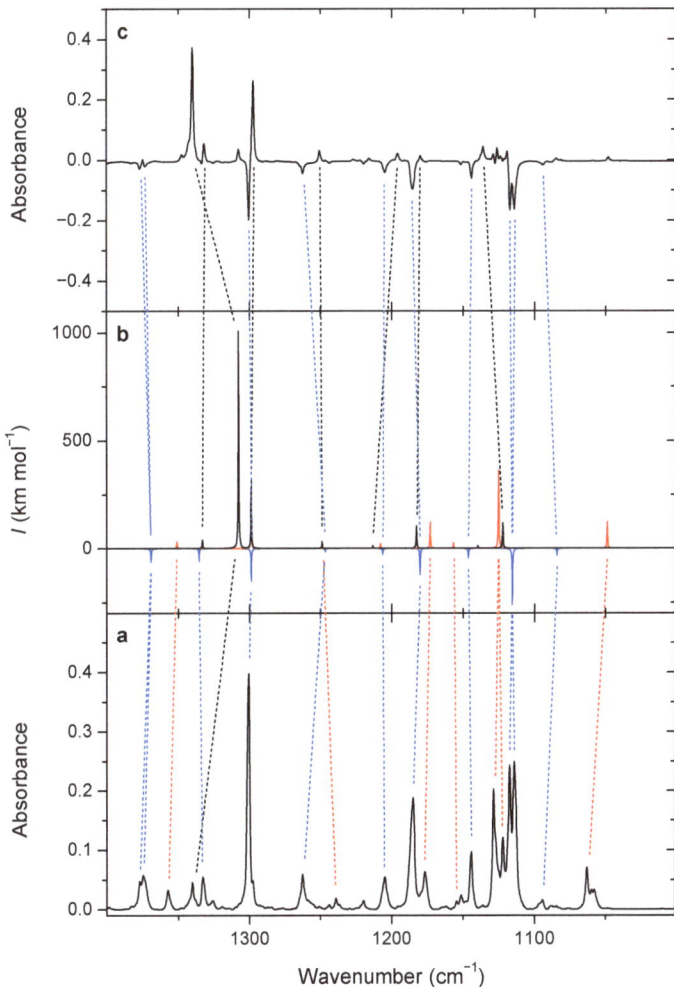

**Figure 3.** (a) Mid-IR spectrum of the 2-F-4-OH benzoic acid in the 1400–1000 cm$^{-1}$ range after deposition; (b) simulated spectrum of conformers **A1** (blue negative trace), **A2** (red), and **D1** (black) based on the anharmonic computations; (c) difference spectrum obtained by subtracting the spectrum taken after deposition from the one collected after 'locally' exciting **A1** at 6952.0 cm$^{-1}$.

**Table 2.** General behavior of the bands of conformers **A1**, **A2**, and **D1** upon near-IR laser excitation of their O–H stretching overtone peaks.

|  | Excitation Wavelength (cm$^{-1}$) [a] | | | |
| --- | --- | --- | --- | --- |
|  | 6952.0 ('Local') | 6994.9 ('Local') | 7093.2 ('Remote') | 7077.1 ('Remote') |
| **A1** | − | + | − | + |
| **A2** | 0 | 0 | 0 | 0 |
| **D1** | + | − | + | − |

+: increasing signal, −: decreasing signal, 0: no change; [a] 'local': excitation of the carboxylic OH, 'remote': excitation of the 4-OH group.

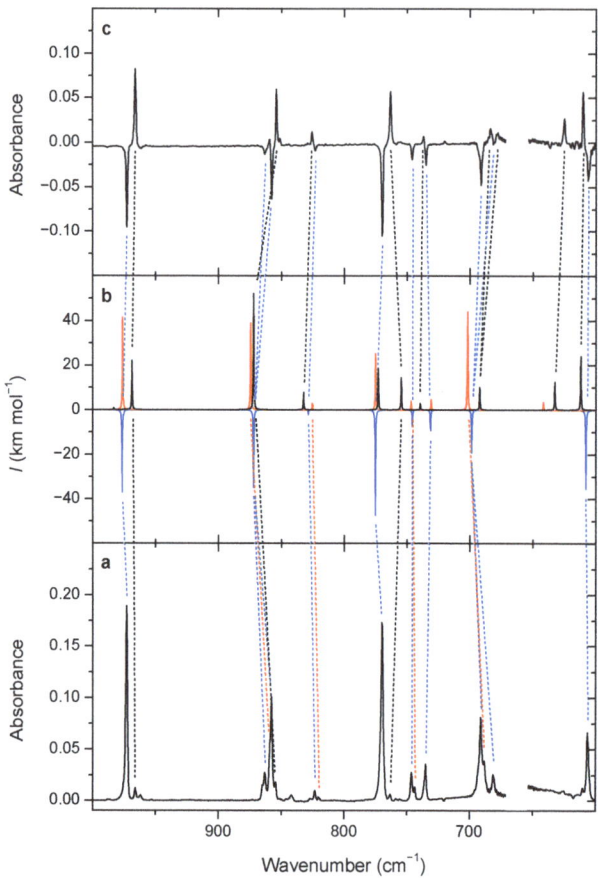

**Figure 4.** (a) Mid-IR spectrum of the 2-F-4-OH benzoic acid in the 1000–600 cm$^{-1}$ range after deposition; (b) simulated spectrum of conformers **A1** (blue negative trace), **A2** (red), and **D1** (black) based on the anharmonic computations; (c) difference spectrum obtained by subtracting the spectrum taken after deposition from the one collected after 'locally' exciting **A1** at 6952.0 cm$^{-1}$. The atmospheric $CO_2$ band is masked in the spectrum.

**Table 3.** Spectral assignment of conformer **A1** for the mid-IR region of 1800–600 cm$^{-1}$.

| Experimental | | Theoretical | | Mode | Description [b] |
| --- | --- | --- | --- | --- | --- |
| $\nu$ (cm$^{-1}$) | $I_{rel.}$ [a] | $\nu$ (cm$^{-1}$) | $I$ (km mol$^{-1}$) | | |
| 1742.5 | 288 | 1746.2 | 329 | $\nu_6$ | $\nu$(C=O) |
| 1625.1 | 80 | 1618.4 | 170 | $\nu_7$ | $\nu$(CC)$_{arom.}$ |
| 1600.3 | 58 | 1586.9 | 52 | $\nu_8$ | $\nu$(CC)$_{arom.}$ |
| 1517.3, 1510.5 | 6.6 | 1498.2 | 11 | $\nu_9$ | $\nu$(CC)$_{arom.}$, $\beta$(CCH) |
| 1466.5, 1461.8 | 119 | 1459.6 | 113 | $\nu_{10}$ | $\nu$(CC)$_{arom.}$, $\beta$(CCH) |
| 1376.9, 1373.3 | 11 | 1369.2 | 49 | $\nu_{11}$ | $\nu$(C–C), $\beta$(COH)$_{COOH}$ |
| 1333.1 | 25 | 1335.4 | 46 | $\nu_{12}$ | $\nu$(CC)$_{arom.}$ |
| 1301.0 | 148 | 1298.8 | 121 | $\nu_{13}$ | $\nu$(C–O)$_{4-OH}$, $\beta$(CCH) |
| 1263.0 | 29 | 1246.5 | 11 | $\nu_{14}$ | $\beta$(CCH) |

Table 3. Cont.

| Experimental | | Theoretical | | Mode | Description [b] |
|---|---|---|---|---|---|
| $\nu$ (cm$^{-1}$) | $I_{rel.}$ [a] | $\nu$ (cm$^{-1}$) | $I$ (km mol$^{-1}$) | | |
| 1243.9 | 1.9 | | | ? | ? |
| 1219.9 | 5.6 | 1217.7 | 21 | $2\nu_{29}$ | $2\delta$(ring) |
| 1204.7 | 36 | 1206.4 | 22 | $\nu_{15}$ | $\beta$(COH)$_{4-OH}$ |
| 1184.9 | 128 | 1180.1 | 98 | $\nu_{16}$ | $\beta$(COH)$_{COOH}$, $\beta$(CCH) |
| 1151.8 | 4.4 | | | ? | ? |
| 1144.2 | 44 | 1146.3 | 38 | $\nu_{17}$ | $\beta$(CCH) |
| 1117.4, 1114.3 | 224 | 1115.6 | 212 | $\nu_{18}$ | $\beta$(CCH) |
| 1094.6 | 9.7 | 1084.7 | 26 | $\nu_{19}$ | $\nu$(C–O)$_{COOH}$, $\nu$(CF) |
| 972.5 | 65 | 976.6 | 29 | $\nu_{21}$ | $\beta$(CCC)$_{arom.}$ |
| 863.5, 858.0 | 44 | 872.8 | 29 | $\nu_{22}$ | $\gamma$(CCH) |
| 823.5 | 2.8 | 829.1 | 1.8 | $\nu_{23}$ | $\gamma$(CCH) |
| 769.7 | 72 | 775.6 | 37 | $\nu_{24}$ | $\tau$(COOH), $\tau$(ring) |
| 746.1 | 7.7 | 746.1 | 6.2 | $\nu_{25}$ | $\delta$(ring) |
| 735.0 | 9.5 | 731.7 | 7.8 | $\nu_{26}$ | $\beta$(O=C–O) |
| 691.1, 680.9 | 33 | 698.9 | 17 | $\nu_{27}$ | $\tau$(COOH), $\tau$(ring) |
| 606.5 | 22 | 608.4 | 28 | $\nu_{29}$ | $\delta$(ring) |

[a] Normalized experimental band areas obtained by multiplying each integrated band area by the sum of the theoretical infrared intensities divided by the sum of the experimental band areas; [b] $\nu$: stretching, $\beta$: in-plane bending, $\gamma$: out-of-plane bending, $\delta$: in-plane deformation, $\tau$: out-of-plane deformation, ?: unassigned peak.

Table 4. Spectral assignment of conformer **D1** for the mid-IR region of 1800–600 cm$^{-1}$.

| Experimental | | Theoretical | | Mode | Description [b] |
|---|---|---|---|---|---|
| $\nu$ (cm$^{-1}$) | $I_{rel.}$ [a] | $\nu$ (cm$^{-1}$) | $I$ (km mol$^{-1}$) | | |
| 1772.0 | 409 | 1778.2 | 257 | $\nu_6$ | $\nu$(C=O) |
| 1628.2sh | 135 | 1619.1 | 140 | $\nu_7$ | $\nu$(CC)$_{arom.}$ |
| 1601.2sh,b | 15 | 1590.5 | 27 | $\nu_8$ | $\nu$(CC)$_{arom.}$ |
| 1508.7, 1484.2 | 20 | 1499.6 | 16 | $\nu_9$ | $\beta$(CCH), $\nu$(CC)$_{arom.}$ |
| 1458.7, 1452.1, 1449.1, 1444.1 | 95 | 1451.6 | 81 | $\nu_{10}$ | $\nu$(CC)$_{arom.}$, $\beta$(CCH) |
| 1347.8, 1340.1 | 465 | 1307.8 | 790 | $\nu_{12}$ | $\beta$(COH)$_{COOH}$ |
| 1331.8 | 44 | 1333.1 | 34 | $\nu_{11}$ | $\nu$(CC)$_{arom.}$, $\beta$(COH) |
| 1308.1 | 43 | | | ? | |
| 1297.8 | 219 | 1298.8 | 247 | $\nu_{13}$ | $\beta$(CCH), $\nu$(C–O)$_{4-OH}$ |
| 1251.0 | 28 | 1248.8 | 26 | $\nu_{14}$ | $\beta$(CCH) |
| 1226.9 | 8.3 | | | ? | |
| 1215.9 | 10 | | | ? | |
| 1196.1 | 31 | 1213.2 | 9.5 | $\nu_{15}$ | $\beta$(CCH), $\beta$(COH)$_{COOH}$ |
| 1179.8 | 15 | 1182.4 | 85 | $\nu_{16}$ | $\beta$(COH)$_{4-OH}$ |
| 1136.2 | 64 | 1122.1 | 97 | $\nu_{18}$ | $\beta$(CCH) |
| 1126.2 | 36 | 1139.6 | 10 | $\nu_{17}$ | $\beta$(CCH) |
| 1085.0, 1081.9 | 15 | 1060.9 | 0.012 | $\nu_{19}$ | $\nu$(C–O)$_{COOH}$, $\nu$(CF) |
| 1048.3 | 14 | | | ? | ? |
| 965.6 | 70 | 968.3 | 18 | $\nu_{21}$ | $\beta$(CCC)$_{arom.}$ |
| 854.4, 851.5 | 45 | 872.4 | 41 | $\nu_{22}$ | $\gamma$(CCH) |
| 825.9 | 10 | 832.5 | 6.3 | $\nu_{23}$ | $\gamma$(CCH) |
| 763.2 | 51 | 754.4 | 11 | $\nu_{25}$ | $\beta$(O=C–O) |

Table 4. Cont.

| Experimental | | Theoretical | | Mode | Description [b] |
|---|---|---|---|---|---|
| $\nu$ (cm$^{-1}$) | $I_{rel.}$ [a] | $\nu$ (cm$^{-1}$) | $I$ (km mol$^{-1}$) | | |
| 736.8 | 5.5 | 739.4 | 2.3 | $\nu_{26}$ | $\delta$(ring) |
| 683.5, 678.1 | 25 | 692.1 | 8.6 | $\nu_{27}$ | $\tau$(COOH), $\tau$(ring) |
| 625.1 | 24 | 632.5 | 10 | $\nu_{28}$ | $\tau$(COOH), $\tau$(ring) |
| 610.5 | 36 | 612.1 | 20 | $\nu_{29}$ | $\delta$(ring) |

[a] Normalized experimental band areas obtained by multiplying each integrated band area by the sum of the theoretical infrared intensities divided by the sum of the experimental band areas; [b] $\nu$: stretching, $\beta$: in-plane bending, $\gamma$: out-of-plane bending, $\delta$: in-plane deformation, $\tau$: out-of-plane deformation, ?: unassigned peak.

Table 5. Spectral assignment of conformer **A2** for the mid-IR region of 1800–600 cm$^{-1}$.

| Experimental | | Theoretical | | Mode | Description [b] |
|---|---|---|---|---|---|
| $\nu$ (cm$^{-1}$) | $I_{rel.}$ [a] | $\nu$ (cm$^{-1}$) | $I$ (km mol$^{-1}$) | | |
| 1763.8, 1757.3 | 378 | 1769.2 | 196 | $\nu_6$ | $\nu$(C=O) |
| 1607.9 | 4.8 | 1616.9 | 89 | $\nu_7$ | $\nu$(CC)$_{arom.}$ |
| 1572.7 | 2.2 | 1587.0 | 98 | $\nu_8$ | $\nu$(CC)$_{arom.}$ |
| 1515.1 | 13 | 1498.0 | 27 | $\nu_9$ | $\beta$(CCH), $\nu$(CC)$_{arom.}$ |
| 1468.4sh | 11 | 1455.3 | 18 | $\nu_{10}$ | $\nu$(CC)$_{arom.}$, $\beta$(CCH) |
| 1357.5 | 16 | 1351.2 | 26 | $\nu_{11}$ | $\nu$(CC)$_{arom.}$, $\beta$(COH)$_{COOH}$ |
| 1301.2sh | 53 | 1298.8 | 162 | $\nu_{13}$ | $\nu$(C–O)$_{4-OH}$, $\beta$(COH)$_{COOH}$ |
| 1239.0 | 8.3 | 1245.5 | 57 | $\nu_{27}+\nu_{31}$ | $\tau$(ring, COOH) + $\gamma$(CCH)$_{COOH}$ |
| 1177.0 | 31 | 1173.0 | 97 | $\nu_{16}$ | $\beta$(COH)$_{COOH}$, $\nu$(C–C) |
| 1154.5 | 1.9 | 1157.0 | 20 | $\nu_{17}$ | $\beta$(CCH) |
| 1128.6, 1122.0 | 123 | 1125.1 | 327 | $\nu_{18}$ | $\beta$(CCH) |
| 1097.2 | 2.8 | | | ? | ? |
| 1063.0, 1059.6, 1057.7 | 58 | 1048.8 | 96 | $\nu_{19}$ | $\nu$(C–O)$_{COOH}$ |
| 859.4sh | 14 | 875.1 | 33 | $\nu_{22}$ | $\gamma$(CCH), $\tau$(ring) |
| 820.3 | 1.1 | 825.9 | 2.5 | $\nu_{23}$ | $\gamma$(CCH), $\tau$(COOH) |
| 770.8sh | 26 | 775.5 | 22 | $\nu_{24}$ | $\tau$(COOH), $\gamma$(CCH) |
| 743.5 | 3.0 | 747.1 | 3.2 | $\nu_{25}$ | $\delta$(ring) |
| 689.1 | 10 | 702.2 | 35 | $\nu_{27}$ | $\tau$(ring), $\tau$(COOH) |

[a] Normalized experimental band areas obtained by multiplying each integrated band area by the sum of the theoretical infrared intensities divided by the sum of the experimental band areas; [b] $\nu$: stretching, $\beta$: in-plane bending, $\gamma$: out-of-plane bending, $\delta$: in-plane deformation, $\tau$: out-of-plane deformation, ?: unassigned peak.

### 3.3. Kinetics of the Near-IR Induced Rotamerization, Quantum Efficiencies

The temporal evolution of the spectrum can be monitored throughout the near-IR irradiation, which provides us with a quantitative means of obtaining the conversion rate. The kinetic curves are plotted in Figure 5 for the 1742.5 and 1628.2 cm$^{-1}$ bands for conformers **A1** and **D1**, respectively. A single exponential function was used to fit the decay profiles:

$$A_t(\mathbf{X}, \tilde{v}) = A_{t=0}(\mathbf{X}, \tilde{v})e^{-kt} + A_{t=\infty}(\mathbf{X}, \tilde{v}) \quad (1)$$

where $A_t(\mathbf{X}, \tilde{v})$ is the integrated area (in cm$^{-1}$) of the band of conformer **X** with a vibrational frequency of $\tilde{v}$ (**X** = **A1** or **D1**; $\tilde{v}$ = 1742.5 cm$^{-1}$ if **X** = **A1** and $\tilde{v}$ = 1628.2 cm$^{-1}$ if **X** = **D1**), $t$ is time passed since the beginning of the irradiation (in sec), and $k$ is the pseudo-first order rate constant (in s$^{-1}$).

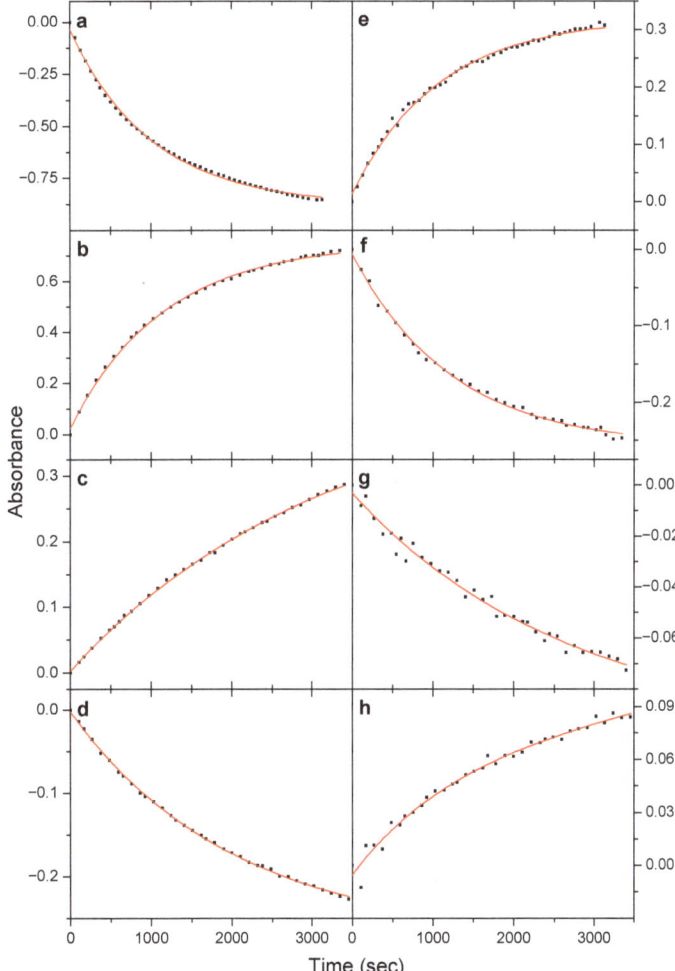

**Figure 5.** Kinetic curves of the 1742.5 cm$^{-1}$ (**a–d**) and 1628.2 cm$^{-1}$ (**e–h**) bands upon near-IR irradiations of 6952.0 cm$^{-1}$ (**a,e**), 6994.9 cm$^{-1}$ (**b,f**), 7093.2 cm$^{-1}$ (**c,g**), and 7077.1 cm$^{-1}$ (**d,h**).

The following equation could be used to fit the growth profiles:

$$A_t(\mathbf{X}, \tilde{v}) = A_{t=\infty}(\mathbf{X}, \tilde{v})(1-e^{-kt}) \quad (2)$$

Moreover, the quantum efficiencies can be estimated using the formula given below [3,6,41]

$$\phi(i) = k(\tilde{v})/\sigma^i(\tilde{v})I(\tilde{v}) \tag{3}$$

where $\phi(i)$ is the quantum yield of the process when exciting the $i$th vibrational mode at the overtone wavenumber $\tilde{v}$; $k(\tilde{v})$ is the rotamerization rate in s$^{-1}$, whereas $\sigma^i(\tilde{v})$ is the absorption cross-section of that particular vibrational mode in cm$^2$; and $I(\tilde{v})$ is the average photon intensity of the laser beam at that wavelength in cm$^2$ s$^{-1}$. We can obtain $\sigma^i(\tilde{v})$ in cm$^2$ when the absorbance of the excited overtone band ($A$, dimensionless value) is divided by the calculated column density of the decaying conformer at the beginning of irradiation ($N$, in cm$^{-2}$). Furthermore, the latter can be deduced from the integrated peak area of one absorption band of the conformer ($A_{int}$, in cm) and by knowing the absorption coefficient ($\alpha$, in cm) and the area of irradiation ($S$ = 1 cm$^2$), respectively:

$$N = \ln(10)\, A_{int}/\alpha S \tag{4}$$

Even though $\alpha$ is not known for our system, a rough estimate can be given based on the computed anharmonic IR intensities (Tables 3 and 4). In order to determine $N$, the strong and well-resolved $\nu$(C=O) stretching vibrational modes were used. Moreover, $I(\tilde{v})$ can be obtained if the output power of the produced near-IR laser light ($P$, in W) is divided by the photon energy ($E_{photon}$, in J) times the surface area $S$. It should be noted that $P$ was measured without the KBr window in the beam path; thus, $P$ and $I(\tilde{v})$ both represent an upper estimate, which also means that the quantum efficiencies deduced here are lower estimates. Table 6 lists all the derived values.

**Table 6.** Rotamerization rates ($k_X$, in sec$^{-1}$, where **X** = **A1** or **D1**, respectively) of the conformers upon near-IR irradiation. Positive values show growth, whereas negative ones indicate decay. Computed quantum yields for the rotamerization of the carboxylic OH group when exciting the OH stretching overtones.

|  | Near-IR Irradiation (cm$^{-1}$) [a] | | | |
| --- | --- | --- | --- | --- |
|  | 6952.0 (Local) | 6994.9 (Local) | 7093.2 (Remote) | 7077.1 (Remote) |
| $k_{A1}$ | $-(9.9 \pm 0.2) \times 10^{-4}$ | $(8.6 \pm 0.2) \times 10^{-4}$ | $-(5.0 \pm 0.1) \times 10^{-4}$ | $(3.2 \pm 0.1) \times 10^{-4}$ |
| $k_{D1}$ | $(9.3 \pm 0.3) \times 10^{-4}$ | $-(8.1 \pm 0.1) \times 10^{-4}$ | $(5.2 \pm 0.5) \times 10^{-4}$ | $-(3.8 \pm 0.5) \times 10^{-4}$ |
| $A_{int}$ (cm) | 1.896 | 1.039 | 1.782 | 1.126 |
| $\alpha$ (cm) | $5.5 \times 10^{-17}$ | $4.3 \times 10^{-17}$ | $5.5 \times 10^{-17}$ | $4.3 \times 10^{-17}$ |
| $N$ (cm$^{-2}$) | $8.0 \times 10^{16}$ | $5.6 \times 10^{16}$ | $7.5 \times 10^{16}$ | $6.0 \times 10^{16}$ |
| $A$ | 0.0020 | 0.0008 | 0.0012 | 0.0006 |
| $\sigma^i(\tilde{v})$ (cm$^2$) | $2.5 \times 10^{-20}$ | $1.4 \times 10^{-20}$ | $1.6 \times 10^{-20}$ | $1.0 \times 10^{-20}$ |
| $P$ (W) | $0.044 \pm 0.008$ | $0.045 \pm 0.001$ | $0.046 \pm 0.001$ | $0.050 \pm 0.001$ |
| $E_{photon}$ (J) | $1.4 \times 10^{-19}$ | $1.4 \times 10^{-19}$ | $1.4 \times 10^{-19}$ | $1.4 \times 10^{-19}$ |
| $I(\tilde{v})$ (cm$^{-2}$ s$^{-1}$) | $(3.1 \pm 0.6) \times 10^{17}$ | $(3.2 \pm 0.1) \times 10^{17}$ | $(3.3 \pm 0.1) \times 10^{17}$ | $(3.6 \pm 0.1) \times 10^{17}$ |
| $\phi(i)$ (E3) | $1.2 \times 10^{-1}$ | $1.9 \times 10^{-1}$ | $9.7 \times 10^{-2}$ | $9.7 \times 10^{-2}$ |
| $N_{iso}$ (s$^{-1}$) | $7.7 \times 10^{13}$ | $4.7 \times 10^{13}$ | $3.8 \times 10^{13}$ | $2.1 \times 10^{13}$ |
| $N_{abs}$ (s$^{-1}$) | $1.4 \times 10^{15}$ | $5.9 \times 10^{14}$ | $9.1 \times 10^{14}$ | $5.0 \times 10^{14}$ |
| $\phi(i)$ (E5) | $5.5 \times 10^{-2}$ | $8.0 \times 10^{-2}$ | $4.2 \times 10^{-1}$ | $4.2 \times 10^{-2}$ |
| $\phi(i)_{average}$ | $(9 \pm 3) \times 10^{-2}$ | $(1.4 \pm 0.5) \times 10^{-1}$ | $(7 \pm 3) \times 10^{-2}$ | $(7 \pm 3) \times 10^{-2}$ |

[a] 'local': excitation of the carboxylic OH, 'remote': excitation of the 4-OH group.

As an alternative approach, $\phi(i)$ can also be estimated as follows:

$$\phi(i) = N_{iso}/N_{abs} \tag{5}$$

where $N_{iso}$ denotes the number of molecules converted, and $N_{abs}$ is equal to the number of photons absorbed per time unit (both in s$^{-1}$).

$$N_{iso} = k(\tilde{v})NS \tag{6}$$

$$N_{abs} = (1 - 10 - A)I(\tilde{v})S \tag{7}$$

The averaged value $\phi(i)_{average}$ varies between $1.4 \times 10^{-1}$ and $7 \times 10^{-2}$ (Table 6), which agree well with, for instance, those of the formic acid ($1.7 \times 10^{-1}$, $7 \times 10^{-2}$) [3], acetic acid ($2.2 \times 10^{-2}$) [6], and propionic acid ($1.4 \times 10^{-2}$) [41], respectively, but they are significantly higher than those of amino acids, such as glycine ($8 \times 10^{-4}$) [18] and alanine ($5 \times 10^{-4}$ and $1 \times 10^{-3}$), respectively [19]. It is also worth noting that the relative uncertainty of $\phi(i)_{average}$ is found to be around 30–40%, which is also in agreement with previous findings [6,41]. By having a look at the $\phi(i)_{average}$ values belonging to the different excitation frequencies, it can be deduced that those of the 'local' processes do not differ significantly from each other. Furthermore, the $\phi(i)_{average}$ values of the 'remote' processes are comparable to the 'local' ones, which is somewhat unexpected, since it would be straightforward to think that energy dissipation is not negligible when exciting a distant functional group that results in lower efficiencies for the 'remote' excitations. Nevertheless, this does not necessarily hold true because, for instance, thioacetamide, which also shows a remote molecular switching property upon near-IR irradiation, has $\phi(i)_{average}$ values of $3.7 \times 10^{-2}$ to $7.2 \times 10^{-2}$ [42]; these are not significantly lower than the quantum efficiencies listed above. It has to be kept in mind, however, that the IVR process in thioacetamide acts through only four bonds instead of the eight in 2-F-4-OH benzoic acid, thus the current findings might still be surprising. A possible explanation for the minor difference between the quantum efficiencies of the 'local' and 'remote' excitations might be the rapid redistribution of the vibrational energy within the molecule, which has a much higher rate than that of the energy dissipation into the surrounding matrix. In this case, the latter would be the one that determines the overall efficiency, i.e., it would matter less which vibrational mode of the molecule is actually excited.

*3.4. Tunneling Decay Kinetics*

After the irradiation experiments, the matrix was irradiated once more with the 6952.0 cm$^{-1}$ laser light to bleach conformer **A1** and generate as much **D1** as possible. Then, the sample was left in the dark overnight while continuously collecting the mid-IR spectra. Furthermore, it is important to recall that a low-pass filter with a cutoff wavelength of 1830 cm$^{-1}$ was installed between the spectrometer and the sample during the experiment. This was done in order to prevent the processes induced by the broadband IR light originating from the spectrometer beam source, which is known to facilitate unwanted rotamerization processes [43–45]. It should be noted that the photon energy at the filter cutoff wavelength is 21.9 kJ mol$^{-1}$, which is significantly below the **A1**←**D1** rotamerization barrier (36.7 kJ mol$^{-1}$, Table S49). However, this cutoff energy is comparable to, or even higher than, the barrier of processes involving the change of other torsional angles. For instance, the 4-OH group almost rotates freely around, and the rotation of the carboxylic group should also be made possible upon exposure to the IR beam. It is also important to note that the 36.7 kJ mol$^{-1}$ barrier (which equals 3068 cm$^{-1}$) should be easily overcome without the presence of the filter by exciting the vibrational modes above this threshold, such as the O–H stretching vibrations; this finding justifies the use of the filter.

Figure 6 shows the kinetic growth/decay of conformers **A1** and **D1**, respectively, when the already irradiated matrix was left in the dark overnight. To fit the experimental data, the same single exponential functions defined by equations E1 and E2 were used as in Section 3.3. The $k$ values, as expected, are significantly (by roughly 2 orders of magnitude) lower than those obtained for the vibrationally induced processes and have a fair agreement with each other. The half-life of **D1** was found to be around $115{,}000 \pm 4000$ s, which is some 32 h. It is interesting to note, however, that the second-order exponentials could also be fitted on these data providing a somewhat improved fit. The same holds true for the irradiation-induced processes, in most cases. This finding implies that there may be matrix sites with different vibrationally induced rotamerization and tunneling rates; thus, they can be classified as 'slow' and 'fast', which is a phenomenon that has been previously described [10,23,46].

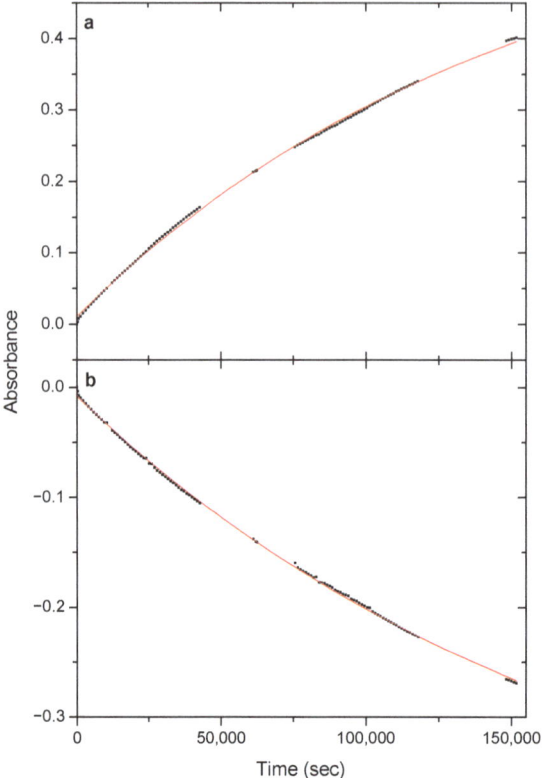

**Figure 6.** Kinetic curves of the (**a**) 1742.5 cm$^{-1}$ (**A1**) and (**b**) 1772.0 cm$^{-1}$ (**D1**) bands after the 6952.0 cm$^{-1}$ near-IR irradiations when the matrix is left in the dark overnight.

*3.5. Estimating the Tunneling Half-Life*

The tunneling rates of the various feasible pathways can be estimated based on computational considerations, which can be used to interpret the experimental findings. The intrinsic reaction paths can be obtained after running IRC computations in Gaussian, which allows for estimation of the width and the height of the barriers that fundamentally determine the tunneling half-lives of the processes. Figures S1–S4 show the computed IRC profiles between the isomers, which also contain the barrier heights and widths used in the calculations. The WKB method mentioned in Section 2.3 was used to predict the tunneling rates using the equation below:

$$P(E) = e^{-\pi^2 w \sqrt{2m(V_0 - E)}/h} \tag{8}$$

where $P(E)$ is the probability of the tunneling process, $m$ stands for the particle mass (1.68 × 10$^{-27}$ kg for H), $V_0$ is the barrier height, $E$ is the particle energy (both in J), $h$ is Planck's constant, and $w$ is the barrier width (in m). The tunneling rate $k$ (in s$^{-1}$) can be obtained by multiplying $P(E)$ by the frequency of attempts ($n$, in s$^{-1}$), which in the latter can be estimated using the anharmonic frequency of the vibrational mode that takes place in the process ($\tilde{v}$, in m$^{-1}$) by means of the formula $\nu = \tilde{v}c$, where c is the speed of light (299,792,458 m s$^{-1}$). The half-life $t_{1/2}$ (in s$^{-1}$) can be derived from $k$ using the equation $t_{1/2} = \ln 2/k$. It is important to note that, due to the inherent uncertainty of the computational results, it is not extraordinary to have a difference of orders of magnitude between the computational and experimental values. What really matters is how the calculated values compare to each other, and this is why the theoretically obtained values can be scaled up with a uniform factor, so that they will be equal to their experimental counterparts [47]. Here, only the **D1→A1** process that can be observed, so the theoretical value was multiplied by a scaling factor of 10,500, and the same was done for all of the other processes as well. Scheme 2 visualizes the tunneling rates of all possible processes, including the rotation of the 4-OH group, the OH group in the COOH moiety, and that of the whole COOH group.

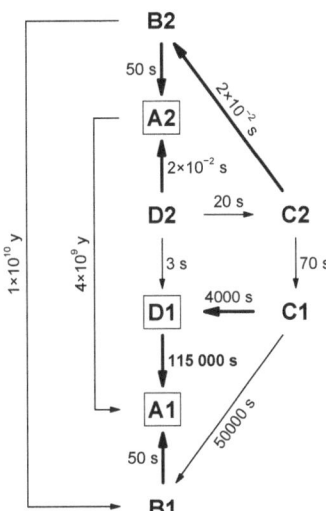

**Scheme 2.** Calculated tunneling half-lives for all conformers. The values are scaled with a uniform factor of 10,500, so that the half-life of the **D1→A1** process is identical to that of the experimentally obtained one. The fastest processes for each conformer are highlighted in bold, whereas the labels of the conformers that can be experimentally observed are outlined.

What can be deduced from Scheme 2 is that all of the higher energy conformers (**C1, C2, D1, D2**), even if they are in the sample in detectable amounts, quickly convert to lower energy ones, such as **A2, B2,** or **D1**. For instance, if there is any **C1** in the matrix at the beginning, most of it should be converted to **D1** by the end of the deposition, and the other processes are even faster than that one. **B2** converts to **A2** in less than a minute, whereas **D1** is shown to slowly transform into **A1**. It is important to note that neither **A2** nor **B2** may convert to their counterparts **A1** or **B1** via tunneling over a reasonable timescale and that the conversion rate of **D2→A2** is some seven orders of magnitude faster than that of **D1→A1**, which may explain why the former process cannot be observed experimentally. Most importantly, the scheme also nicely shows why only three forms may be expected to be present in the matrix after deposition, namely **A1, A2**, and **D1**, which is in accordance with the experimental results.

## 4. Conclusions

In this study, 2-F-4-OH benzoic acid was isolated in Ar matrices at low temperatures while collecting its mid-IR spectra. The molecule has eight different conformers based on the position of three fundamental torsional angles, which are the orientation of the H atom in the 4-OH group and that of the O and H atoms in the –COOH group at the other end of the aromatic ring. The spectral assignment can be made by comparing the experimental spectrum with the theoretically predicted spectra of the conformers. Based on the vibrational analysis, only the three most stable conformers (**A1, A2,** and **D1**) are expected to be present in the matrix after deposition, which are supported by their calculated Boltzmann populations at the deposition temperature. When exciting the $2\nu$(OH) stretching overtones of the conformers, it can be deduced that **A1** converts to **D1** and vice versa, meaning that rotation of the H atom on the COOH group is induced. In contrast to this, no similar effect could be observed for conformer **A2**. The response of **A1** or **D1** depends only marginally on whether the irradiation occurred 'locally' (i.e., by the excitation of the OH of the COOH group) or 'remotely' (by the excitation of the 4-OH moiety). The behavior of the bands upon excitation further confirms the vibrational assignment. The kinetic rates of the rotamerization, as well as the quantum efficiencies of the 'local' and 'remote' excitations, were also determined. According to this, the efficiency is similar in all cases independently on the excitation wavelength. The spontaneous conversion of the higher energy conformer **D1** to the more stable **A1** via tunneling was also examined, and, as such, its rate was determined and found to be roughly two orders of magnitude slower than that of the vibrationally induced one.

This molecule represents a great example, where both the local and remote switching could be studied simultaneously, thus allowing for their direct comparison. Therefore, the molecular system presented here further extends our understanding of intramolecular vibrational energy transfer processes.

**Supplementary Materials:** The following supporting information can be downloaded at: https://www.mdpi.com/article/10.3390/photochem2010009/s1, Tables S1–S20: the optimized geometries as well as harmonic and anharmonic vibrational frequencies and their intensities of **A1**, **A2**, **B1**, **B2**, **C1**, **C2**, **D1**, and **D2** are listed. Tables S21–S48: the optimized geometries of transition states **A1–A2**, **(A1–A2)***, **A1–B1**, **(A1–B1)***, **A1–D1**, **(A1–D1)***, **A2–B2**, **(A2–B2)***, **A2–D2**, **(A2–D2)***, **B1–B2**, **(B1–B2)***, **B1–C1**, **(B1–C1)***, **B2–C2**, **(B2–C2)***, **C1–C2**, **(C1–C2)***, **C1–D1**, **(C1–D1)***, **C2–C2***, **(C2–D2)$_1$**, **(C2–D2)$_2$**, **(C2*–D2*)$_1$**, **(C2*–D2*)$_2$**, **D1–D2**, **(D1–D2)***, and **D2–D2***. Table S49: torsional angles and harmonic zero-point energy corrected relative energy values of the transition states. Figures S1–S4: IRC profiles of the different rotamerization processes.

**Author Contributions:** Conceptualization, G.T.; methodology, G.T. and S.G.; formal analysis, S.G. and M.B.; investigation, S.G. and M.B.; resources, G.T.; writing–original draft, S.G. and M.B.; writing–review and editing, G.T. and S.G.; visualization, S.G.; supervision, G.T.; project administration, G.T.; funding acquisition, G.T. All authors have read and agreed to the published version of the manuscript.

**Funding:** This reseach was funded by the Lendület program of the Hungarian Academy of Sciences and by the ELTE Institutional Excellence Program, grant number TKP2020-IKA-05, financed by the Hungarian Ministry of Human Capacities.

**Institutional Review Board Statement:** Not applicable.

**Informed Consent Statement:** Not applicable.

**Data Availability Statement:** The data are available from the authors upon request.

**Acknowledgments:** The support of the Lendület program of the Hungarian Academy of Sciences is acknowledged. This work was completed in the ELTE Institutional Excellence Program (TKP2020-IKA-05), financed by the Hungarian Ministry of Human Capacities.

**Conflicts of Interest:** The authors declare no conflict of interest.

## References

1. Pettersson, M.; Lundell, J.; Khriachtchev, L.; Räsänen, M. IR Spectrum of the Other Rotamer of Formic Acid, cis-HCOOH. *J. Am. Chem. Soc.* **1997**, *119*, 11715–11716. [CrossRef]
2. Pettersson, M.; Maçôas, E.M.S.; Khriachtchev, L.; Lundell, J.; Fausto, R.; Räsänen, M. Cis → trans conversion of formic acid by dissipative tunneling in solid rare gases: Influence of environment on the tunneling rate. *J. Chem. Phys.* **2002**, *117*, 9095–9098. [CrossRef]
3. Maçôas, E.M.S.S.; Khriachtchev, L.; Pettersson, M.; Juselius, J.; Fausto, R.; Räsänen, M. Reactive vibrational excitation spectroscopy of formic acid in solid argon: Quantum yield for infrared induced trans → cis isomerization and solid state effects on the vibrational spectrum. *J. Chem. Phys.* **2003**, *119*, 11765–11772. [CrossRef]

4. Maçôas, E.M.S.S.; Khriachtchev, L.; Pettersson, M.; Fausto, R.; Räsänen, M. Rotational Isomerism in Acetic Acid: The First Experimental Observation of the High-Energy Conformer. *J. Am. Chem. Soc.* **2003**, *125*, 16188–16189. [CrossRef] [PubMed]
5. Maçôas, E.M.S.S.; Khriachtchev, L.; Fausto, R.; Räsänen, M. Photochemistry and Vibrational Spectroscopy of the Trans and Cis Conformers of Acetic Acid in Solid Ar. *J. Phys. Chem. A* **2004**, *108*, 3380–3389. [CrossRef]
6. Maçôas, E.M.S.S.; Khriachtchev, L.; Pettersson, M.; Fausto, R.; Räsänen, M. Rotational isomerism of acetic acid isolated in rare-gas matrices: Effect of medium and isotopic substitution on IR-induced isomerization quantum yield and cis → trans tunneling rate. *J. Chem. Phys.* **2004**, *121*, 1331–1338. [CrossRef]
7. Apóstolo, R.F.G.F.G.; Bazsó, G.; Bento, R.R.F.R.F.; Tarczay, G.; Fausto, R. The first experimental observation of the higher-energy trans conformer of trifluoroacetic acid. *J. Mol. Struct.* **2016**, *1125*, 288–295. [CrossRef]
8. Apóstolo, R.F.G.G.; Bazsó, G.; Ogruc-Ildiz, G.; Tarczay, G.; Fausto, R. Near-infrared in situ generation of the higher-energy trans conformer of tribromoacetic acid: Observation of a large-scale matrix-site changing mediated by conformational conversion. *J. Chem. Phys.* **2018**, *148*, 044303. [CrossRef]
9. Maçôas, E.M.S.; Khriachtchev, L.; Pettersson, M.; Fausto, R.; Räsänen, M.; Maças, E.M.S.; Khriachtchev, L.; Pettersson, M.; Fausto, R.; Räsänen, M. Internal rotation in propionic acid: Near-infrared-induced isomerization in solid Argon. *J. Phys. Chem. A* **2005**, *109*, 3617–3625. [CrossRef]
10. Bazsó, G.; Góbi, S.; Tarczay, G. Near-Infrared Radiation Induced Conformational Change and Hydrogen Atom Tunneling of 2-Chloropropionic Acid in Low-Temperature Ar Matrix. *J. Phys. Chem. A* **2012**, *116*, 4823–4832. [CrossRef]
11. Halasa, A.; Lapinski, L.; Reva, I.; Rostkowska, H.; Fausto, R.; Nowak, M.J. Near-Infrared Laser-Induced Generation of Three Rare Conformers of Glycolic Acid. *J. Phys. Chem. A* **2014**, *118*, 5626–5635. [CrossRef] [PubMed]
12. Reva, I.; Nunes, C.M.; Biczysko, M.; Fausto, R. Conformational switching in pyruvic acid isolated in Ar and $N_2$ matrixes: Spectroscopic analysis, anharmonic simulation, and tunneling. *J. Phys. Chem. A* **2015**, *119*, 2614–2627. [CrossRef] [PubMed]
13. Halasa, A.; Lapinski, L.; Rostkowska, H.; Reva, I.; Nowak, M.J. Tunable Diode Lasers as a Tool for Conformational Control: The Case of Matrix-Isolated Oxamic Acid. *J. Phys. Chem. A* **2015**, *119*, 2203–2210. [CrossRef] [PubMed]
14. Maçôas, E.M.S.S.; Fausto, R.; Pettersson, M.; Khriachtchev, L.; Räsänen, M. Infrared-induced rotamerization of oxalic acid monomer in argon matrix. *J. Phys. Chem. A* **2000**, *104*, 6956–6961. [CrossRef]
15. Maçôas, E.M.S.S.; Fausto, R.; Lundell, J.; Pettersson, M.; Khriachtchev, L.; Räsänen, M. Conformational Analysis and Near-Infrared-Induced Rotamerization of Malonic Acid in an Argon Matrix. *J. Phys. Chem. A* **2000**, *104*, 11725–11732. [CrossRef]
16. Maçôas, E.M.S.S.; Fausto, R.; Lundell, J.; Pettersson, M.; Khriachtchev, L.; Räsänen, M. A matrix isolation spectroscopic and quantum chemical study of fumaric and maleic acid. *J. Phys. Chem. A* **2001**, *105*, 3922–3933. [CrossRef]
17. Bazsó, G.; Magyarfalvi, G.; Tarczay, G. Near-infrared laser induced conformational change and UV laser photolysis of glycine in low-temperature matrices: Observation of a short-lived conformer. *J. Mol. Struct.* **2012**, *1025*, 33–42. [CrossRef]
18. Bazsó, G.; Magyarfalvi, G.; Tarczay, G. Tunneling Lifetime of the ttc/VIp Conformer of Glycine in Low-Temperature Matrices. *J. Phys. Chem. A* **2012**, *116*, 10539–10547. [CrossRef]
19. Bazsó, G.; Najbauer, E.E.; Magyarfalvi, G.; Tarczay, G. Near-Infrared Laser Induced Conformational Change of Alanine in Low-Temperature Matrixes and the Tunneling Lifetime of Its Conformer VI. *J. Phys. Chem. A* **2013**, *117*, 1952–1962. [CrossRef]
20. Nunes, C.M.; Lapinski, L.; Fausto, R.; Reva, I. Near-IR laser generation of a high-energy conformer of L-alanine and the mechanism of its decay in a low-temperature nitrogen matrix. *J. Chem. Phys.* **2013**, *138*, 125101. [CrossRef]
21. Najbauer, E.E.; Bazsó, G.; Góbi, S.; Magyarfalvi, G.; Tarczay, G. Exploring the Conformational Space of Cysteine by Matrix Isolation Spectroscopy Combined with Near-Infrared Laser Induced Conformational Change. *J. Phys. Chem. B* **2014**, *118*, 2093–2103. [CrossRef] [PubMed]
22. Najbauer, E.E.; Bazsó, G.; Apóstolo, R.; Fausto, R.; Biczysko, M.; Barone, V.; Tarczay, G. Identification of Serine Conformers by Matrix-Isolation IR Spectroscopy Aided by Near-Infrared Laser-Induced Conformational Change, 2D Correlation Analysis, and Quantum Mechanical Anharmonic Computations. *J. Phys. Chem. B* **2015**, *119*, 10496–10510. [CrossRef] [PubMed]
23. Lapinski, L.; Reva, I.; Rostkowska, H.; Halasa, A.; Fausto, R.; Nowak, M.J. Conformational transformation in squaric acid induced by near-IR laser light. *J. Phys. Chem. A* **2013**, *117*, 5251–5259. [CrossRef] [PubMed]
24. Halasa, A.; Lapinski, L.; Reva, I.; Rostkowska, H.; Fausto, R.; Nowak, M.J. Three Conformers of 2-Furoic Acid: Structure Changes Induced with Near-IR Laser Light. *J. Phys. Chem. A* **2015**, *119*, 1037–1047. [CrossRef]
25. Kuş, N.; Fausto, R. Effects of the matrix and intramolecular interactions on the stability of the higher-energy conformers of 2-fluorobenzoic acid. *J. Chem. Phys.* **2017**, *146*, 124305. [CrossRef]
26. Halasa, A.; Lapinski, L.; Rostkowska, H.; Nowak, M.J. Intramolecular Vibrational Energy Redistribution in 2-Thiocytosine: SH Rotamerization Induced by Near-IR Selective Excitation of $NH_2$ Stretching Overtone. *J. Phys. Chem. A* **2015**, *119*, 9262–9271. [CrossRef]
27. Lopes Jesus, A.J.; Reva, I.; Araujo-Andrade, C.; Fausto, R. Conformational Switching by Vibrational Excitation of a Remote NH Bond. *J. Am. Chem. Soc.* **2015**, *137*, 14240–14243. [CrossRef]
28. Halasa, A.; Reva, I.; Lapinski, L.; Rostkowska, H.; Fausto, R.; Nowak, M.J. Conformers of Kojic Acid and Their Near-IR-Induced Conversions: Long-Range Intramolecular Vibrational Energy Transfer. *J. Phys. Chem. A* **2016**, *120*, 2647–2656. [CrossRef]
29. Lopes Jesus, A.J.; Nunes, C.; Fausto, R.; Reva, I.; Nunes, M.C.; Fausto, R.; Reva, I.; Nunes, C.; Fausto, R.; Reva, I. Conformational control over an aldehyde fragment by selective vibrational excitation of interchangeable remote antennas. *Chem. Commun.* **2018**, *54*, 4778–4781. [CrossRef]

30. Kovács, B.; Kuş, N.; Tarczay, G.; Fausto, R. Experimental Evidence of Long-Range Intramolecular Vibrational Energy Redistribution through Eight Covalent Bonds: NIR Irradiation Induced Conformational Transformation of E-Glutaconic Acid. *J. Phys. Chem. A* **2017**, *121*, 3392–3400. [CrossRef]
31. Ogruc Ildiz, G.; Fausto, R. Structural Aspects of the Ortho Chloro- and Fluoro-Substituted Benzoic Acids: Implications on Chemical Properties. *Molecules* **2020**, *25*, 4908. [CrossRef] [PubMed]
32. Frisch, M.J.; Trucks, G.W.; Schlegel, H.B.; Scuseria, G.E.; Robb, M.A.; Cheeseman, J.R.; Scalmani, G.; Barone, V.; Mennucci, B.; Petersson, G.A.; et al. *Gaussian 09*; revision D.01; Gaussian, Inc.: Wallingford, CT, USA, 2013.
33. Becke, A.D. Density-functional thermochemistry. III. The role of exact exchange. *J. Chem. Phys.* **1993**, *98*, 5648–5652. [CrossRef]
34. Lee, C.; Yang, W.; Parr, R.G. Development of the Colle-Salvetti correlation-energy formula into a functional of the electron density. *Phys. Rev. B* **1988**, *37*, 785–789. [CrossRef] [PubMed]
35. Vosko, S.H.; Wilk, L.; Nusair, M. Accurate spin-dependent electron liquid correlation energies for local spin density calculations: A critical analysis. *Can. J. Phys.* **1980**, *58*, 1200–1211. [CrossRef]
36. Dunning, T.H. Gaussian basis sets for use in correlated molecular calculations. I. The atoms boron through neon and hydrogen. *J. Chem. Phys.* **1989**, *90*, 1007–1023. [CrossRef]
37. Irikura, K.K. *SYNSPEC*; National Institute of Standards and Technology: Gaithersburg, MD, USA, 1995.
38. Peng, C.; Ayala, P.Y.; Schlegel, H.B.; Frisch, M.J. Using redundant internal coordinates to optimize equilibrium geometries and transition states. *J. Comput. Chem.* **1996**, *17*, 49–56. [CrossRef]
39. Fukui, K. The Path of Chemical Reactions—The IRC Approach. *Acc. Chem. Res.* **1981**, *14*, 363–368. [CrossRef]
40. Borden, W.T. Reactions that involve tunneling by carbon and the role that calculations have played in their study. *Wiley Interdiscip. Rev. Comput. Mol. Sci.* **2016**, *6*, 20–46. [CrossRef]
41. Maçôas, E.M.S.S.; Khriachtchev, L.; Pettersson, M.; Fausto, R.; Räsänen, M.M. Rotational isomerization of small carboxylic acids isolated in argon matrices: Tunnelling and quantum yields for the photoinduced processes. *Phys. Chem. Chem. Phys.* **2005**, *7*, 743–749. [CrossRef]
42. Góbi, S.; Reva, I.; Tarczay, G.; Fausto, R. Amorphous and crystalline thioacetamide ice: Infrared spectra as a probe for temperature and structure. *J. Mol. Struct.* **2020**, *1220*, 128719. [CrossRef]
43. Reva, I.; Nowak, M.J.; Lapinski, L.; Fausto, R. Spontaneous tunneling and near-infrared-induced interconversion between the amino-hydroxy conformers of cytosine. *J. Chem. Phys.* **2012**, *136*, 064511. [CrossRef] [PubMed]
44. Lapinski, L.; Reva, I.; Rostkowska, H.; Fausto, R.; Nowak, M.J. Near-IR-Induced, UV-Induced, and Spontaneous Isomerizations in 5-Methylcytosine and 5-Fluorocytosine. *J. Phys. Chem. B* **2014**, *118*, 2831–2841. [CrossRef] [PubMed]
45. Lopes Jesus, A.J.; Nunes, C.M.; Reva, I.; Pinto, S.M.V.V.; Fausto, R. Effects of Entangled IR Radiation and Tunneling on the Conformational Interconversion of 2-Cyanophenol. *J. Phys. Chem. A* **2019**, *123*, 4396–4405. [CrossRef] [PubMed]
46. Schreiner, P.R.; Wagner, J.P.; Reisenauer, H.P.; Gerbig, D.; Ley, D.; Sarka, J.; Császár, A.G.; Vaughn, A.; Allen, W.D. Domino Tunneling. *J. Am. Chem. Soc.* **2015**, *137*, 7828–7834. [CrossRef] [PubMed]
47. Góbi, S.; Nunes, C.M.; Reva, I.; Tarczay, G.; Fausto, R. S–H rotamerization via tunneling in a thiol form of thioacetamide. *Phys. Chem. Chem. Phys.* **2019**, *21*, 17063–17071. [CrossRef]

Article

# Ar-Matrix Studies of the Photochemical Reaction between $CS_2$ and ClF: Prereactive Complexes and Bond Isomerism of the Photoproducts

Michelle T. Custodio Castro [1], Carlos O. Della Védova [1], Helge Willner [2] and Rosana M. Romano [1,*]

[1] CEQUINOR (UNLP, CCT-CONICET La Plata, Associated with CIC), Departamento de Química, Facultad de Ciencias Exactas, Universidad Nacional de La Plata, Blvd. 120 N° 1465, La Plata 1900, Argentina
[2] Anorganische Chemie, Bergische Universität Wuppertal, Gaußstraße 20, D-42097 Wuppertal, Germany
* Correspondence: romano@quimica.unlp.edu.ar

**Abstract:** In this work, prereactive complexes, reaction products, and conformational preferences derived from the photochemical reaction between $CS_2$ and ClF were analyzed following the codeposition of the reactants trapped in argon matrices at cryogenic temperatures. After codeposition of $CS_2$ and ClF diluted in Ar, the formation of van der Waals complexes is observed. When the mixture is subsequently irradiated by means of broad-band UV-visible light ($225 \leq \lambda \leq 800$ nm), fluorothiocarbonylsulfenyl chloride (FC(S)SCl) and chlorothiocarbonylsulfenyl fluoride (ClC(S)SF) are produced. These species exist as two stable planar anti- and syn-conformers (anti- and syn- of the C=S double bond with respect to the S–Cl or S–F single bond, respectively). For both novel molecules, anti-FC(S)SCl and anti-ClC(S)SF are the lowest-energy computed rotamers. As expected due to the photochemical activity of these species, additional reaction products due to alternative or subsequent photochannels are formed during this process.

**Keywords:** matrix-isolation; photochemistry; carbon disulfide; chloromonofluoride

## 1. Introduction

The preparation and study of properties of new covalent compounds has been and will be a central challenge for chemists of all time. Through this knowledge, chemists and scientists from related branches can design their work with novelty and unconventionality. In this context, the synthesis of small and new covalent compounds, as a linking piece between disciplines such as inorganic and organic chemistry, biology, biochemistry, medicine, physics, materials science, different spectroscopies, and photochemistry, is one of the goals of the present work. An emerging edge to be approached with the systematized information obtained is the one referring to the world of conformations and their equilibria.

The matrix-isolation technique in combination with photochemistry is particularly suitable for the isolation of novel small molecules for which no other alternative synthetic route was found, and also the understanding of the reaction pathways that may lead to more efficient control of the reactions [1–3]. Our research group has been able to isolate and study different families of novel molecules by matrix-isolation photochemistry coupled with FTIR spectroscopy (see, for example, Refs. [4–7] and references cited therein). A prerequisite for a photochemical reaction in matrix conditions to occur is that the reactants are held in the same matrix site. Due to the isolation conditions, the reaction is favored when a prereactive molecular complex is formed between the reactants. Furthermore, the geometry of the prereactive complex often determines the course of the photochemical reaction [7].

In this work, we explored the Ar-matrix photochemical reaction between carbon disulfide and chlorine monofluoride. The codeposition of $CS_2$ with ClF, both diluted in Ar, gives rise to the formation of van der Waals complexes, which are stable species from a thermodynamic point of view. Several molecular complexes between ClF and different Lewis bases were previously studied by a combination of matrix-isolation technique with IR spectroscopy [6,8–16]. A T-shaped structure for 1:1 complexes of ClF with a series of alkynes and alkenes was inferred from the Ar-matrix IR spectra, with the interhalogen molecule interacting with the π electron density of the alkyne or alkene [8]. It was reported that complexes with alkenes produced larger shifts than complexes formed with alkynes, and the wavenumber shifts increased with increasing methyl substitution near the carbon-carbon multiple bond. A redshift of the ClF stretching vibrational mode was also observed for 1:1 molecular complexes of $H_2Se$ and $H_3As$ with the interhalogen molecule isolated in solid Ar, interpreted as an electron density transfer from the Se or As atom to the lowest unoccupied antibonding molecular orbital of ClF [9]. Since the base subunit is donating nonbonding electron density, only slight perturbations were observed in the base modes of the complex. Comparable results were obtained for complexes of ClF with sulfur and nitrogen-containing macrocycles [11].

After irradiation, the formation of the novel FC(S)SCl and ClC(S)SF species were detected by FTIR spectra of the matrices. The relative stabilities of the syn- and anti-rotamers of each of these molecules are discussed, and compared with analogue molecules. Their formation could be related to roaming mechanisms or frustrated dissociation occurring during the photoisomerization process of these penta-atomic species. The identification of the van der Waals complexes and the two conformers of both novel molecules, FC(S)SCl and ClC(S)SF, were assisted by the predictions of DFT calculations.

## 2. Materials and Methods

CAUTION: Handling pure fluorine implies that pertinent precautions should be observed. The reactor and the vacuum lines have to be adequately pretreated with fluorine prior to use. ClF was obtained by the reaction of stoichiometric amounts of $F_2$ (Solvay, Germany) and $Cl_2$ at 250 °C in a Monel vessel. The reaction mixture was subsequently distilled to separate ClF from $Cl_2$, $ClF_3$ and $F_2$. The interhalogen ClF was transferred into a 1 L stainless-steel container at a vacuum line and diluted with Ar in a ClF:Ar = 2:100 proportion. Separately, a sample of $CS_2$ was mixed with argon in a 0.5 L glass container in 1:100 ratio. Both containers were connected via needle valves and stainless-steel capillaries to the spray-on nozzle of the matrix support. About 0.5–1.0 mmol of the gas mixtures were codeposited within 10–20 min on the mirror support at 15 K (a mirror plane of a rhodium-plated copper block). A 150 W high-pressure Xe lamp (Heraeus, Hanau, Germany) in combination with a 225 nm cut-off filter were used for the matrix irradiation. The light was directed through water-cooled quartz lenses onto the matrix for 1 to a maximum of 90 min. The photolysis process was followed by IR spectroscopy. Details of the matrix apparatus are given elsewhere [17]. Matrix IR spectra were recorded on a Bruker IFS 66v/S spectrometer with a resolution of 0.5 cm$^{-1}$ in absorption/reflection mode. The IR spectra were processed by curve-fitting analysis using the OPUS Program and the intensities were determined integrating the areas of the individual peaks.

Quantum chemical calculations were performed using either Gaussian 03 [18] or Gaussian 09 [19] program. Density Functional Theory (B3LYP and B3LYP-D3) method was tried in combination with the 6-311+G(d,p) basis sets. Relaxed two-dimensional scans for the 1:1 $CS_2$:ClF complexes were performed in order to find the energy minima. Geometry optimizations were sought using standard gradient techniques by simultaneous relaxation of all the geometrical parameters. The calculated vibrational properties correspond in all cases to potential energy minima with no imaginary frequencies.

The binding energies of the molecular complexes were calculated using the correction proposed by Nagy et al. [20]. The basis set superposition errors (BSSE) have been calculated by applying the counterpoise procedure developed by Boys and Bernardi [21]. The electronic transitions for the previously optimized structure of the molecular complexes were calculated using the TD-DFT formalisms, with a maximum of 100 states and S = 1 [22,23].

## 3. Results and Discussion

### 3.1. Codeposition of $CS_2$ + ClF in Ar Matrix

The reactants were codeposited simultaneously in the mirror plane cooled to about 15 K. The FTIR spectra of the matrix obtained before irradiation were analyzed and compared with the experimental spectra of the monomers taken in similar conditions. The signal corresponding to the $CS_2$ antisymmetric vibration, $\nu_{as}CS_2$, of the carbon disulfide molecule appears in the spectrum at 1527.9 cm$^{-1}$ (Figure 1). In addition, there are also absorptions at 1524.4 cm$^{-1}$ assigned to the $^{34}$SCS isotopologue and at 1533.5 cm$^{-1}$ originating by the dimer $(CS_2)_2$. This dimer was also formed in the reaction between $CS_2$ and $F_2$ achieved under similar conditions to the present work, where the corresponding band appears at 1533.6 cm$^{-1}$ [4]. Figure 1 also shows a band at 2169.5 cm$^{-1}$, assigned to a combination mode of $CS_2$ ($\nu_{as}CS_2 + \nu_s CS_2$). The ClF interhalogen presents FTIR signals at 767.0 and 759.8 cm$^{-1}$, due to $\nu^{35}$ClF and $\nu^{37}$ClF vibrations, and also mirror bands of lower intensity at 769.9 and 762.7 cm$^{-1}$, attributed to matrix effects, and at 755.5 and 748.7 cm$^{-1}$ due to molecular aggregation (Figure 2) [6].

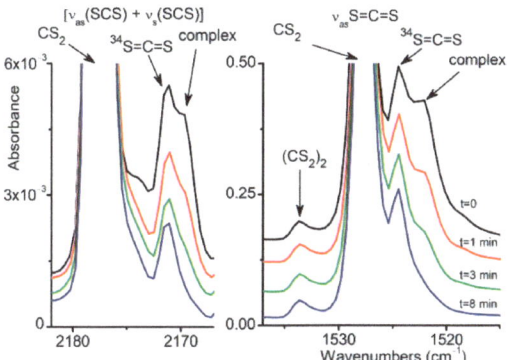

**Figure 1.** FTIR spectra of the $CS_2$/ClF/Ar matrix ($CS_2$:ClF:Ar = 1:2:200) at about 15 K after deposition (**top**, black trace) and, from **top** to **bottom**, after 1 (red trace), 3 (green trace) and 8 min (blue trace) of irradiation with broad-band UV-visible light (225 ≤ λ ≤ 800 nm) in the 2182–2167 and 1537–1515 cm$^{-1}$ regions.

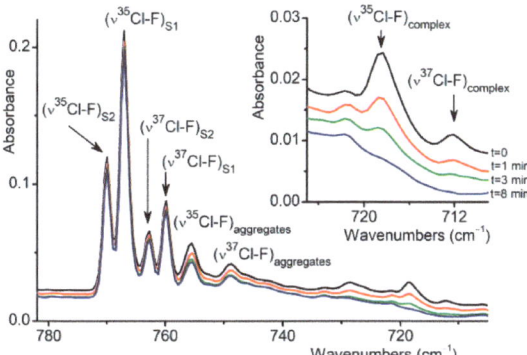

**Figure 2.** FTIR spectra of the $CS_2$/ClF/Ar matrix ($CS_2$:ClF:Ar = 1:2:200) at about 15 K after deposition (**top**, black trace) and, from **top** to **bottom**, after 1 (red trace), 3 (green trace) and 8 min (blue trace) of irradiation with broad-band UV-visible light (225 ≤ λ ≤ 800 nm) in the 782–705 cm$^{-1}$ region.

At this stage of the experiment, that is, when the matrix was not yet irradiated, four new bands were observed at 2169.5, 1522.2, 718.5 and 712.2 cm$^{-1}$ (see Figures 1 and 2).

These bands are not present in the experimental spectrum of the isolated monomers of $CS_2$ and ClF and their subsequent behavior during the photochemical irradiation of the matrix; that is, the tendency to decrease their intensities with the irradiation time allows for them to be assigned to signals originated by the formation of van der Waals complexes between $CS_2$ and ClF. In order to dispose of photochemical kinetic data that allow for relevant information to be provided to the present study, the matrix formed with a 1:2:200 $CS_2$:ClF:Ar concentration was irradiated in the UV-visible range ($\lambda > 225$ nm) for 1, 3, 8, 15, 45, and 90 min. As can be observed in Figures 1 and 2, these new absorptions completely vanish after 8 min of irradiation. Figure 3 shows the variation of the FTIR band intensities assigned to the monomers, $CS_2$ and ClF, and that to the $CS_2$:ClF van der Waals complex as a function of the irradiation time. As can be observed in Figure 3, while the intensities of the IR bands of the monomers show a slight decrease (their intensities after 45 min of irradiation are approximately 80% of the initial values), the IR features assigned to the complex follow a different kinetic behavior, disappearing after 8 min of exposition to broad-band radiation. Figure 3 also reveals that the absorptions attributed to the complex follow the same pattern as a function of irradiation time, a necessary condition to be assigned to the same species.

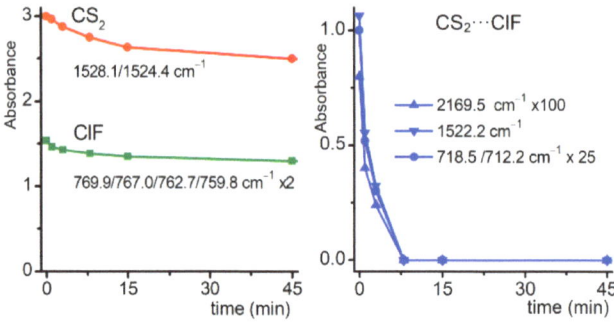

**Figure 3.** Plots of the intensities of the bands in the FTIR spectra of the $CS_2$/ClF/Ar matrix ($CS_2$:ClF:Ar = 1:2:200) at about 15 K vs irradiation times with broad-band UV-visible light ($225 \leq \lambda \leq 800$ nm). **Left**: bands assigned to $CS_2$ and ClF; **Right**: bands assigned to the 1:1 $CS_2$:ClF van der Walls complex.

Computational calculations are especially valuable for reproducing experimental results. The geometry of the stable systems formed by $CS_2$ and ClF were determined searching for the energy minima of the complexes varying the S···Cl and S···F distances in 0.1 Å steps and the C=S···Cl and C=S···F intermolecular angles in 10° steps. Figure 4 shows the contour map of the potential energy surface for the $CS_2$···ClF complex calculated with the B3LYP/6-311G+(d,p) approximation. On the right part of Figure 4, the molecular model corresponding to the optimized structure of the energy minimum calculated with the same theoretical approximation is also represented in Figure 4. Equivalently, Figure 5 shows the molecular complex that links the $CS_2$ molecule with the fluorine atom of the ClF interhalogen. As in many branches of chemistry, fluorine again shows its particularity. When the van der Waals complex is formed using it as a bridge, the geometry of the complex is now linear.

The calculated geometrical parameters of the optimized structures of the two complexes are presented as Supporting Information (Table S1). An intermolecular distance S···Cl of 2.9411 Å, which corresponds to a van der Waals penetration distance of 0.62 Å, and an intermolecular C=S···Cl angle of 97.9° are predicted for the $CS_2$···ClF complex. On the other hand, for the $CS_2$···FCl structure, a value of 3.2531 Å is obtained for the S···F distance, that gives a 0.05 Å van der Waals penetration distance. The predicted intermolecular C=S···F angle for the optimized structure is 180.0°, as shown in Figure 5. According to the B3LYP/6-311G+(d,p) approximation, the $CS_2$···FCl complex is 1.83 kcal/mol (7.66 kJ/mol) higher in energy than the CS$_2$···ClF form. The former structure presents almost the same energy that the isolated monomers (ΔE = E(CS$_2$···FCl)–E(CS$_2$)–E(FCl) = −0.06 kcal/mol), while the latter complex is predicted to be energetically favored with respect to the monomers by 1.89 kcal/mol. The Supporting Information contains detailed data of the calculated energies and the corresponding corrections (Table S2). In agreement with this stability difference, a value of −9.79 and −0.71 kcal/mol were obtained for the most important contribution to the orbital stabilization energies of the CS$_2$···ClF and CS$_2$···FCl complexes, respectively. The charge transfer resulting in the complex formation occurring mainly from the free-electron pair of the sulfur atom to the σ antibonding orbital of the ClF molecule (lpS→σ*$_{ClF}$) acquires a value of −0.0894 e for the CS$_2$···ClF complex. On the other hand, for the CS$_2$···FCl structure, a much lower charge transfer of −0.0013 e was predicted, while the main orbital interactions occur from the ClF unit to the CS$_2$ molecule: σClF→RyS (−0.71 kcal/mol) and lpF→σ*C=S (−0.29 kcal/mol). Figure 6 shows a schematic representation of these two orbital interactions. The geometries of the complexes are consistent with the main contributions to the orbital interactions in each case. In the CS$_2$···ClF adduct, carbon disulfide acts as a donor molecule and chlorine monofluoride as the acceptor unit, favoring the angular geometry. The donor and acceptor roles are inverted in the CS$_2$···FCl complex, determining the collinear geometry.

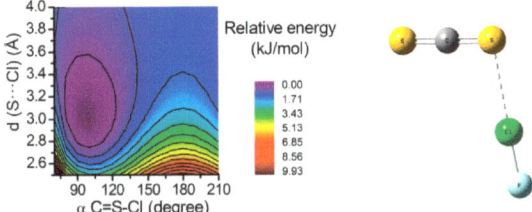

**Figure 4.** Contour map of the potential energy surface of the S=C=S···ClF molecular complex calculated by the variation of the S···Cl intermolecular distance from 2.5 to 4.0 Å in 0.1 Å steps and the C=S···Cl intermolecular angle from 70 to 210° in 10° steps with the B3LYP/6-311G+(d,p) (**left**) and molecular model of the optimized minimum (**right**).

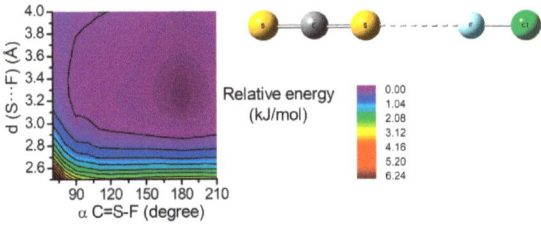

**Figure 5.** Contour map of the potential energy surface of the S=C=S···FCl molecular complex calculated by the variation of the S···Cl intermolecular distance from 2.5 to 4.0 Å in 0.1 Å steps and the C=S···F intermolecular angle from 70 to 210° in 10° steps with the B3LYP/6-311G+(d,p) (**left**) and molecular model of the optimized minimum (**right**).

The vibrational spectra of the complexes were also calculated using the B3LYP/6-311G+(d,p) approximation, and the predicted wavenumbers shifts with respect to the monomers were compared with the experimental findings. A complete list of the calculated frequencies is presented as Supporting Information in Table S3, while Table 1 compiles the vibrational wavenumbers that were observed in the experimental FTIR spectra.

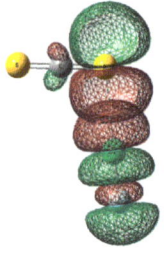

**Figure 6.** Schematic representations of the main contributions to the orbital interactions in the CS$_2$⋯ClF (**left**) and CS$_2$⋯FCl (**right**) complexes calculated with the B3LYP/6-311G+(d,p).

**Table 1.** Selected wavenumbers for the different complexes formed between CS$_2$ and ClF computed with the B3LYP/6-311+G(d,p) approximation (wavenumbers are in cm$^{-1}$ and relative IR intensities are given between parentheses) and comparison with the experimental value.

| Ar Matrix | | B3LYP/6-311+G(d,p) | | | | Tentative Assignment |
| --- | --- | --- | --- | --- | --- | --- |
|  |  | S=C=S⋯Cl−F | | S=C=S⋯F−Cl | |  |
| ν (cm$^{-1}$) | Δν (cm$^{-1}$) [a] | ν (cm$^{-1}$) [b] | Δν (cm$^{-1}$) [a] | ν (cm$^{-1}$) [b] | Δν (cm$^{-1}$) [a] |  |
| 2169.5 | −8.3 |  |  |  |  | ν$_{as}$(SCS) + ν$_s$(SCS) |
| 1522.2 | −5.7 | 1547.1 (100) | −6.2 | 1552.2 (100) | −1.9 | ν$_{as}$(SCS) |
|  |  | 671.8 (<1) | −2.1 | 673.3 (<1) | −0.6 | ν$_s$(SCS) |
| 718.5 | −48.6 | 646.5 (31.8) | −93.3 | 740.3 (3.8) | +0.7 | ν($^{35}$Cl−F) |
| 712.2 | −47.7 | 640.8 (10.2) | −92.2 | 733.2 (1.2) | +0.5 | ν($^{37}$Cl−F) |

[a] Δν = ν$_{complex}$ − ν$_{monomer}$. [b] Relative intensities between parentheses.

The simulated IR spectrum of the CS$_2$⋯ClF complex is the one that best reproduces the experimental results. The calculated ν$_{as}$(S=C=S) is shifted by −6 cm$^{-1}$, in complete agreement with the experimental value (see Figure 1). The direction of the shift in the Cl–F stretching vibrational mode is what allows the structure to be assigned with confidence. Meanwhile, in the complex that interacts through the Cl atom of the diatomic molecule, a shift towards lower wavenumbers of approximately 90 cm$^{-1}$ is predicted, in accordance with the redshift observed in the FTIR spectra depicted in Figure 2; for the CS$_2$⋯FCl adduct, the shift is less than 1 cm$^{-1}$ and in the opposite direction. This difference in the ν(Cl–F) shift is closely related to the main components of the orbital interactions of each of the structures, discussed previously. As mentioned in the Introduction, the interactions between ClF and several N, O, S, Se, and As- containing compounds, as well as with different alkyne or alkene molecules studied by IR matrix-isolation spectroscopy [6,8–13,16], were previously interpreted as an electron density transfer from the corresponding Lewis base to the lowest unoccupied antibonding molecular orbital of ClF, in accordance with the results presented here.

From the comparison between the experimental and calculated IR spectra described above, it can be concluded that the additional bands in the FTIR spectrum taken immediately after deposition with respect to the FTIR spectra of the monomers, and whose intensities decrease until they disappear after 8 min of irradiation, belong to the CS$_2$⋯ClF structure. This is also in agreement with the lower energy predicted for this species. Previous studies on other Lb⋯ClF complexes (Lb is one of several Lewis bases), either by FTIR matrix-isolation spectroscopy, OCS⋯ClF [16] and OCSe⋯ClF [6], or by microwave spectroscopy, H$_2$S⋯ClF [14] and H$_2$O⋯ClF [14], also conclude that the structures interacting through the chlorine atom of the ClF molecules are the lower-energy forms. Regarding the CS$_2$⋯FCl adduct, its presence cannot be completely discarded, since the expected small wavenumber changes would be overlapped to monomer bands. Furthermore, a close inspection of Figure 1 reveals a shoulder of the [ν$_{as}$(SCS) + ν$_s$(SCS)] combination band blueshift by ~2 cm$^{-1}$ that might be tentatively attributed to this complex, taking into account the calculated vibrational data presented in Table 1.

TD-DFT calculations with the B3LYP/6-311+G(d,p) approximation were performed for the optimized CS$_2$···ClF structure, to simulate the electronic transitions of this species in the energy range of the irradiation source used for the photochemical experiments, in order to predict the possibility of photolysis of this species under the experimental conditions. The calculated wavelengths for the one-electron transitions are listed in Table 2, together with the theoretical oscillator strength (f) and a tentative approximate assignment. Only transitions with λ > 200 nm and f ≥ 0.002 are included in the Table. A schematic representation of the molecular orbitals involved in the electronic transitions presented in Table 2 is presented in Figure 7.

**Table 2.** Electronic transitions calculated with the TD-DFT (B3LYP/6-311+G(d,p)) approximation for the CS$_2$···ClF complex [a].

| λ (nm) | Oscillator Strength | Transition | Tentative Approximate Assignment [b] |
|---|---|---|---|
| 374.6 | 0.0002 | HOMO → LUMO | $\pi_z$ (C=S1) → $\pi_z$* (S=C=S) |
| 340.1 | 0.0010 | (HOMO−2) → LUMO | lp$_z$ Cl → $\pi_z$* (S=C=S) |
| 333.9 | 0.0003 | (HOMO−3) → LUMO | $\pi_z$ (C=S2) → $\pi_z$* (S=C=S) |
| 311.7 | 0.1381 | (HOMO−1) → LUMO | $\pi_{i.p.}$ (C=S1) → $\pi_z$* (S=C=S) |
| 271.7 | 0.0002 | HOMO → (LUM0+2) | $\pi_z$ (C=S1) → σ* (Cl–F) |
| 263.7 | 0.2342 | (HOMO−1) → (LUM0+2) | $\pi_{i.p.}$ (C=S1) → σ* (Cl–F) |
| 220.7 | 0.0138 | (HOMO−3) → (LUM0+1) | $\pi_z$ (C=S2) → $\pi_{i.p.}$* (S=C=S) |
| 215.2 | 0.0236 | (HOMO−2) → (LUM0+2) | lp$_z$ Cl → σ*(Cl–F) |

[a] Only transition with oscillator strength ≥ 0.002 are included. [b] S1 and S2 correspond to the interacting and noninteracting sulfur atom, respectively. The z axis is perpendicular to the molecular plane.

### 3.2. Photochemistry of CS$_2$ + ClF in Ar Matrix

The examination of the FTIR spectra taken after irradiation of the matrix allows for the observation that whereas intensities corresponding to the set of bands assigned to the CS$_2$:ClF molecular complex decrease, another group of bands arises, evidencing the formation of species that were not originally present in the codeposited Ar matrix of CS$_2$ and ClF at cryogenic temperatures. Some of the IR absorptions appearing after irradiation that could not be associated with any known compound were assigned to novel pentatomic molecules with XC(S)SY general formula, with X, Y = F, Cl, based on (i) the comparison with previous results on similar systems and (ii) the comparison with the IR spectra predicted by DFT methods for these species. Selected spectral regions of the FTIR spectra of the matrix at different irradiation time are presented in Figure 8. A complementary way to evaluate this information is displayed in Figure 9, which shows the variation of band intensities as a function of irradiation time for the signals assigned to the proposed photoproducts.

The region between 1250 and 950 cm$^{-1}$ in the FTIR spectra of the irradiated matrix is particularly rich in information concerning the formation of the new molecules reported in this work. The absorptions at 1228.9/1226.0 and 1213.5/1208.2 cm$^{-1}$, which grow and then decay on continued irradiation (Figures 8 and 9), were assigned to the ν(C=S) stretching vibration of syn-FC(S)SCl and anti-FC(S)SCl, respectively. The positions of these signals are in very good agreement with the computed wavenumbers at 1220.9 and 1214.0 cm$^{-1}$ using the B3LYP/6-311G+(d,p) level of approximation, as presented in Table 3. According to this computational calculation, the anti-form of FC(S)SCl is the one with the lowest energy. The syn-rotamer is calculated to be 0.76 kcal/mol higher in energy than the anti-form. The IR absorption coefficients values for the ν(C=S) vibrational mode are 383 and 288 km/mol for the syn- and anti-conformer, respectively. Even taking into account the higher absorption coefficient of the syn-rotamer, the experimental abundance of the syn-form is greater than that of the anti-form. Unlike what was observed in other similar photochemical reactions, a process called randomization was not observed in this case [7].

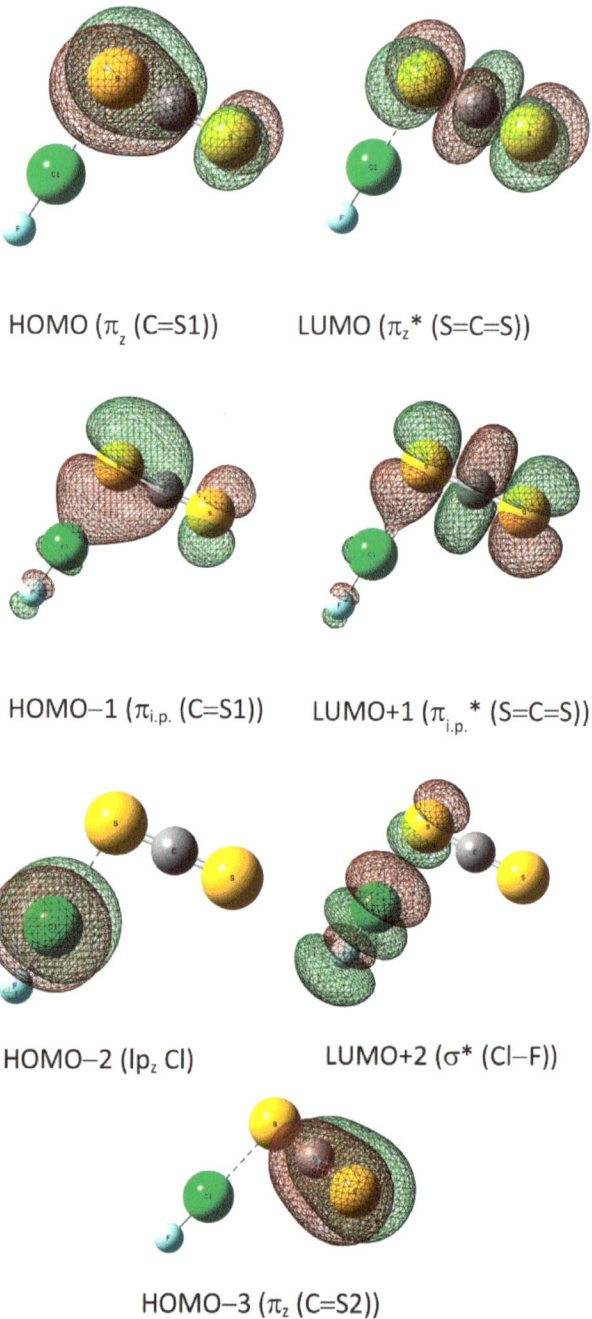

**Figure 7.** Schematic representation and approximate assignment of the molecular orbitals of the $CS_2\cdots ClF$ complex involved in the electronic transitions with $\lambda > 200$ nm and $f \geq 0.002$, calculated with the B3LYP/6-311+G(d,p) approximation.

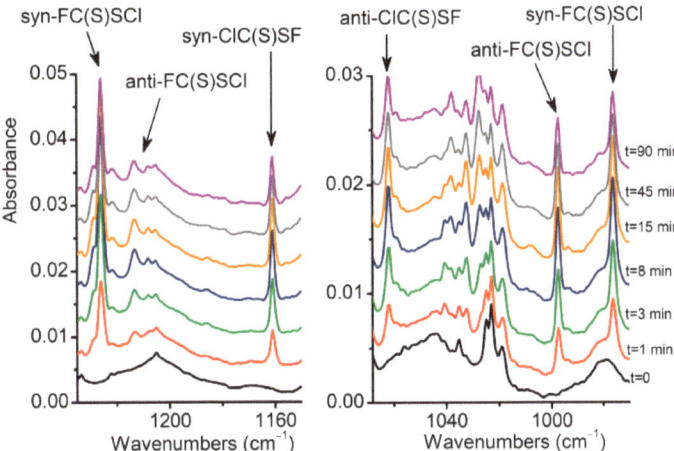

**Figure 8.** FTIR spectra of the $CS_2/ClF/Ar$ matrix ($CS_2$:ClF:Ar = 1:2:200) at about 15 K after deposition (**bottom**, black trace) and from **bottom** to **top** after 1 (red trace), 3 (green trace), 8 min (blue trace), 15 min (orange trace), 45 min (grey trace), and 90 min (purple trace) of irradiation with broad-band UV-visible light ($225 \leq \lambda \leq 800$ nm) in the 1235–1150 and 1068–970 $cm^{-1}$ regions.

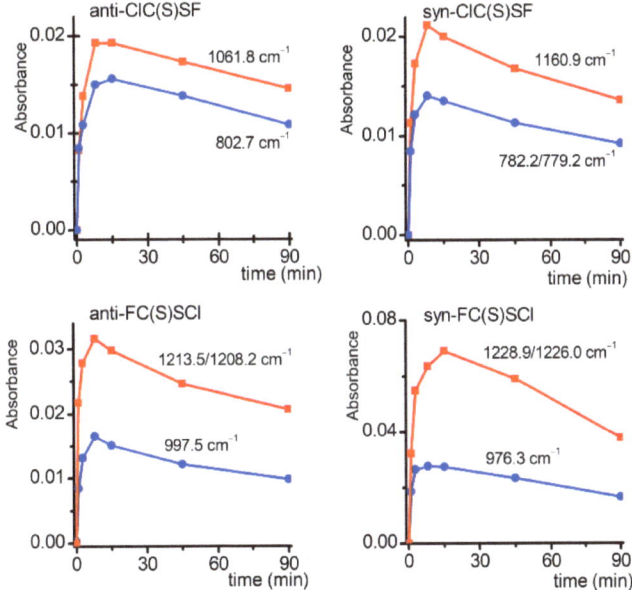

**Figure 9.** Plots of the intensities of the bands assigned to anti-ClC(S)SF (**top left**), syn-ClC(S)SF (**top right**), anti-FC(S)SCl (**bottom left**) and syn-FC(S)SCl (**bottom right**) in the FTIR spectra of the $CS_2/ClF/Ar$ matrix ($CS_2$:ClF:Ar = 1:2:200) at about 15 K vs. irradiation times with broad-band UV-visible light ($225 \leq \lambda \leq 800$ nm).

**Table 3.** Wavenumbers (in cm$^{-1}$) assigned to syn-FC(S)SCl, anti-FC(S)SCl, syn-ClC(S)SF and anti-ClC(S)SF after irradiation of CS$_2$/ClF/Ar matrix (CS$_2$:ClF:Ar = 1:2:200) and computed values with the B3LYP/6-311+G(d,p) approximation (relative IR intensities are given between parentheses).

| syn-FC(S)SCl | | anti-FC(S)SCl | | syn-ClC(S)SF | | anti-ClC(S)SF | | Tentative |
| --- | --- | --- | --- | --- | --- | --- | --- | --- |
| Ar Matrix | Calculated | Ar Matrix | Calculated | Ar Matrix | Calculated | Ar Matrix | Calculated | Assignment |
| 1228.9 1226.0 (100) | 1220.9 (100) | 1213.5 1208.2 (100) | 1214.0 (100) | 1160.9 (100) | 1159.5 (100) | 1061.8 (100) | 1084.3 (100) | ν (C=S) |
| 976.3 (40) | 965.5 (66) | 997.5 (50) | 1040.6 (78) | - | 743.1 (4) | 802.7 (78) | 893.6 (56) | ν (C–X) [a] |
| 718.5 | 589.6 (3) | - | 628.9 (2) | - | 514.4 (8) | - | 481.4 (12) | ν (C–S) [b] |
| 712.2 | 526.4 (7) | - | 464.2 (63) | 782.2 779.2 (70) | 728.3 (93) | 668.8 (50) | 671.4 (63) | ν (S–Y) |

[a] ν$_{as}$ (Cl–C–S) for syn- and anti-ClC(S)SF. [b] ν$_s$ (Cl–C–S) for syn- and anti-ClC(S)SF.

Two absorptions at 1160.9 and 1061.8 cm$^{-1}$ (Figures 8 and 9) can be attributed to the presence of the bond isomer ClC(S)SF in its two planar syn-ClC(S)SF and anti-ClC(S)SF conformations, respectively. The experimental values compare fairly well with those computed at 1159.5 and 1084.3 cm$^{-1}$ using the B3LYP/6-311G+(d,p) level of approximation (Table 3). The values of the absorption coefficients corresponding to the C=S stretching vibrations of the two rotamers are very similar. According to the performed calculations, the syn-conformer presents an absorption coefficient of 316 km/mol, while a value of 253 km/mol is predicted for the anti-form. The formation kinetics for the two conformers is also similar; the maximum intensity of these absorptions is reached at around 8 min of UV-visible irradiation for both rotamers. As in the case of its FC(S)SCl constitutional isomer, anti-ClC(S)SF was found to be the lowest-energy conformer, being the syn-ClC(S)SF rotamer 2.16 kcal/mol higher in energy than the anti-form. The compared photoevolution of the anti- and syn-ClC(S)SF forms is in this case indicative of a randomization process with approximately equal experimental concentration of each form after irradiation (see Figure S1 in the Supporting Information).

Figure 10 shows the molecular models of the two novel photochemically obtained species, each with its two rotamers, optimized with the B3LYP/6-311+(d,p) approximation. Table S4 of the Supporting Information compiles the theoretical geometric parameters of these species.

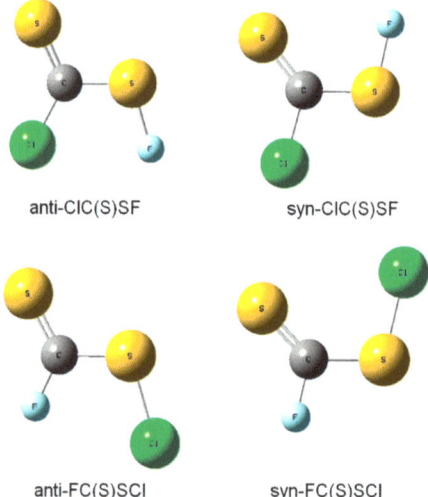

**Figure 10.** Molecular models of anti-ClC(S)SF, syn-ClC(S)SF, anti-FC(S)SCl and syn-FC(S)SCl calculated with the B3LYP/6-311+G(d,p) approximation.

The presence of anti- and syn-ClC(S)SF rotamers was also reconfirmed experimentally due the additional absorptions appearing when the Ar matrix of codeposited $CS_2$ and ClF at cryogenic temperatures was irradiated. For the syn-rotamer the computed band at 743.1 cm$^{-1}$ ($\nu_{as}$ Cl–C–S) was not observed in the experimental spectra, in accordance with its predicted absorption coefficient of only 9.2 km/mol. The absorption at 782.2/779.2 cm$^{-1}$ could in principle be attributed to the ~778 cm$^{-1}$ IR band of the triatomic species ClSF, isolated by the photolysis of FC(O)SCl in Ar matrix [24]. Although the formation of this molecule cannot be completely discarded, the absence of the other band expected in the recorded region of the spectra, at 543/537 cm$^{-1}$, allows us to conclude that the major contribution to the signals around 780 cm$^{-1}$ is originated by syn-ClC(S)SF.

In addition to the bands assigned to the XC(S)SY (X, Y = F, Cl) molecules, other features were observed to appear in the IR spectra taken after irradiation. Figure 11 shows the intensity vs. irradiation time plots of the most intense of these IR absorptions. A complete list of the wavenumbers appearing during the irradiation of the $CS_2$:ClF mixture in Ar matrix, as well as their tentative assignment, is presented in the Supporting Information (Table S5). A weak band at 1481.9 cm$^{-1}$ is developed in the spectra after irradiation. Figure 9 shows the time photoevolution of this absorption, assigned to the $\nu_{as}$(S=C=S) vibrational mode of the Cl•···SCS complex, by the comparison with the reported value at 1481.5 cm$^{-1}$ [5]. Two signals can be observed at 1353.6 and 1346.0 cm$^{-1}$ in the FTIR spectra, attributed to the $\nu$(S=C) vibrational mode of the $SCF_2$ species [25].

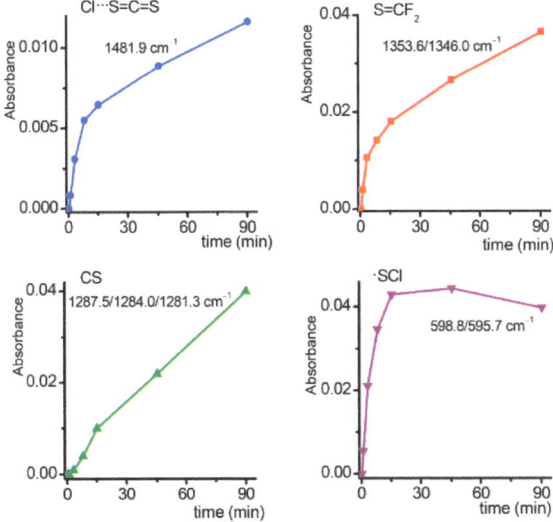

**Figure 11.** Plots of the intensities of the bands assigned to Cl•···S=C=S (**top left**), S=CF$_2$ (**top right**), CS (**bottom left**) and •SCl (**bottom right**) in the FTIR spectra of the $CS_2$/ClF/Ar matrix ($CS_2$:ClF:Ar = 1:2:200) at about 15 K vs. irradiation times with broad-band UV-visible light (225 ≤ λ ≤ 800 nm).

A set of bands at 1185.3/1182.2/1179.3 cm$^{-1}$ that follows a similar behavior against the irradiation time and presents the expected intensity ratio with respect to the $\nu$(S=C) mode was attributed to the $\nu_{as}$(F–C–F) of $SCF_2$, reinforcing the proposed assignment. For its formation, the migration of a F radical from the original matrix cage containing $CS_2$ and ClF, an already-documented process [26,27], is mandatory. The formation of S=CClF can be ruled out since the most intense FTIR bands of this compound, reported at 1257.4, 1014.9, and 612.4 cm$^{-1}$ [28], are not present in the corresponding vibrational spectra. The absorptions observable at 1287.5, 1284.0, and 1281.3 cm$^{-1}$ can be attributed to complexed CS species, by comparison with its 1276 cm$^{-1}$ reported value for the isolated molecule [29]. A signal with a characteristic $^{35}$Cl/$^{37}$Cl pattern, at 598.8/595.7 cm$^{-1}$, is tentatively assigned to the SCl• radical, presumably perturbed by the other species present in the same matrix cage (reported values are 574.4 and 566.9 cm$^{-1}$) [30].

The features observed to develop at 961.5 and 953.2 cm$^{-1}$ (see Figure S2 in the Supporting Information) could not be assigned to any product arising from the $CS_2$:ClF mixture. Instead, they are originated by the ClClO molecule, formed through the photolysis of $Cl_2O$ [31]. As described in the literature, $Cl_2O$ is a by-product in the preparation of ClF [32]. In fact, the IR absorptions of $Cl_2O$ are present in the initial spectrum of the $CS_2$:ClF Ar matrix with very low intensities, at 677.1/674.8 and 638.9/636.2/633.5 cm$^{-1}$ [33], and decreased as the bands of ClClO increased.

## 4. Conclusions

The codeposition of $CS_2$ and ClF, both diluted in Ar, at about 15 K conducts to the formation of the $CS_2 \cdots$ClF complex, as revealed by the comparison of the IR spectrum taken immediately after deposition with the one calculated with DFT methods. Although this complex is predicted to be energetically favorable with respect to the $CS_2 \cdots$FCl adduct, the presence of the latter cannot be completely discarded, since if the signals corresponding to this species were present, they would probably be overlapped with those of the monomers due to the expected small wavenumbers shifts. The photolysis of the $CS_2 \cdots$ClF species with broad-band UV-visible light (225 $\leq \lambda \leq$ 800 nm) is completed after 8 min of irradiation, in agreement with the predicted electronic transitions of this complex by TD-DFT calculations. The first step in the photolysis of the complex isolated in solid Ar is the formation of the constitutional isomers, FC(S)SCl and ClC(S)SF, each of them in their two anti- and syn-rotamers. Although no conclusive evidence exists so far, roaming mechanisms [34] involving either F or Cl atoms could not be ruled out to explain the formation of these two molecules. The novel pentatomic molecules isolated in this work are also photoactive, and evolve inside the Ar-matrix cage, forming CS, the SCl radical, the $CS_2 \cdots$Cl complex, and the thiofluorophosgene molecule, $F_2CS$, in a mechanism that involve the migration of fluorine atoms in the matrix. The proposed photoproducts are summarized in Scheme 1.

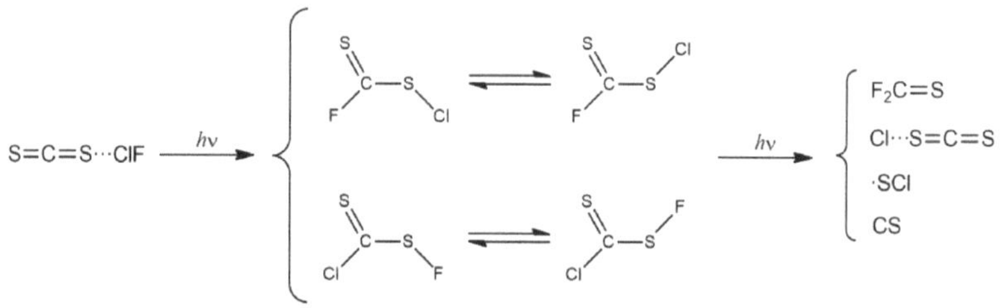

**Scheme 1.** Proposed photoproducts detected in the FTIR spectra of the $CS_2$/ClF/Ar matrix ($CS_2$:ClF:Ar = 1:2:200) at about 15 K after irradiation with broad-band UV-visible light (225 $\leq \lambda \leq$ 800 nm).

**Supplementary Materials:** The following supporting information can be downloaded at: https://www.mdpi.com/article/10.3390/photochem2030049/s1, Figure S1: Plot of the relative intensities of the ν(C=S) absorptions of syn- and anti-ClC(S)SF as a function of the broad-band UV-visible irradiation time; Figure S2: FTIR spectra of the $CS_2$/ClF/Ar matrix ($CS_2$:ClF:Ar = 1:2:200) at about 15 K after deposition (bottom, black trace) and (from bottom to top) 1 (red trace), 3 (green trace), 8 min (blue trace), 15 min (orange trace), 45 min (grey trace) and 90 min (purple trace) of irradiation with broad-band UV-visible light (225 $\leq \lambda \leq$ 800 nm) in the 970–940 and 655–635 cm$^{-1}$ regions; Figure S3: Plot of the relative intensities of the ν(C=S) absorptions of syn- and anti-ClC(S)SF as a function of the broad-band UV-visible irradiation time; Figure S4: FTIR spectra of the $CS_2$/ClF/Ar matrix ($CS_2$:ClF:Ar = 1:2:200) at about 15 K after deposition (bottom, black trace) and (from bottom to top) 1 (red trace), 3 (green trace), 8 min (blue trace), 15 min (orange trace), 45 min (grey trace) and 90 min (purple trace) of irradiation with broad-band UV-visible light (225 $\leq \lambda \leq$ 800 nm) in the 970–940 and 655–635 cm$^{-1}$ regions; Table S1: Geometric parameters for the different complexes formed between $CS_2$ and ClF (distances in Å, angles in degrees) calculated using the B3LYP/6-

311+G(d,p) approximation; Table S2: $\Delta E^{(SCF)}$, $\Delta E^{CP}$, BSSE and GEOM corrections (in kcal.mol$^{-1}$), transferred charge (q), orbital stabilization energy ($\Delta E^{(2)}$ in kcal.mol$^{-1}$) for the different complexes formed between $CS_2$ and ClF computed using the B3LYP/6-311+G(d,p) approximation; Table S3: Wavenumbers for the different complexes formed between $CS_2$ and ClF computed with the B3LYP/6-311+G(d,p) approximation (wavenumbers are in cm$^{-1}$ and relative IR intensities are given between parentheses) and comparison with the experimental values; Table S4: Geometrical parameters of anti-ClC(S)SF, syn-ClC(S)SF, anti-FC(S)SCl, and syn-FC(S)SCl calculated with the B3LYP/6-311+G(d,p) approximation (distances in Å and angles in degrees); Table S5: FTIR wavenumber and proposed assignment of the photoproducts formed by UV-visible irradiation of $CS_2$ and ClF co-deposited in an Ar matrix ($CS_2$:ClF:Ar 1:2:200) at cryogenic temperatures. Refs. [5,25,29,30,35] have been cited in supplementary materials.

**Author Contributions:** Conceptualization, C.O.D.V. and R.M.R.; formal analysis, M.T.C.C., C.O.D.V. and R.M.R.; investigation, C.O.D.V. and H.W.; writing—original draft preparation, C.O.D.V. and R.M.R.; writing—review and editing, C.O.D.V. and R.M.R.; funding acquisition, C.O.D.V. and R.M.R. All authors have read and agreed to the published version of the manuscript.

**Funding:** This research was funded by Consejo Nacional de Investigaciones Científicas y Técnicas (CONICET), PIP 0352, Agencia Nacional de Promoción Científica y Tecnológica (ANPCyT), PICT-2014-3266 and PICT-2018-04355, Universidad Nacional de La Plata, UNLP-X822.

**Institutional Review Board Statement:** Not applicable.

**Informed Consent Statement:** Not applicable.

**Data Availability Statement:** Not applicable.

**Conflicts of Interest:** The authors declare no conflict of interest.

# References

1. Ault, B.S. Matrix isolation spectroscopic studies: Thermal and soft photochemical bimolecular reactions. *Front. Adv. Mol. Spectrosc.* **2018**, *20*, 667–712.
2. Wu, Z.; Shao, X.; Zhu, B.; Wang, L.; Lu, B.; Trabelsi, T.; Francisco, J.S.; Zeng, X. Spectroscopic characterization of two peroxyl radicals during the $O_2$-oxidation of the methylthio radical. *Commun. Chem.* **2022**, *5*, 19. [CrossRef]
3. Young, N.A. Main group coordination chemistry at low temperatures: A review of matrix isolated Group 12 to Group 18 complexes. *Coord. Chem. Rev.* **2013**, *257*, 956–1010. [CrossRef]
4. Bava, Y.B.; Cozzarín, M.V.; Della Védova, C.O.; Willner, H.; Romano, R.M. Preparation of FC(S)SF, FC(S)SeF and FC(Se)SeF through matrix photochemical reactions of $F_2$ with $CS_2$, SCSe, and $CSe_2$. *Phys. Chem. Chem. Phys.* **2021**, *23*, 20892–20900. [CrossRef]
5. Tobón, Y.A.; Romano, R.M.; Della Védova, C.O.; Downs, A.J. Formation of new halogenotiocarbonylsulfenyl halides, XC(S)SY, through photochemical matrix reactions starting from $CS_2$ and dihalogen molecule XY (XY = $Cl_2$, $Br_2$, or BrCl). *Inorg. Chem.* **2007**, *46*, 4692–4703. [CrossRef]
6. Gomez Castaño, J.A.; Romano, R.M.; Della Védova, C.O.; Willner, H. Photochemical reaction of OCSe with ClF in argon matrix: A light-driven formation of XC(O)SeY (X, Y = F or Cl) species. *J. Phys. Chem. A* **2017**, *121*, 2878–2887. [CrossRef]
7. Gómez Castaño, J.A.; Picone, A.L.; Romano, R.M.; Willner, H.; Della Védova, C.O. Early barriers in the matrix photochemical formation of syn–anti randomized FC(O)SeF from the OCSe:$F_2$ complex. *Chem. Eur. J.* **2007**, *13*, 9355–9361. [CrossRef]
8. Ault, B.S. Matrix isolation investigation of the interaction of ClF and $Cl_2$ with carbon-carbon multiple bonds. *J. Phys. Chem.* **1987**, *91*, 4723–4727. [CrossRef]
9. Machara, N.P.; Ault, B.S. Infrared spectroscopy studies of the interactions of ClF and $Cl_2$ with $H_2Se$, $(CH_3)_2Se$ and $AsH_3$ in argon matrices. *Inorg. Chem.* **1988**, *27*, 2383–2385. [CrossRef]
10. Bai, H.; Ault, B.S. Infrared spectroscopy investigation of complexes of ClF and $Cl_2$ with crown ethers and related cyclic polyethers in argon matrices. *J. Mol. Struct.* **1989**, *196*, 47–56. [CrossRef]
11. Bai, H.; Ault, B.S. Infrared matrix isolation of the 1 1 complexes of HCl and ClF with sulfur and nitrogen-containing macrocycles. *J. Mol. Struct.* **1990**, *238*, 223–230. [CrossRef]
12. Bai, H.; Ault, B.S. Infrared matrix isolation investigation of the molecular complexes of ClF with benzene and its derivatives. *J. Phys. Chem.* **1990**, *94*, 199–203. [CrossRef]
13. Bai, H.; Ault, B.S. Matrix isolation study of complexation and with terf-butyl halides. *J. Phys. Chem.* **1991**, *95*, 3080–3084. [CrossRef]
14. Bloemink, H.I.; Hinds, K.; Holloway, J.H.; Legon, A.C. Isolation of $H_2S$...ClF in a pre-reactive mixture of $H_2S$ and ClF expanded in a coaxial jet and characterisation by rotational spectroscopy. *Chem. Phys. Lett.* **1995**, *242*, 113–120. [CrossRef]

15. Cooke, S.A.; Cotti, G.; Evans, C.M.; Holloway, J.H.; Kisiel, Z.; Legon, A.C.; Thumwood, J.M.A. Pre-reactive complexes in mixtures of water vapour with halogens: Characterisation of $H_2O \cdots ClF$ and $H_2O \cdots F_2$ by a combination of rotational spectroscopy and ab initio calculations. *Chem. Eur. J.* **2001**, *7*, 2295–2305. [CrossRef]
16. Picone, A.L.; Della Védova, C.O.; Willner, H.; Downs, A.J.; Romano, R.M. Experimental and theoretical characterization of molecular complexes formed between OCS and XY molecules (X, Y = F, Cl and Br) and their role in photochemical matrix reactions. *Phys. Chem. Chem. Phys.* **2010**, *12*, 563–571. [CrossRef]
17. Schnöckel, H.; Willner, H. Matrix-isolated molecules. In *Infrared and Raman Spectroscopy, Methods and Applications*; Schrader, B., Ed.; VCH: Weinheim, Germany, 1995; pp. 297–313.
18. Frisch, M.J.; Trucks, G.W.; Schlegel, H.B.; Scuseria, G.E.; Robb, M.A.; Cheeseman, J.R.; Montgomery, J.A., Jr.; Vreven, T.; Kudin, K.N.; Burant, J.C.; et al. *Gaussian 03, Rev. B.04*; Gaussian, Inc.: Pittsburgh, PA, USA, 2003.
19. Frisch, M.J.; Trucks, G.W.; Schlegel, H.B.; Scuseria, G.E.; Robb, M.A.; Cheeseman, J.R.; Scalmani, G.; Barone, V.; Mennucci, B.; Petersson, G.A.; et al. *Gaussian 09, Rev. D.01*; Gaussian, Inc.: Wallingford, CT, USA, 2013.
20. Nagy, P.I.; Smith, D.A.; Alagona, G.; Ghio, C. Ab initio studies of free and monohydrated carboxylic acids in the gas phase. *J. Phys. Chem.* **1994**, *98*, 486–493. [CrossRef]
21. Boys, S.F.; Bernardi, F. The calculation of small molecular interactions by the differences of separate total energies. Some procedures with reduced errors. *Mol. Phys.* **1970**, *19*, 553–566. [CrossRef]
22. Bauernschmitt, R.; Ahlrichs, R. Treatment of electronic excitations within the adiabatic approximation of time dependent density functional theory. *Chem. Phys. Lett.* **1996**, *256*, 454–464. [CrossRef]
23. Stratmann, R.E.; Scuseria, G.E.; Frisch, M.J. An efficient implementation of time-dependent density-functional theory for the calculation of excitation energies of large molecules. *J. Chem. Phys.* **1998**, *109*, 8218–8224. [CrossRef]
24. Willner, H. Das Infrarotspektrum von matrixisoliertem SFCl. *Z. Für Nat. B* **1984**, *39*, 314–316. [CrossRef]
25. Haas, A.; Willner, H.; Bürger, H.; Pawelke, G. Matrix-infrarot-Spektren und Kraftkonstanten von $SCF_2$ und $SeCF_2$. *Spectrochim. Acta* **1977**, *33*, 937–945. [CrossRef]
26. Alimi, R.; Gerber, R.B.; Apkarian, V.A. Dynamics of molecular reactions in solids: Photodissociation of $F_2$ in crystalline Ar. *J. Chem. Phys.* **1990**, *92*, 3551–3558. [CrossRef]
27. Feld, J.; Kunttu, H.; Apkarian, V.A. Photodissociation of $F_2$ and mobility of F atoms in crystalline argon. *J. Chem. Phys.* **1990**, *93*, 1009–1020. [CrossRef]
28. Hamm, R. Schwingungsspektrum von CSFCl. *Z. Naturforsch.* **1979**, *34 A*, 325–332. [CrossRef]
29. Jacox, M.E.; Milligan, D.E. Matrix isolation study of the infrared spectrum of thioformaldehyde. *J. Mol. Spectrosc.* **1975**, *58*, 142–157. [CrossRef]
30. Willner, H. Die Infrarotabsorption des matrixisolierten SCl-Radikal. *Spectrochim. Acta* **1981**, *37*, 405–406. [CrossRef]
31. Johnsson, K.; Engdahl, A.; Nelander, B. The UV and IR Spectra of the ClClO Molecule. *J. Phys. Chem.* **1995**, *99*, 3965–3968. [CrossRef]
32. Ruff, O.; Ascher, E.; Lass, F. Das Chlorfluorid. *Z. Anorg. Allg. Chem.* **1928**, *176*, 258–270. [CrossRef]
33. Andrews, L.; Raymond, J.I. Argon matrix infrared spectrum of the ClO radical. *J. Chem. Phys.* **1971**, *55*, 3087–3094. [CrossRef]
34. Townsend, D.; Lahankar, S.A.; Lee, S.K.; Chambreau, S.D.; Suits, A.G.; Zhang, X.; Rheinecker, J.; Harding, L.B.; Bowman, J.M. The roaming atom: Straying from the reaction path in formaldehyde decomposition. *Science* **2004**, *306*, 1158–1161. [CrossRef] [PubMed]
35. Bondi, A. van der Waals Volumes and Radii. *J. Phys. Chem.* **1964**, *68*, 441–451.

Article

# Conformational-Dependent Photodissociation of Glycolic Acid in an Argon Matrix

Jussi Ahokas [1,2], Timur Nikitin [3], Justyna Krupa [4], Iwona Kosendiak [4], Rui Fausto [3], Maria Wierzejewska [4] and Jan Lundell [2,*]

[1] Financial Services, University of Jyväskylä, P.O. Box 35, FI-40014 Jyväskylä, Finland
[2] Department of Chemistry, University Jyväskylä, P.O. Box 35, FI-40014 Jyväskylä, Finland
[3] CQC-IMS, Department of Chemistry, University of Coimbra, P-3004-535 Coimbra, Portugal
[4] Faculty of Chemistry, University of Wrocław, Joliot-Curie 14, PL-50-383 Wrocław, Poland
* Correspondence: jan.c.lundell@jyu.fi

**Abstract:** Ultraviolet-induced photodissociation and photo-isomerization of the three most stable conformers (SSC, GAC, and AAT) of glycolic acid are investigated in a low-temperature solid argon matrix using FTIR spectroscopy and employing laser radiation with wavelengths of 212 nm, 226 nm, and 230 nm. The present work broadens the wavelength range of photochemical studies of glycolic acid, thus extending the understanding of the overall photochemistry of the compound. The proposed kinetic model for the photodissociation of glycolic acid proceeds from the lowest energy conformer (SSC). The model suggests that ultraviolet light induces isomerization only between the SSC and GAC conformers and between the SSC and AAT conformers. The relative reaction rate coefficients are reported for all proposed reactions. These results suggest that the direct photodissociation of GAC and AAT conformer does not occur in an argon matrix. The main photodissociation channel via the SSC conformer produces formaldehyde–water complexes. The proposed photodissociation mechanism emphasizes that the conformers' relative abundancies can significantly affect the photodissociation rate of the molecule. For example, in the case of high relative GAC and AAT concentrations, the ultraviolet photodissociation of glycolic acid requires the proceeding photo-isomerization of GAC and AAT to SSC.

**Keywords:** glycolic acid; hydroxy acid; dissociation; isomerization; chemical kinetics; reaction mechanism; matrix isolation; vibrational spectroscopy; photochemistry

## 1. Introduction

Glycolic acid (GA), the smallest α-hydroxy acid, is a suitable prototype molecule for photochemical studies. The molecular structure of GA, with functional OH groups at both ends of the molecule, and its ability to form inter- and intramolecular hydrogen bonds gives versatile possibilities to investigate how vibrational energy relaxes and redistributes in the molecule and how hydrogen bonding affects these relaxation processes [1,2]. The three lowest energy conformers of GA are populated at room temperature, and they offer a set of probe conformers for both dissociative and non-dissociative photochemical studies.

Most photochemical studies have focused on the vibrational excitations on the electronic ground state, both in the gas phase and in low temperature matrices [3–8]. The reported light-induced processes reported in the literature for glycolic acid isolated in low-temperature matrix experiments are shown in Figure 1. Broadband infrared excitation of GA initiates conformational photo-isomerization reactions [3–5]. Selective conformational photo-isomerization reactions between the conformers can be initiated by near-infrared excitation of first and second OH stretching overtones [6,7]. At shorter wavelengths, the excitation of high OH stretching overtones by visible light, at 532 nm, is also followed by conformational photo-isomerization reactions [8]. Photons from visible to infrared light were then found to be capable of vibrationally exciting GA in the electronic ground state and, thus, capable of initiating isomerization processes. Recently, Krupa et al. [8] widened the wavelength range of the photochemical studies of GA to ultraviolet radiation. They investigated the photochemistry of GA in an argon matrix using laser light at 212 and 226 nm. The ultraviolet excitation led to the photo-isomerization and photodissociation of GA. The main dissociation channels produced molecular complexes of formaldehyde and water with carbon monoxide [9].

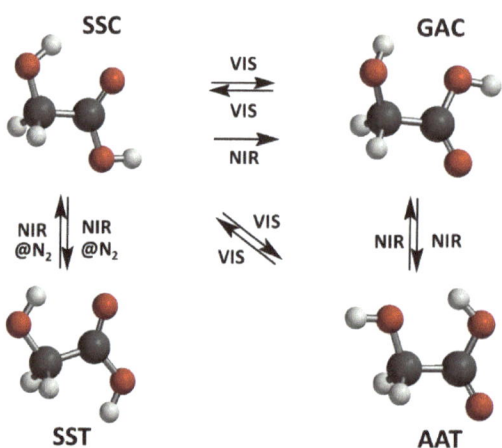

**Figure 1.** The experimentally identified photo-induced processes of GA in a low-temperature argon matrix. The photo-induced reactions between SSC and SST have been observed only in a nitrogen matrix and denoted as @N$_2$ in the Figure. NIR stands for near-infrared radiation, and VIS for visible light radiation. The Figure is adapted from information in Ref. [8].

Previous photochemical studies show that the conformational photo-isomerization of GA can be initiated by both vibrational excitations on the electronic ground state and electronic excitations, while only the latter also leads to photodissociation. The competing isomerization and dissociation reactions raise an exciting research question: what are the photodissociation and photo-isomerization kinetics of individual conformers? The kinetics of individual conformers is essential information to estimate how changes in relative concentrations of the conformers affect the ultraviolet photokinetics of GA. For example, in some instances, such as atmospheric or interstellar environments, longer wavelengths can also be available for vibrational excitation. GA is present in the atmosphere [10], where the wavelength intensity distribution repeatedly changes so that a wide variety of excitations can happen depending on the sunlight spectrum and the conditions of a specific atmospheric region [11–13]. The mechanisms in narrow wavelength ranges are of prime importance for understanding the overall photochemical mechanisms in the case of broad spectral range excitations. So far, only a kinetic model for the high overtone excitation-induced isomerization reactions of GA has been proposed [8].

The present study aims to develop a kinetic model to describe the ultraviolet-induced photochemistry of GA. The model provides a better understanding of the processes that follow the electronic excitation of the molecule. Together with the previous studies on the electronic ground state photochemistry of GA, the present work promotes an understanding of how different mechanisms could affect the photochemistry of a single molecule when photons from infrared to ultraviolet are available for excitations.

## 2. Materials and Methods

The GA/argon matrices were prepared by passing an argon gas (Messer 5.0) flow over solid GA (Janssen Chimica, Warsaw, Poland, purity 99% or Sigma-Aldrich, Lisbon,

Portugal, Reagent Plus®, purity 99%) at room temperature. This method yields suitable monomeric matrices, even though the guest-to-host ratio can be controlled only by the gas flow rate and temperature of the substrate. A precise guest-to-host ratio is unspecified throughout the study. The typical deposition temperature was 15 K, and measurements were performed at 10–15 K.

Ultraviolet photolysis at 212, 226, and 230 nm was performed in Wrocław, and near-infrared photolysis with subsequent 230 nm photolysis experiments were carried out in Coimbra. The experimental setup in Wrocław consisted of a closed-cycle helium cryostat APD-Cryogenics (ARS-2HW), equipped with CsI windows; a Bruker IFS 66 FTIR spectrometer, equipped with a liquid $N_2$ cooled MCT detector; a Scientific Instruments 9700 temperature controller, equipped with a silicon diode; a UV radiation source; and the frequency-doubled signal beam of a pulsed (7 ns, repetition rate of 10 Hz) optical parametric oscillator Vibrant (Opotek Inc., Carlsbad, CA, USA), pumped with an Nd:YAG laser (Quantel, Lannion, France). For more details, refer to Ref. [9].

The experimental setup in Coimbra consisted of a closed-cycle helium cryostat APD-Cryogenics (DE-202A), equipped with external KBr windows and with a CsI substrate mounted at the cold tip of the cryostat; a Thermo Nicolet 6700 FTIR spectrometer, purged through the optical path with dry, $CO_2$-filtered air to avoid interference of atmospheric $H_2O$ and $CO_2$, equipped with a deuterated triglycine sulfate detector (DTGS) and a KBr beam splitter; and a Scientific Instruments, model 9650-1 temperature controller, equipped with a silicon diode; NIR radiation source: tunable narrowband ($\sim$0.2 cm$^{-1}$ spectral width) light generated by the idler beam of a Quanta-Ray MOPO-SL optical parametric oscillator (OPO), pumped by a pulsed Nd:YAG laser (pulse duration: 10 ns; repetition rate: 10 Hz). UV radiation source: same as for the NIR irradiation, but in this case, the frequency-doubled signal beam of the OPO has been used.

Mathematica software was used for the analytical solutions of the differential equations and data fitting [14]. The built-in functions DSolve and NonLinearModelFit of the software were utilised.

## 3. Results

We studied the UV photochemistry of the three most abundant conformers of GA (SSC, GAC, and AAT) in low-temperature argon matrices. Three different laser wavelengths (212, 226, and 230 nm) were used in the photodissociation studies. The initial relative concentrations of the conformers were varied before ultraviolet photodissociation using NIR irradiation with the photon energies of 6954 cm$^{-1}$.

The first set of photodissociation experiments consisted of UV photodissociation of freshly deposited samples of GA in an argon matrix. After deposition, three conformers of GA were identified in the IR spectra (Figure 2) [3,6]. The SSC conformer shows the strongest absorbance, reflecting its higher relative concentration than the other conformers used under the experimental conditions. The relative populations of the SSC, GAC, and AAT conformers at room temperature, obtained from a previous computational study, are 94.8, 3.7, and 1.3%, respectively [8]. The photolysis of GA in an argon matrix at 212 and 226 nm yielded similar results, in agreement with a recent report [9]. Figure 2 shows the main changes in the IR spectra of GA upon UV irradiation of the sample at 212 nm. The increase in the absorbances of the GAC and AAT conformers follows the photo-induced decrease of those ascribed to the SSC conformer, which suggests a significant conformational photo-isomerization channel of the SSC conformer.

The generation of the photoproducts upon photolysis at 212 nm is illustrated in Figure 3. The main photoproducts were the molecular complexes $H_2O$–CO, HCHO–$H_2O$, HCHO–$CO_2$, and HCHO–CO. There was no qualitative difference between the photolysis at 212 nm and 226 nm, in agreement with a precedent study [9].

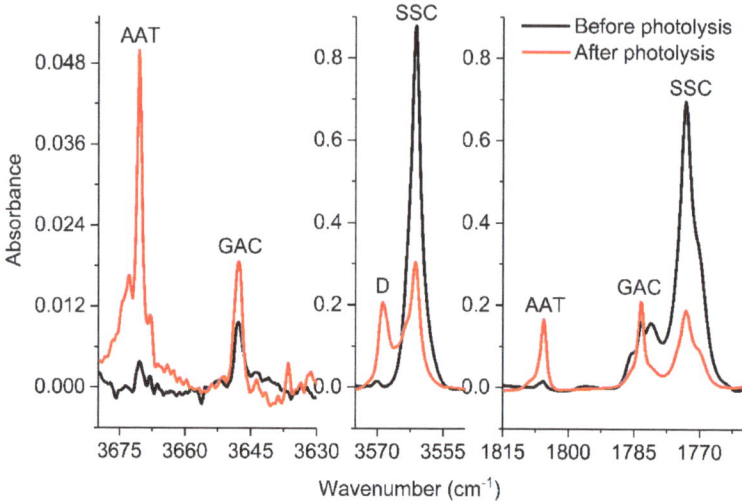

**Figure 2.** IR spectra of GA in argon matrix before and after irradiation with λ = 212 nm. Three GA conformers and HCHO–H$_2$O photoproduct (D) are labeled in the graphs.

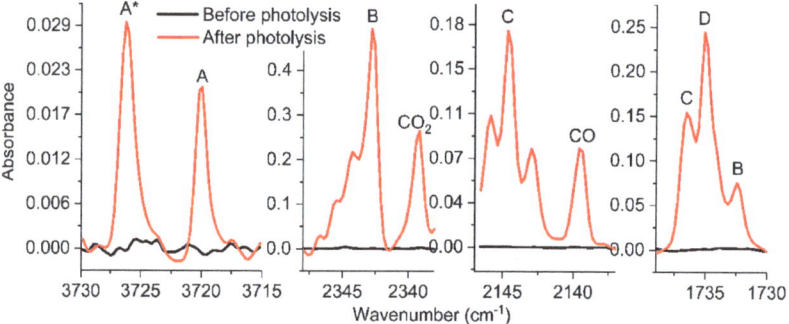

**Figure 3.** The IR spectra of the photodissociation products of GA (λ = 212 nm). (**A**) is H$_2$O–CO; (**B**) is HCHO–CO$_2$; (**C**) is HCHO–CO; (**D**) is HCHO–H$_2$O. An asterisk indicates a thermally unstable site. Monomeric CO$_2$ and CO perturbed by neighboring species are also marked in the spectra.

In the present study, we extended the wavelength range used in the photolysis to 230 nm to study the photo-induced kinetics in more detail. Qualitatively, the same products appeared in the photolysis at 230 nm as at 212 nm and 226 nm. Figure 4 shows the relative abundance kinetic curves for the SSC, GAC, and ATT conformers upon irradiation at different wavelengths. The initial relative concentration of each conformer was set according to the above-mentioned computational values. The computational abundances have been successfully used in the kinetic analysis of high overtone-induced isomerization of GA isolated in solid argon [8]. The main differences observed in the photodissociation and photo-isomerization kinetic curves along an increasing wavelength of irradiation are: (i) The lowering of the relative concentrations of both GAC and ATT conformers, (ii) The approach of the decay curves of the SSC conformer to the total decay curve of GA (see Figure 4). The former observation suggests that the possible photo-equilibrium between the SSC and the other conformers changes as the wavelength changes. The latter observation suggests that the total dissociation of GA proceeds via the SSC conformer. This also indicates that the photodissociation of GAC and AAT conformers could be hindered.

However, these findings do not fully exclude the existence of possible photodissociation channels also for GAC and AAT conformers.

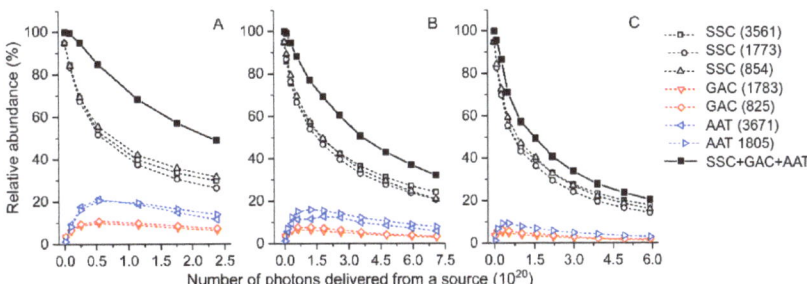

**Figure 4.** The relative abundance kinetic curves of the SSC, GAC, and ATT conformers in the irradiation at 212 nm (**A**), 226 nm (**B**), and 230 nm (**C**). The initial relative abundances of the conformers are set to 94.8, 3.7, and 1.3 % for SSC, GAC, and AAT, respectively [8]. Numbers in parenthesis are the wavenumbers of the vibrational bands from where the kinetic data were obtained. The total population (SSC + GAC + AAT) is calculated from the average values of the measured relative population of each conformer.

Inspection of the distributions of photodissociation products was expected to shed more light on the photo-induced kinetics of the conformers. Remarkably, the IR spectra of the photoproducts were found to be almost identical when about 50 percent of the GA was dissociated, despite using different wavelengths in the photolysis (Figure 5). The similarity of the spectra suggests that the main photodissociation channel remains almost the same when the photolysis wavelength is varied. At this point, the main dissociation channel seems to proceed mainly via the SSC conformer. The kinetics of the SSC conformer in the 230 nm photolysis supports this assumption. These findings suggest that the photodissociation of the GAC and AAT conformers is indeed hindered in an argon matrix.

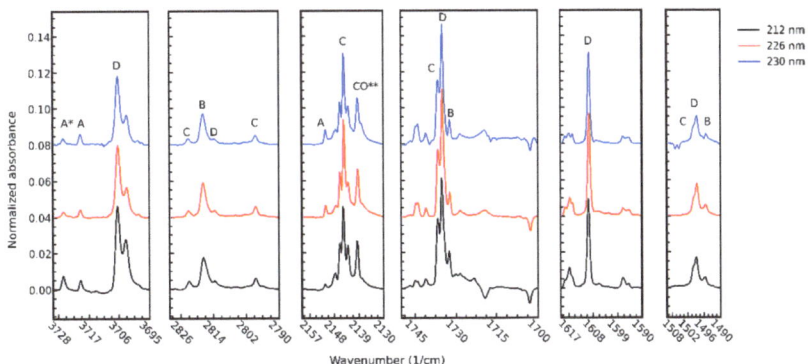

**Figure 5.** The IR difference spectrum of the photodissociation products of GA at different photodissociation wavelengths. The intensity of each spectrum is normalized by the intensity of the SSC band at 3561 cm$^{-1}$ after deposition. The dissociation degree of GA is about 50 percent in each spectrum. A is $H_2O$–CO; B is HCHO–$CO_2$; C is HCHO–CO; D is HCHO–$H_2O$. An asterisk indicates a thermally unstable site. CO** stands for monomeric CO that is perturbed by the neighboring species.

In the next series of experiments, we varied the relative populations of the conformers via NIR-induced conformational isomerization. The NIR excitation band centered at 6954 cm$^{-1}$ overlaps the first OH stretching overtones of the SSC and GAC conformers at 6954 cm$^{-1}$ and 6958 cm$^{-1}$, respectively. The excitation at 6954 cm$^{-1}$ led to SSC → GAC and

GAC → AAT isomerizations [6]. Figure 6 shows the variation of the relative populations of the conformers in the NIR-induced isomerization and subsequent UV photolysis experiments. Upon NIR irradiation, SSC isomerizes to GAC, which subsequently isomerizes to AAT. Therefore, the prolonged NIR exposure yields a high AAT concentration.

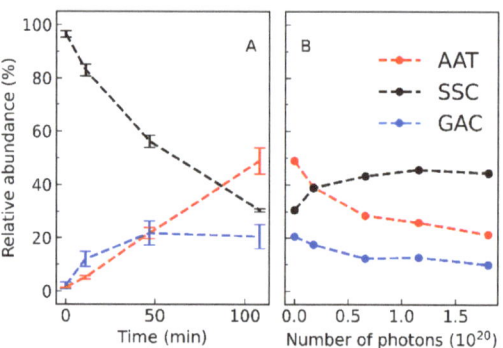

**Figure 6.** (**A**): The change of the conformer populations upon NIR irradiation at 6954 cm$^{-1}$. Error bars indicate the error in the determined relative populations of the conformers. (**B**): The change of the conformer populations in UV photolysis at 230 nm, subsequent to the NIR irradiation.

The NIR irradiation experiments allowed us to test our assumption of the initial concentrations of the conformers. There is strong evidence that the NIR irradiation does not cause photodissociation of GA or the formation of other conformers than SSC, GAC, and AAT in the argon matrix [6]. Therefore, one may expect the total GA concentration to remain unchanged, and only the relative concentrations of the SSC, GAC, and AAT vary. At each phase of the NIR irradiation, the sum of the relative concentrations of the conformers must obey the equation:

$$c_{\text{SSC}} I_{\text{SSC}}^{(i)} + c_{\text{GAC}} I_{\text{GAC}}^{(i)} + c_{\text{AAT}} I_{\text{AAT}}^{(i)} = 100\% \tag{1}$$

where $c$ is the conversion coefficient to convert the IR intensity, $I$, of a certain IR band to the relative concentration of the conformer, and $i$ is the phase of the photolysis. Here, the three conformers (SSC, GAC and AAT) are assumed to exist in the present experimental conditions. The system of linear equations was solved for the three cases where different NIR doses were applied to the samples. The coefficients are also solved for different IR band combinations of the conformers. The experimental errors for the concentrations were calculated as a standard deviation of the results from the different combinations. The experimental concentrations agree well with the computational values (Table 1). Experimentally, the SSC concentration is only about 2 percent higher, and the GAC concentration is about 2 percent lower than the computational values suggest [4,8]. The AAT concentration does not show significant deviation from the computational values. The computational values refer to 298 K, whereas the present experiments were carried out at ca. 15 K. Lower temperature favors the lowest energy SSC conformer, and conformational cooling can *a priori* be expected to take place to some extent during deposition. However, the good agreement between the calculated and observed relative populations demonstrates that the population of the conformers practically did not change upon deposition, allowing us to conclude that the conformational cooling during matrix deposition was, in fact, negligible, justifying the use of the calculated relative conformational concentrations in Figure 6.

Table 1. Estimated initial relative populations of the GA conformers in the Ar matrix and literature values obtained from quantum chemical calculations. For the experimental data, the standard deviations are shown in the parenthesis.

| Experiment | SSC | GAC | AAT | Other Conformers | Method |
|---|---|---|---|---|---|
| 1 | 97.4 (0.74) | 1.2 (0.97) | 1.1 (0.29) | 0 | |
| 2 | 96.4 (0.17) | 2.92 (0.17) | 0.69 (0.07) | 0 | Expt. |
| 3 | 96.5 (1.29) | 2.08 (1.32) | 1.35 (0.13) | 0 | Ar matrix |
| Average | 96.8 (0.73) | 2.07 (0.82) | 1.05 (0.16) | | |
| Literature | 94.8 | 3.7 | 1.3 | 0.2 | Calc. [8] |
| | 94.98 | 4.09 | 0.73 | 0.86 | Calc. [5] |

After different NIR exposures, the samples were UV-irradiated ($\lambda$ = 230 nm). The increase in the concentration of the SSC conformer was evident at the early stages of the 230 nm photolysis (see Figure 6). This finding supports the assumption of the existence of significant GAC/AAT → SSC isomerization channels in the UV irradiation of matrix-isolated GA. Remarkably, the UV irradiation performed after the NIR conformational conversions led to similar amounts of the photodissociation products as obtained in the UV photolysis carried out without preceding NIR irradiation (Figure 7). These results suggest that only one dissociation channel exists since the dissociation of different conformers is supposed to yield a somewhat different distribution of the photodissociation products. However, as mentioned above, the distribution of the products seems to be very similar in the different experiments. These findings strongly support the significance of the SSC products' photodissociation channel within the 212–230 nm wavelength range.

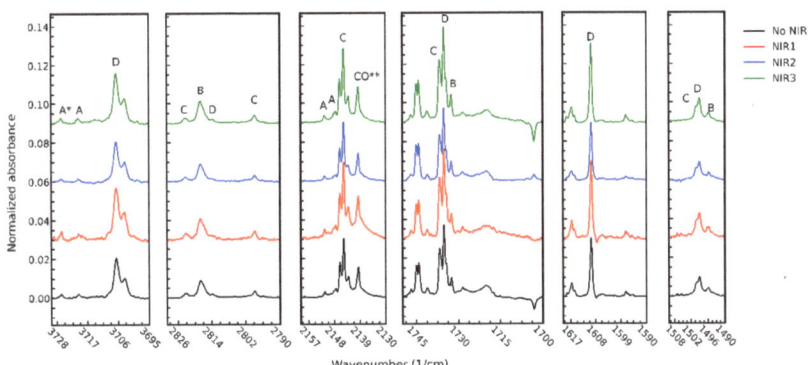

Figure 7. The IR difference spectrum of the photodissociation products of GA at 230 nm. The NIR irradiation of the sample was used to vary the initial relative concentrations of the precursor conformers (refer to Figure 6 and text for more details). The intensities are normalized by the intensity of the SSC band at 3561 cm$^{-1}$ after deposition. The dissociation degree of GA is about 25% in each spectrum. A is $H_2O$–CO; B is HCHO–$CO_2$; C is HCHO–CO; D is HCHO–$H_2O$. An asterisk indicates a thermally unstable site. CO** stands for monomeric CO that is perturbed by the neighboring species. The initial relative concentrations prior to the UV photolysis (SSC/GAC/AAT, %): No NIR: 96.9/2.1/0.9; NIR1: 62.1/17.8/20.3; NIR2: 36.0/33.2/30.8; NIR3: 30.5/20.5/49.0.

Combining all the findings, we may assume that the photo-induced kinetics proceeds via five channels: SSC ⇌ GAC, SSC ⇌ AAT, and SSC → products. Nevertheless, the possible existence of the GAC ⇌ AAT channels cannot be neglected here, as there is no direct evidence for their non-existence. The direct photodissociation of the GAC and AAT conformers is also possible a priori. The simplest reaction scheme, including all these reaction channels, can be described with the following equations, where $k_i$ represents a reaction rate constant:

$$\text{products} \overset{k_6}{\leftarrow} \text{AAT} \underset{k_1}{\overset{k_4}{\rightleftharpoons}} \text{SSC} \underset{k_5}{\overset{k_2}{\rightleftharpoons}} \text{GAC} \overset{k_7}{\rightarrow} \text{products} \quad (2)$$

$$k_3 \downarrow \text{products}$$

and

$$\text{GAC} \underset{k_9}{\overset{k_8}{\rightleftharpoons}} \text{AAT} \quad (3)$$

These reaction equations lead to the reaction rate equations:

$$\frac{d[\text{SSC}]}{dt} = -(k_1 + k_2 + k_3)[\text{SSC}] + k_4[\text{GAC}] + k_5[\text{AAT}] \quad (4)$$

$$\frac{d[\text{GAC}]}{dt} = -k_4[\text{GAC}] + k_1[\text{SSC}] \underbrace{-k_7[\text{GAC}]}_{\text{optional photodissociation}} \underbrace{-k_8[\text{GAC}] + k_9[\text{AAT}]}_{\text{optional equilibrium}} \quad (5)$$

$$\frac{d[\text{AAT}]}{dt} = -k_5[\text{AAT}] + k_2[\text{SSC}] \underbrace{-k_6[\text{AAT}]}_{\text{optional photodissociation}} \underbrace{+k_8[\text{GAC}] - k_9[\text{AAT}]}_{\text{optional equilibrium}}. \quad (6)$$

The variable $t$ in the above equations is a general (time) variable representing the number of photons delivered from a source.

The analytical solutions of the differential equations were obtained with the Mathematica software with and without the optional photodissociation and equilibrium channels of the GAC and ATT conformers. The solution functions were fitted simultaneously to the experimental kinetic data of the conformers with shared parameters $k_1$ to $k_9$. The model without the optional photodissociation and equilibrium channels of the GAC and AAT conformers fits well with the experimental kinetic curves of 212 nm and 226 nm photodissociation experiments. The 230 nm photodissociation curves modeled slightly deviate from the model without pre-NIR photolysis. However, the model fits well to 230 nm data when the relative concentrations of the conformers are varied prior to the 230 nm irradiation (NIR3 in Figure 7 was used in the fitting). The results of the data fitting are shown in Figure 8.

When all the optional reactions in the reaction rate equations were used in the data fitting procedure, the solutions became more uncertain, and the problem seemed to have approached over-parametrisation. The increasing complexity of the model, with more parameters, did not, in fact, show any improvement in the data fitting. To avoid overparametrisation, we sought acceptable solutions with the minimum number of parameters. The best solutions were obtained with the model, including the reactions described by the rate coefficients $k_1$ to $k_5$, i.e., when the photodissociation and equilibrium channels of the GAC and AAT conformers were not accounted for. The results supported the above assumption of the conformer-dependent photodissociation of GA.

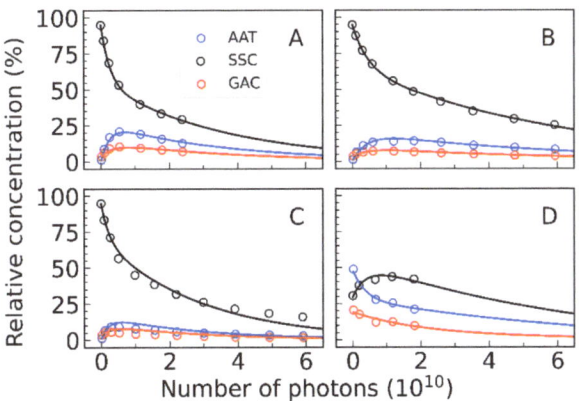

**Figure 8.** The results of the kinetic model fitting to the experimental data. (**A**): 212 nm photodissociation. (**B**): 226 nm photodissociation. (**C**): 230 nm photodissociation. (**D**): 230 nm photodissociation after NIR irradiation (NIR3, see Figure 6).

It needs to be mentioned that some experiments suffered from fluctuating laser power during the photolysis. We controlled the laser power during the experiments, but we believe that in some cases (specifically the experiments designated as "No NIR", "NIR1", and "NIR2", see Figure 7), the laser power fluctuations were the main problem for the poorer data fitting. The small number of data points also increased the risk of overparameterization in these cases. The obtained parameters are collected in Table 2.

**Table 2.** The reaction rate coefficients of the photodissociation and photo-isomerization reactions of the GA conformers and the photoproducts.

| | $k_i$ | | | | | | | | | Initial Concentrations (%) | | |
|---|---|---|---|---|---|---|---|---|---|---|---|---|
| Wavelength (nm) | $k_1$ | $k_2$ | $k_3$ | $k_4$ | $k_5$ | $H_2O$–CO | HCHO–$H_2O$ | HCHO–$CO_2$ | HCHO–CO | $[SSC]_0$ | $[GAC]_0$ | $[AAT]_0$ |
| 212 | 1.06 | 0.36 | 0.51 | 2.52 | 1.70 | 0.098 | 0.33 | 0.018 | 0.034 | 95.9 | 4.3 | 2.0 |
| 226 | 0.43 | 0.19 | 0.25 | 1.52 | 1.44 | 0.051 | 0.14 | 0.022 | 0.031 | 94.0 | 3.4 | 1.9 |
| 230 | 0.65 | 0.31 | 0.42 | 3.24 | 2.49 | 0.066 | 0.28 | 0.043 | 0.080 | 92.0 | 1.0 | 0.9 |
| 230 (NIR3) [1] | 0.81 | 0.02 | 0.32 | 1.67 | 0.46 | 0.11 | 0.22 | 0.0093 | 0.40 | 29.6 | 19.1 | 47.1 |

[1] The 230 nm experiments are carried out with different laser systems, line widths and wavelength accuracies, so the values are not directly comparable.

The final test of the chosen kinetic model was the data fitting of the kinetic curves of the photodissociation products. According to the observations and analyses described above, the photodissociation products were assumed to form from the SSC conformer. We have not yet discussed the possible photodissociation of the primary photoproducts of the SSC conformer, but it must be included optionally in the kinetic model of the photoproducts. Then, the primary ($P_i^{primary}$) and secondary ($P_j^{secondary}$) photoproducts are formed as

$$\text{SSC} \xrightarrow{k_i} P_i^{primary} \xrightarrow{k_i'} P_j^{secondary} \tag{7}$$

The rate equation of the primary photoproducts can be written as

$$\frac{d\left[P_i^{primary}\right]}{dt} = k_i[\text{SSC}] - k_i'\left[P_i^{primary}\right] \tag{8}$$

where $k_i$ and $k_i'$ are the rate coefficients of the growth and decay of the photodissociation product $i$. The [SSC] function is obtained from the kinetic analysis of the precursor conformer. Before the analysis, the IR intensities of the photoproducts were converted to percentages. We used a similar approach here as we used with the GA conformers (Equation (1)). The total concentration of the photoproducts was assumed to be equal to the loss of the GA concentration. At each experimental phase, the total concentration of photoproducts must obey the equation

$$\sum_i c_i I_i^{(k)} = [GA]_0 - [GA]^{(k)} \qquad (9)$$

where the $c_i$ is the conversion coefficient from the IR intensity, $I_i$, to concentration for the product $i$ at the phase of the experiment $k$. At each phase, the lefthand side of the equation must equal the total loss of the GA. For this, we analysed a large number of IR bands of the photoproducts. The IR bands were assigned according to Ref. [9]. However, most bands overlap with the bands of other absorbers, and spectral simulations were needed to extract each component from the bands. All the acceptable solutions for Equation (9), with all possible peak combinations, suggested that HCHO–H$_2$O is the main photoproduct. Thus, we filtered the solutions that agreed with the assumption of the main photoproduct. The concentrations were taken as an average over the satisfactory results. Then, the solution of the rate Equation (8) was fitted to the experimental data. The results are shown in Figure 9 and Table 2. The best fits were obtained by setting $k'_i$, as the experiments did not show further photodissociation of the primary photoproducts. The validity of the results was supported by $\sum_i k_i \approx k_3$. This analysis supports the assumption of the single photodissociation channel of the GA via the SSC conformer and the formation of four primary photoproducts with branching ratios of $k_i/k_3$.

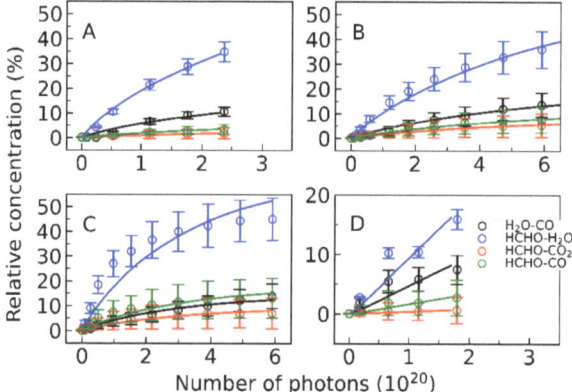

**Figure 9.** Relative concentrations of the photoproducts. The concentrations obtained from the data are shown with the error bars representing the standard deviation of the averaged experimental values. Solid lines represent the results of the data-fitting procedure. (**A**): 212 nm photodissociation. (**B**): 226 nm photodissociation. (**C**): 230 nm photodissociation. (**D**): 230 nm photodissociation after NIR photolysis (NIR3, see Figure 7).

## 4. Discussion

The present study introduces a tentative photochemical kinetic model for the UV photolysis of GA in a low-temperature argon matrix. The main findings obtained by fitting the kinetic model to the experimental data are the existence of a single dissociation channel via the SSC conformer and the simultaneous occurrence of the SSC $\rightleftharpoons$ GAC and SSC $\rightleftharpoons$ AAT conformational photo-isomerization reactions. For comparison, the same photo-isomerization reactions were suggested by the kinetic model developed for the high overtone-induced photochemistry of GA [8]. The first OH overtone induced by NIR irradiation of matrix-isolated GA was found to lead to the isomerization reaction between the GAC and AAT conformers, but no isomerization reaction between SSC and AAT [6]. The generation of the AAT conformer is favored over the GAC conformer in ultraviolet irradiation, while the GAC conformer is favored in visible light irradiation [8].

The main result of the present study is the conformer-dependent photodissociation of the matrix-isolated GA. The electronic excitations of the GAC and AAT at 212, 226, and 230 nm are followed by relaxation processes that result in their conversion into the SSC conformer. The observed conformer-dependent isomerization for GA appears in line with the previously reported conformer-dependent photodissociation of formic acid [15] and N–C bonded bicyclic azoles [16].

**Author Contributions:** Conceptualisation, J.A., J.L. and M.W.; methodology, J.A.; software, J.A.; validation, J.A., M.W., R.F. and J.L.; formal analysis, J.A.; investigation, J.A., I.K. and T.N.; resources, J.L., M.W. and R.F.; data curation, J.A. and J.L.; writing—original draft preparation, J.A.; writing—review and editing, J.A., J.K., I.K., T.N., M.W., R.F. and J.L.; visualization, J.A.; supervision, J.L., M.W. and R.F.; project administration, J.L., R.F. and M.W.; funding acquisition, M.W., R.F. and J.L. All authors have read and agreed to the published version of the manuscript.

**Funding:** This research was funded by the Academy of Finland, grant numbers 286844 and 332023. The CQC-IMS is financially supported by the Portuguese Science Foundation ("Fundação para a Ciência e a Tecnologia"—FCT)—Projects CQC UIDB/00313/2020 and UIDP/00313/2020 (National Funds).

**Data Availability Statement:** Data is contained within the article.

**Acknowledgments:** Access to instruments from Laser-Lab Coimbra is gratefully acknowledged for the work performed in Portugal.

**Conflicts of Interest:** The authors declare no conflict of interest.

# References

1. Godfrey, P.D.; Rodgers, F.M.; Brown, R.D. Theory versus Experiment in Jet Spectroscopy: Glycolic Acid. *J. Am. Chem. Soc.* **1997**, *119*, 2232–2239. [CrossRef]
2. Havey, D.K.; Feierabend, K.J.; Takahashi, K.; Skodje, R.T.; Vaida, V. Experimental and Theoretical Investigation of Vibrational Overtones of Glycolic Acid and Its Hydrogen Bonding Interactions with Water. *J. Phys. Chem. A* **2006**, *110*, 6439–6446. [CrossRef] [PubMed]
3. Hollenstein, H.; Ha, T.-K.; Günthard, H. IR induced conversion of rotamers, matrix spectra, AB initio calculation of conformers, assignment and valence force field of trans glycolic acid. *J. Mol. Struct.* **1986**, *146*, 289–307. [CrossRef]
4. Reva, I.; Jarmelo, S.; Lapinski, L.; Fausto, R. First experimental evidence of the third conformer of glycolic acid: Combined matrix isolation, FTIR and theoretical study. *Chem. Phys. Lett.* **2004**, *389*, 68–74. [CrossRef]
5. Reva, I.D.; Jarmelo, S.; Lapinski, L.; Fausto, R. IR-Induced Photoisomerization of Glycolic Acid Isolated in Low-Temperature Inert Matrices. *J. Phys. Chem. A* **2004**, *108*, 6982–6989. [CrossRef]
6. Halasa, A.; Lapinski, L.; Reva, I.; Rostkowska, H.; Fausto, R.; Nowak, M.J. Near-Infrared Laser-Induced Generation of Three Rare Conformers of Glycolic Acid. *J. Phys. Chem. A* **2014**, *118*, 5626–5635. [CrossRef] [PubMed]
7. Ahokas, J.M.; Kosendiak, I.; Krupa, J.; Wierzejewska, M.; Lundell, J. High vibrational overtone excitation-induced conformational isomerization of glycolic acid in solid argon matrix. *J. Raman Spectrosc.* **2018**, *49*, 2036–2045. [CrossRef]
8. Nunes, C.M.; Reva, I.; Fausto, R. Conformational isomerizations triggered by vibrational excitation of second stretching overtones. *Phys. Chem. Chem. Phys.* **2019**, *21*, 24993–25001. [CrossRef] [PubMed]
9. Krupa, J.; Kosendiak, I.; Wierzejewska, M.; Ahokas, J.; Lundell, J. UV laser induced photolysis of glycolic acid isolated in argon matrices. *J. Photochem. Photobiol. A Chem.* **2021**, *412*, 113236. [CrossRef]
10. Souza, S. Low molecular weight carboxylic acids in an urban atmosphere: Winter measurements in Sao Paulo City, Brazil. *Atmos. Environ.* **1999**, *33*, 2563–2574. [CrossRef]
11. Vaida, V.; Donaldson, D.J. Red-light initiated atmospheric reactions of vibrationally excited molecules. *Phys. Chem. Chem. Phys.* **2014**, *16*, 827–836. [CrossRef] [PubMed]
12. Donaldson, D.J.; Tuck, A.F.; Vaida, V. Atmospheric Photochemistry via Vibrational Overtone Absorption. *Chem. Rev.* **2003**, *103*, 4717–4730. [CrossRef] [PubMed]
13. Vaida, V.; Feierabend, K.J.; Rontu, N.; Takahashi, K. Sunlight-Initiated Photochemistry: Excited Vibrational States of Atmospheric Chromophores. *Int. J. Photoenergy* **2008**, *2008*, 138091. [CrossRef]
14. Wolfram Research Inc. *Mathematica*; Version 13.0.1.0; Wolfram Research Inc.: Champaign, IL, USA, 2022. Available online: http://www.wolfram.com/mathematica (accessed on 10 August 2020).

15. Khriachtchev, L.; Pettersson, M.; Räsänen, M. Conformational Memory in Photodissociation of Formic Acid. *J. Am. Chem. Soc.* **2002**, *124*, 10994–10995. [CrossRef] [PubMed]
16. Pagacz-Kostrzewa, M.; Reva, I.D.; Bronisz, R.; Giuliano, B.M.; Fausto, R.; Wierzejewska, M. Conformational Behavior and Tautomer Selective Photochemistry in Low Temperature Matrices: The Case of 5-(1*H*-Tetrazol-1-yl)-1,2,4-triazole. *J. Phys. Chem. A* **2011**, *115*, 5693–5707. [CrossRef] [PubMed]

**Disclaimer/Publisher's Note:** The statements, opinions and data contained in all publications are solely those of the individual author(s) and contributor(s) and not of MDPI and/or the editor(s). MDPI and/or the editor(s) disclaim responsibility for any injury to people or property resulting from any ideas, methods, instructions or products referred to in the content.

Article

# Conformational Structure, Infrared Spectra and Light-Induced Transformations of Thymol Isolated in Noble Gas Cryomatrices

António Jorge Lopes Jesus [1,*], Cláudio M. Nunes [2] and Igor Reva [3]

1 Faculty of Pharmacy, University of Coimbra, CQC-IMS, 3004-295 Coimbra, Portugal
2 Department of Chemistry, University of Coimbra, CQC-IMS, 3004-535 Coimbra, Portugal; cmnunes@qui.uc.pt
3 Department of Chemical Engineering, University of Coimbra, CIEPQPF, 3030-790 Coimbra, Portugal; reva@eq.uc.pt
* Correspondence: ajorge@ff.uc.pt

**Abstract:** The conformational space of the natural product thymol (2-isopropyl-5-methylphenol) was investigated using quantum chemical calculations at the B3LYP and MP2 levels, which revealed the existence of four types of conformers differing in the orientation of the isopropyl and hydroxyl groups. Thymol monomers were isolated in noble gas (Ar and Xe) matrices (at 15 K) and characterized by IR spectroscopy. With the support of B3LYP harmonic vibrational calculations, the two most stable trans-OH-conformers, differing in the isopropyl orientation, were identified in the cryomatrices. The two less stable cis-OH conformers were not detected as they shall undergo fast tunneling to the most stable ones. Annealing experiments in a Xe matrix up to 75 K did not lead to any conversion between the two isolated conformers, which is in accordance with the significative energy barrier computed for rotamerization of the bulky isopropyl group (~24 kJ mol$^{-1}$). Vibrational excitation promoted by broadband or by narrowband irradiation, at the 2ν(OH) frequencies of the isolated conformers, did not lead to any conversion either, which was interpreted in terms of a more efficient energy transfer to the hydroxyl rotamerization (associated with a lower energy barrier and a light H-atom) than to the isopropyl rotamerization coordinate. Broadband UV irradiation experiments (λ > 200 nm) led to a prompt transformation of matrix isolated thymol, with spectroscopic evidence suggesting the formation of isomeric alkyl-substituted cyclohexadienones, Dewar isomers and open-chain conjugated ketenes. The photochemical mechanism interpretation concords with that reported for analogous phenol derivatives.

**Keywords:** thymol; photochemistry; matrix-isolation; IR spectra; conformational isomerization; H-tunneling

Citation: Lopes Jesus, A.J.; Nunes, C.M.; Reva, I. Conformational Structure, Infrared Spectra and Light-Induced Transformations of Thymol Isolated in Noble Gas Cryomatrices. *Photochem* **2022**, *2*, 405–422. https://doi.org/10.3390/photochem2020028

Academic Editors: Gulce Ogruc Ildiz and Licinia L.G. Justino

Received: 19 May 2022
Accepted: 3 June 2022
Published: 7 June 2022

**Publisher's Note:** MDPI stays neutral with regard to jurisdictional claims in published maps and institutional affiliations.

**Copyright:** © 2022 by the authors. Licensee MDPI, Basel, Switzerland. This article is an open access article distributed under the terms and conditions of the Creative Commons Attribution (CC BY) license (https://creativecommons.org/licenses/by/4.0/).

## 1. Introduction

Thymol (2-isopropyl-5-methylphenol, Scheme 1) is a naturally occurring monoterpenoid phenol derivative of *p*-cymene (1-isopropyl-4-methylbenzene), being one of the main chemical constituents of essential oils extracted from thyme plants. This compound is reported to exhibit, among others, antioxidant, analgesic, anti-inflammatory, antimicrobial, and anticancer activities [1–5]. Besides its pharmacological activities, thymol is also used as a flavoring and preservative ingredient in the food and cosmetic industries [6], as well as an insecticide, acaricide, and antiseptic agent [7,8].

Despite the wide range of bioactivities, the number of studies on the structure of thymol available in literature is relatively scarce [9–11]. X-ray data show that in the crystalline state, the thymol molecules are held together through O–H···O hydrogen bonds forming hexamers, which in turn are linked to each other by van der Waals contacts, while the crystal conformation is characterized by having the O–H group pointing away from the isopropyl fragment and the tertiary C–H bond of this fragment pointing towards the hydroxyl oxygen [9]. In gas phase, combination of microwave spectroscopy with theoretical calculations led to the experimental identification of three conformers for thymol, with one of them being much more populated than the other two [10].

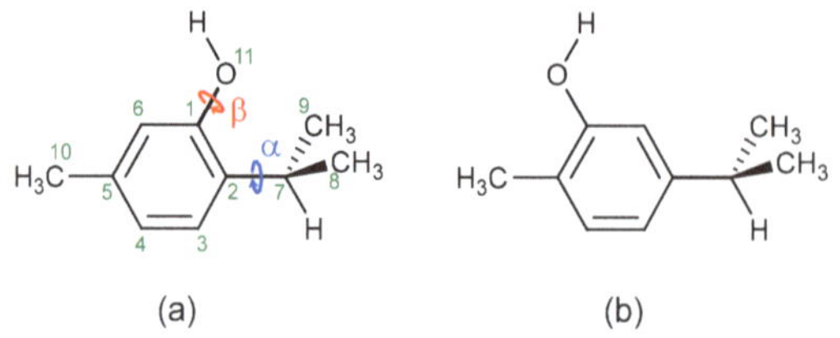

**Scheme 1.** Molecular structures of thymol (**a**) and carvacrol (**b**), including numbering of the heavy atoms and identification of the two conformationally relevant dihedral angles for the thymol structure. The H atoms connected to the aromatic ring are implicit.

Some infrared (IR) spectroscopic studies have been performed for thymol in the solid state [12,13] and in solution [14]. However, as far as we are aware, IR spectra of the monomeric compound have never been reported, nor has its light-induced transformations. Therefore, one of the objectives of the present paper is to fill this gap and to present, for the first time, a detailed analysis of the IR spectra of thymol isolated in low-temperature Ar and Xe matrices, and of the transformations triggered by exposing the matrix-isolated compound to IR- and UV-radiation.

Thymol is a positional isomer of carvacrol (5-isopropyl-2-methylphenol, see Scheme 1). The two molecules only differ in the relative position of the OH and isopropyl groups. In thymol, they are in *ortho*-position, while in carvacrol, they are in *meta*-position. Very recently, we have studied the conformational, vibrational and photochemical properties of carvacrol isolated in a low-temperature Ar matrix [15]. In that study, it has been shown that the molecule exists as two conformers in the cryogenic environment, differing from each other by the orientation of the isopropyl fragment. Exposition of the matrix-isolated carvacrol to broadband IR radiation was found to be effective in changing the relative population of the two conformers, by internal rotation of the bulky isopropyl fragment. A more specific goal of the current work is to verify whether this channel of vibrational energy dissipation is still effective in thymol, where the OH and isopropyl groups are closer together, making the barrier for internal rotation of the isopropyl group higher than in carvacrol.

## 2. Methods

### 2.1. Experimental Methods

Solid thymol (melting point of 49–51 °C) was supplied by Fluka with a purity degree of 99.5%. The matrix host gases (Ar N60 or Xe N48) were provided by Air Liquide. The low-temperature matrices were prepared by placing a small amount of thymol crystals into a glass tube, which was then connected to a cryostat (APD Cryogenics, Allentown, PA, USA, closed-cycle He refrigerator, with a DE-202A expander) through a needle valve, kept at ~298 K. The vapor of the sublimating compound was co-deposited with an excess of the matrix gas onto a CsI window cooled to 15 K. The temperature of the CsI window was measured directly at the sample holder by using a silicon diode temperature sensor connected to a digital controller (Scientific Instruments, Allentown, PA, USA, model 9650-1), providing the stabilization of temperature with an accuracy of 0.1 K. IR spectra of the matrix-isolated compound were recorded in the mid-IR range (4000–400 $cm^{-1}$) with 0.5 $cm^{-1}$ resolution using a Nicolet 6700 Fourier-transform infrared (FTIR) spectrometer (Porto, Portugal), equipped with a deuterated triglycine sulphate (DTGS) detector (Porto, Portugal) and a KBr beam splitter (Porto, Portugal). For the experiments carried out in xenon, spectra in the near-IR (NIR) range (7500–4000 $cm^{-1}$, 1 $cm^{-1}$ resolution) were also recorded using the same spectrometer equipped with an Indium Gallium Arsenide (InGaAs) detector and a $CaF_2$ beamsplitter.

The thymol/Xe matrix was irradiated through the outer quartz window of the cryostat by using narrowband frequency-tunable NIR light provided by the idler beam of a Quanta-

Ray MOPO-SL optical parametric oscillator (Mountain View, CA, USA) (pulse energy 10 mJ, duration 10 ns, repetition rate 10 Hz) pumped with a pulsed Nd:YAG laser. Alternatively, the sample was irradiated with broadband IR light provided by a kanthal wire electrically heated to a reddish-orange glow. Broadband UV-irradiations were also performed on the thymol/Ar matrix by using a 200 W output from a medium pressure Xe/Hg lamp (Oriel, Newport, UK).

*2.2. Computational Methods*

The thymol molecule has two conformational degrees of freedom, corresponding to the H–C7–C2–C1 ($\alpha$) and H–O11–C1–C2 ($\beta$) dihedral angles (see Scheme 1). For identification of the conformers exhibited by this molecule, a two-dimensional (2D) potential energy surface (PES) was computed by increasing both dihedral angles in steps of 20° over the 0–360° range, in all combinations, while all the remaining internal coordinates were optimized by using the B3LYP [16–18] hybrid DFT method in conjunction with the Pople-type 6-311++G(d,p) basis set. Then, the equilibrium geometries of the thymol conformers, as well as of the putative photoproducts resulting from the UV-photolysis of the matrix-isolated compound, were fully optimized followed by harmonic vibrational calculations at the same DFT level. For the thymol conformers, calculations were also carried out by combining the second-order Møller–Plesset (MP2) perturbation method [19] with the aug-cc-pVDZ Dunning's correlation consistent basis set. Single-point energy calculations were further carried out on the MP2 optimized geometries using the QCISD method [20] combined with the same aug-cc-pVDZ basis set. The Cartesian coordinates of the B3LYP and MP2 optimized geometries of the thymol conformers are provided in Tables S1 and S2.

The B3LYP/6-311++G(d,p) harmonic wavenumbers were corrected by using the following scale factors [15,21]: 0.950 (>3200 cm$^{-1}$), 0.960 (3200–2000 cm$^{-1}$) and 0.980 (<2000 cm$^{-1}$). A full list of the B3LYP/6-311++G(d,p) calculated wavenumbers (scaled) and IR intensities computed for the thymol conformers are provided in Table S3. For graphical comparison with the experimental spectra, the scaled wavenumbers and IR intensities (in km mol$^{-1}$) were convoluted with Lorentzian functions with a full width at half-maximum (fwhm) of 2 cm$^{-1}$ and peak heights matching the calculated IR intensities by using the Chemcraft software [22].

All quantum-mechanical computations reported in this work were performed with the Gaussian 09 software package [23], while the vibrational assignments were carried out with the vibAnalysis software [24].

## 3. Results and Discussion

*3.1. Conformers and Barriers to Internal Interconversion*

The B3LYP/6-311++G(d,p) 2D-PES calculated for thymol, as a function of the H–C7–C2–C1 ($\alpha$) and H–O11–C1–C2 ($\beta$) dihedrals, is represented in Figure 1 as a contour map. Two of the four minima found on the PES exist as pairs of mirror-image structures lying within relatively broad and shallow potential energy valleys along the $\alpha$ coordinate, extending from $-40$ to $+40°$. One of those doubly degenerate minima is found at ($\alpha$, $\beta$) = ($\pm 35°$, 0°) and is designated as **gc** ($g^+c/g^-c$), while the other is located at ($\alpha$, $\beta$) = ($\pm 35°$, 180°) and is labeled as **gt** ($g^+t/g^-t$), where $g^\pm$, **c** and **t** are abbreviations of gauche plus/minus, cis and trans orientations, respectively. The first and second letters refer to the orientation around the $\alpha$ and $\beta$ dihedrals, respectively. The $g^+c/g^-c$ and $g^+t/g^-t$ pairs of mirror-image structures are separated by first-order saddle points characterized by $\alpha = 0°$. Two additional unique minima are found on the PES at ($\alpha$, $\beta$) = (180°, 180°) and (180°, 0°) and are respectively called **tt** and **tc**. In Figure 2 are represented the geometries of the four identified thymol conformers, while in Table 1 are given their electronic ($\Delta E_{el}$), zero-point corrected ($\Delta E_0$) and Gibbs energies ($\Delta G$) at 298.15 K (temperature of the gaseous compound before the matrix deposition), calculated with the B3LYP/6-311++G(d,p) and MP2/aug-cc-pVDZ model chemistries on geometries fully optimized at the respective levels. In Table 1 are included the QCISD/aug-cc-pVDZ electronic energies calculated for the MP2/aug-cc-pVDZ optimized geometries. The results of our computations are concordant with those published earlier by Schmitz et al. [10].

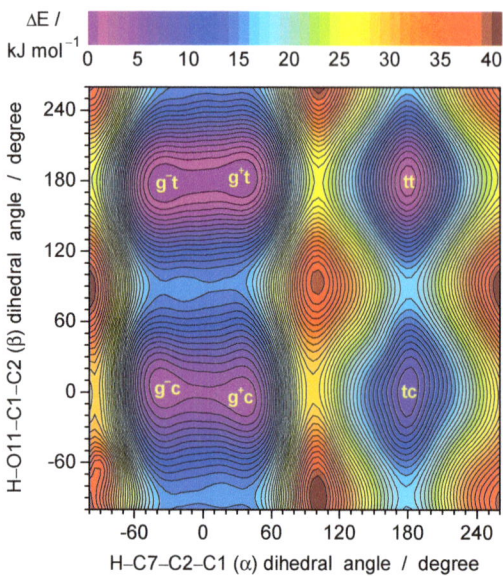

**Figure 1.** Contour plot of the potential energy surface computed for the thymol molecule as a function of the H−C7−C2−C1 (α) and H−O11−C1−C2 (β) dihedral angles (specified in Scheme 1). Both angles were incrementally fixed in steps of 20°, in all combinations, while all the remaining parameters were optimized at the B3LYP/6-311++G(d,p) level of theory. The minima are indicated in the plot as **gt** ($g^+t/g^-t$), **gc** ($g^+c/g^-c$), **tt** and **tc** (geometries are shown in Figure 2). The color bar on the top represents the relative energy scale (zero is assigned to the energy of the global minimum, **gt**). The isoenergy lines are traced with steps of 1 kJ mol$^{-1}$.

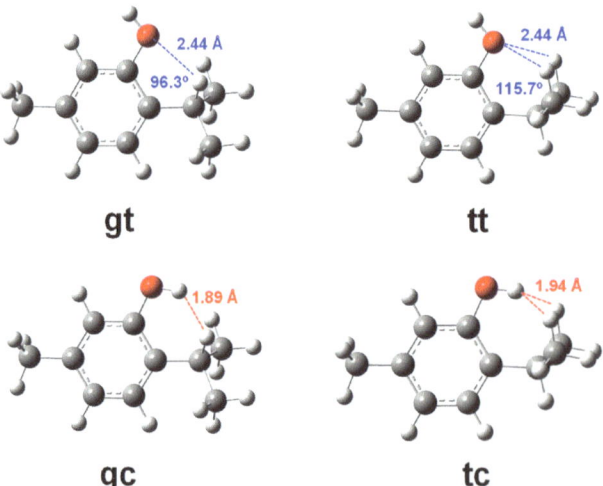

**Figure 2.** Geometries of the thymol conformers, including some geometrical parameters calculated at the MP2/aug-cc-pVDZ. For the doubly generated **gt** and **gc** conformers, only one of the mirror-image structures ($g^+t$ and $g^+c$) is shown.

Table 1. Relative energies and Boltzmann populations calculated for the thymol conformers at different levels of theory [a].

| Level of Theory | Conformer | | | |
|---|---|---|---|---|
| | gt | tt | gc | tc |
| | B3LYP/6-311++G(d,p) | | | |
| $\Delta E_{el}$ | 0.00 | 1.37 | 3.24 | 7.34 |
| $\Delta E_0$ | 0.00 | 1.11 | 3.54 | 6.07 |
| $\Delta G$ (298 K) | 0.00 | 1.27 | 4.96 | 5.18 |
| Pop. (%) | 66.9 | 20.0 | 9.0 | 4.1 |
| | (75.9) | (24.1) | | |
| | MP2/aug-cc-pVDZ | | | |
| $\Delta E_{el}$ | 0.00 | 2.51 | 2.54 | 7.94 |
| $\Delta E_0$ | 0.00 | 1.94 | 2.43 | 7.01 |
| $\Delta G$ (298 K) | 0.00 | 1.52 | 2.12 | 5.43 |
| Pop. (%) | 57.0 | 15.5 | 24.3 | 3.2 |
| | (81.3) | (18.7) | | |
| | QCISD/aug-cc-pVDZ [b] | | | |
| $\Delta E_{el}$ | 0.00 | 2.29 | 2.56 | 7.30 |

[a] $\Delta E_{el}$—relative electronic energy; $\Delta E_0$—relative electronic energy corrected for zero-point vibrational energy; $\Delta G$—relative Gibbs energy at 298 K; all relative energies are expressed in kJ mol$^{-1}$ and were calculated with respect to the energy of the most stable conformer **gt**. Equilibrium populations (Pop.) were estimated by means of the Boltzmann distribution based on the $\Delta G$ values at 298 K (weighting factors of 2/2 or 1/1 were used for conformers **gt/gc** or **tt/tc**, respectively). Values in parentheses correspond to the expected populations of conformers **gt** and **tt** with the contributions of the populations of conformers **gc** and **tc**, respectively. [b] Single-point calculations performed on the MP2/aug-cc-pVDZ optimized geometries.

At all levels of theory, **gt** is the lowest-energy conformer. It has the hydroxyl group pointing away from of the isopropyl fragment, while the C7–H bond is ± gauche with respect to the C2–C1 bond (see Figure 2), with α taking values of ±35.6° (B3LYP) or ±39.0° (MP2). The deviation of the C7–H bond from the aromatic ring plane should be attributed to the presence of the hydroxyl group in *ortho*-position relative to the isopropyl fragment, a feature also observed in a structural microwave spectroscopic study of the 2,6-diisopropylphenol analog [25]. Note that in carvacrol or *p*-isopropylphenol, where the OH and isopropyl substituents occupy *meta*- or *para*-positions, respectively, all conformers identified for these molecules were found to have the tertiary C–H bond of the isopropyl group lying in the aromatic ring plane [15,26,27]. The geometry of **gt** is not too different from that adopted by the thymol molecules in the crystal, although in the latter case the C7–H bond is less deviated from the ring plane (α = ±13.3°) [9]. The second most stable conformer, **tt**, also has the hydroxyl group pointing away from the isopropyl fragment, but the C7–H and C2–C1 bonds are trans to each other, bringing the two methyl groups of the isopropyl close to the oxygen atom (Figure 2). This conformational arrangement results in an increase of the electronic, zero-point corrected, or Gibbs energies by 1.1–2.5 kJ mol$^{-1}$ relative to the global minimum.

As shown in Figure 2, the (C7)H···O distance in **gt** and the (C8)H···O (equivalent of (C9)H···O) distance in **tt** are smaller than the sum of the van der Waals radii of the oxygen and hydrogen atoms (2.72 Å). This suggests the formation of C–H···O interactions in both trans-OH-conformers, as has been claimed in previous studies carried out for thymol [10] and 2-isopropylphenol [28]. These interactions, however, should be very weak because of the significant deviation of the C–H···O bonding angles (96–116°) from the linearity: it is well known that the strongest nonbonded interactions are observed for the linear arrangements (optimal value of 180°) [29,30]. In the **gc** and **tc** conformers, the hydroxyl group is directed towards the isopropyl fragment. Such an orientation does not allow the formation of stabilizing C–H···O interactions. Instead, it originates one (**gc**) or two (**tc**) repulsive CH···HO close contacts (see Figure 2). Both factors explain the lower stability of these cis-OH-conformers, and in particular of **tc**, relative to the **gt** and **tt** ones.

To characterize the conformational composition of the gaseous sample of thymol immediately before the matrix deposition, the populations of the thymol conformers in the gaseous phase were estimated from the values of $\Delta G$ at 298.15 K, through the application of the Boltzmann statistics, and the obtained values have been included in Table 1. According to these estimates, **gt** is by far the most abundant conformer in the gaseous phase, with a population of 57–67%. Conformer **tt** is also predicted to be present in the gaseous phase, but in a smaller amount (15–20%), while the population of conformer **gc** is of only 9% when it is estimated from the B3LYP Gibbs energies, increasing to 24% when obtained from the MP2 Gibbs energies. Nevertheless, as will be shown below, the difference of population predicted for this conformer by the two theoretical methods is not relevant from the experimental viewpoint. Finally, a very small the gaseous phase abundance (3–4%) is predicted for the less-stable **tc** conformer.

For the interpretation of the results of the matrix-isolation experiments, it is also relevant to have an idea of the energy barriers corresponding to the conversion of the higher into the lower energy conformers [31–33]. In Figure 3 are represented the OH-torsional barriers interconnecting the pair of conformers (**gc** and **gt**), and the pair of conformers (**tc** and **tt**). The heights of these barriers in the direction of conformational relaxation were calculated to be 10–12 kJ mol$^{-1}$. These values are of the same order as those computed for the conversion of higher into lower energy OH-rotamers in various compounds containing the phenol moiety, such as 2-cyanophenol (15 kJ mol$^{-1}$) [34], 2-isocyanophenol (12 kJ mol$^{-1}$) [35], hydroquinone (11 kJ mol$^{-1}$) [36], tetrachlorohydroquinone (19 kJ mol$^{-1}$) [37], and carvacrol (16 kJ mol$^{-1}$) [21]. Although these energy barriers are too high to be surmounted in a low-temperature matrix environment, for all these previously studied derivatives of phenol compounds, the higher energy OH-rotamers escaped from spectroscopic detection in matrices of inert gases because of their fast depopulation in favor of the corresponding lower energy OH-rotamers by means of quantum tunneling of hydrogen atom [38,39]. The same behavior has been reported in various matrix-isolation studies carried out for other compounds containing conformers differing from each other by 180° rotation of an OH group, namely carboxylic acids [40–44] and aminoacids [45]. Therefore, it is expected that in the case of thymol, the **gc** and **tc** cis-OH-conformers shall also undergo fast tunneling OH rotamerization to the respective **gt** and **tt** trans-OH-forms in the low temperature Ar and Xe matrices, thus preventing experimental identification of the cis-OH-forms.

To theoretically address this issue, we carried out computations of reaction paths for the flip of the OH group in the (**gc**, **gt**) and (**tc**, **tt**) pairs of conformers. Initially, the first-order transition states for OH-torsion were optimized, and the respective force constants were computed analytically. The intrinsic reaction paths were then followed from the transition states in both directions. The intrinsic reaction coordinate (IRC) was set using the "IRC = Cartesian" option of the Gaussian software and is expressed in units of Bohr. The computed reaction paths for the OH torsion in the (**gc**, **gt**) and (**tc**, **tt**) pairs are presented in Figure 3.

The probability of tunneling through a parabolic barrier was estimated by using the equation

$$P(E) = e^{-4\pi w \sqrt{2mE}/h} \qquad (1)$$

where a particle with mass $m$ tunnels through a barrier with height $E$ and width $w$, and $h$ is the Planck constant [38,39]. Using the calculated barrier heights $E$ of 10.32 and 8.57 kJ mol$^{-1}$ and widths $w$ of 2.915 and 2.715 Bohr for the **gc** $\rightarrow$ **gt** and **tc** $\rightarrow$ **tt** conversions, respectively, one can estimate the tunneling probability (transmission coefficient) of the light H-atom ($m = 1$) in **gc** as $2.97 \times 10^{-8}$, and in **tc** as $4.08 \times 10^{-7}$. The tunneling rate is a product of the transmission coefficient and the frequency of attempts. Assuming that the OH group is vibrating at an OH torsional frequency of 317 cm$^{-1}$ in **gc** and 275 cm$^{-1}$ in **tc** (computed values), tunneling rates of $2.82 \times 10^5$ s$^{-1}$ (half-life time of ~2.5 microseconds) and $3.36 \times 10^6$ s$^{-1}$ (half-life time of ~0.2 microseconds) were estimated for the **gc** $\rightarrow$ **gt** and **tc** $\rightarrow$ **tt** rotamerizations, respectively. According to these results, there is no doubt that the

**gc** and **tc** conformers are unlikely to survive long enough in low temperature Ar and Xe matrices to allow their spectroscopic detection.

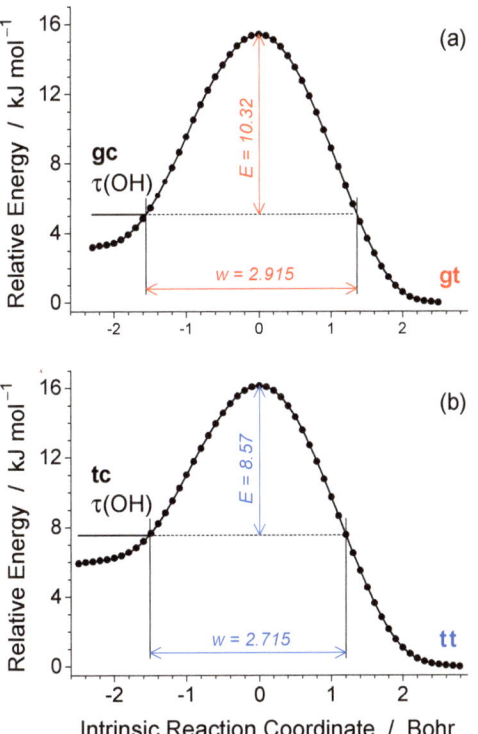

**Figure 3.** IRC profiles for the flip of OH-group in (**gc**, **gt**) (**a**) and (**tc**, **tt**) (**b**) pairs of thymol conformers, computed at the B3LYP/6-311++G(d,p) level of theory in Cartesian (non-mass-weighted) coordinates. The vertical arrows ($E$ = 10.32 or 8.57 kJ mol$^{-1}$) designate the computed ZPE-corrected energies of the transition states with respect to the reactants (**gc** or **tc**, left in each frame). The horizontal arrows designate the estimated barrier widths (w = 2.915 or 2.715 Bohr) of the barrier at the ZPE level of the reactants. The relative zero energies correspond to the products (**gt** or **tt**, right in each frame).

Conversion of the **tt** into the **gt** conformer involves the internal rotation of a bulky isopropyl group and should only occur through an over-the-barrier mechanism. The torsional barrier for this isomerization was computed to be ~24 kJ mol$^{-1}$ (see Figures 1 and S1). This value is too high to allow the occurrence of a **tt** → **gt** relaxation in a matrix environment at a temperature as low as 15 K [46,47]. Accordingly, both conformers are expected to be trapped in the as-deposited Ar and Xe matrixes with populations of 76–81% (**gt**) and 24–19% (**tt**). These values were obtained by considering the conversion of the estimated gas phase populations of **gc** to **gt** and of **tc** to **tt** (see Table 1).

It is interesting to compare the potential energy profile for the rotation of the isopropyl group in thymol with that computed for the closely related carvacrol and *p*-cymene molecules (see Figure S1). The heights of the barriers separating the conformers of these last two molecules differing from each other by internal rotation of the isopropyl group are ~10 kJ mol$^{-1}$, which is 2.5 times lower than the one computed for thymol. For these three molecules, the first-order transition state separating the conformers is characterized by having the tertiary C-H bond of the isopropyl group nearly perpendicular to the aromatic ring plane. In *p*-cymene and carvacrol, this gives rise to steric repulsions between the two methyl groups of the isopropyl fragment and their neighboring aromatic C-H bonds, whereas in thymol, one of the methyl groups stays too close to the OH group, leading to stronger steric repulsions, which is responsible for the significant increase in the barrier height associated with the isopropyl group rotamerization. As it will be shown below, this result has important implications for the outcomes of the experiments of thermal and IR excitations carried out for matrix-isolated thymol.

### 3.2. Infrared Spectra of Matrix-Isolated Thymol, Annealing and IR Irradiations

Monomers of thymol were trapped in Ar and Xe matrixes at 15 K, as described in Section 2.2, and the mid-IR spectra recorded for the freshly deposited samples are shown in Figure 4a,b (3675–3580 and 1650–700 cm$^{-1}$ regions), while the spectral positions and relative intensities (expressed qualitatively) of the observed bands are collected in Table 2. The two experimental spectra are well reproduced by the population-weighted theoretical spectrum of the matrix-isolated compound, which is displayed in Figure 4c. This spectrum was obtained by summing the individual spectra calculated for the conformers present in the matrix, shown in Figure 4d, with the computed IR intensities weighted by their expected populations in the noble gas matrices: 0.79 for **gt** and 0.21 for **tt** (mean values obtained from the B3LYP and MP2 computations, see Table 1). From the comparison of the experimental and theoretical spectra, an assignment of the experimental bands was carried out and has been included in Table 2.

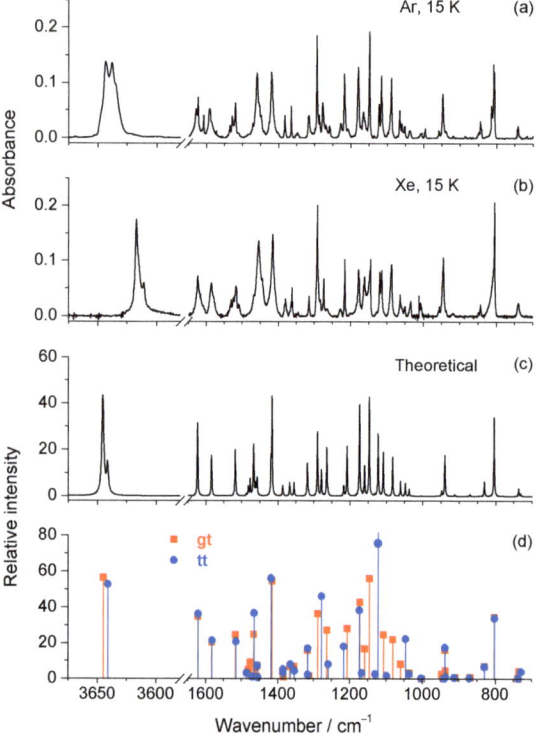

**Figure 4.** Experimental IR spectra of thymol isolated in (**a**) Ar and (**b**) Xe matrices at 15 K; (**c**) Theoretical IR spectrum of thymol, obtained as a sum of the B3LYP/6-311++G(d,p) spectra calculated for the two experimentally relevant conformers with their intensities scaled by the respective expected populations in the noble gas matrices. This spectrum was simulated by applying Lorentzian functions (fwhm = 2 cm$^{-1}$) centered at the scaled wavenumbers; (**d**) Scaled wavenumbers and unscaled IR intensities extracted from the harmonic vibrational calculations carried out at the B3LYP/6-311++G(d,p) level for the **gt** and **tt** conformers.

Table 2. Bands observed in the experimental IR spectra of thymol isolated in Ar and Xe matrices, compared with the harmonic wavenumbers ($\tilde{\nu}$/cm$^{-1}$) and absolute IR intensities (A$^{th}$/km mol$^{-1}$) calculated for the **gt** and **tt** conformers at the B3LYP/6-311++G(d,p) level, and vibrational assignment.

| Ar, 15 K [a] | | Xe, 15 K [a] | | Calc. gt [b] | | Calc. tt [b] | | Assignment [c] |
|---|---|---|---|---|---|---|---|---|
| $\tilde{\nu}$ | Int. | $\tilde{\nu}$ | Int. | $\tilde{\nu}$ | A$^{th}$ | $\tilde{\nu}$ | A$^{th}$ | |
| 3643/3638/3635 (sh) | vs | 3617/3611 | vs | 3645.5 | 56.6 | 3641.5 | 52.8 | ν(OH) |
| 1629 | m | 1624 | m | 1623.9 | 34.9 | 1622.9 | 36.3 | ν(CC)$_{ring}$ |
| 1591 | m | 1586 | m | 1584.8 | 20.6 | 1583.5 | 21.5 | ν(CC)$_{ring}$ |
| 1519 (split) | m | 1517 (split) | m | 1518.2 | 24.7 | 1516.5 | 20.5 | ν(CC)$_{ring}$; δ(CH)$_{ring}$ |
| | | | | 1481.9 | 5.1 | 1487.3 | 3.3 | δ(CH$_3$)$_{iso,as'}$ (+) |
| | | | | 1476.9 | 9.2 | 1471.4 | 1.1 | δ(CH$_3$)$_{iso,as''}$ (+) |
| 1460 (split) | s | 1455 (split) | s | 1467.7 | 24.8 | 1466.3 | 36.7 | δ(CH$_3$)$_{as'}$ |
| | | | | 1461.7 | 6.0 | 1456.6 | 0.8 | δ(CH$_3$)$_{iso,as''}$ (−) |
| | | | | 1459.3 | 1.3 | 1469.3 | 1.4 | δ(CH$_3$)$_{iso,as'}$ (−) |
| | | | | 1457.6 | 7.3 | 1458.0 | 7.3 | δ(CH$_3$)$_{as''}$ |
| 1420 | s | 1416 | s | 1417.0 | 54.5 | 1420.0 | 56.6 | δ(CH)$_{ring}$; ν(CC)$_{ring}$; ν(CO) |
| 1383 | w | 1380 | w | 1387.4 | 3.8 | 1386.3 | 3.4 | δ(CH$_3$)$_{iso,s}$ (+) |
| | | | | 1385.5 | 1.1 | 1386.5 | 5.3 | δ(CH$_3$)$_s$ |
| 1365 | w | 1362 | w | 1367.4 | 6.8 | 1365.8 | 8.0 | δ(CH$_3$)$_{iso,s}$ (−) |
| 1347 | vw | 1344 | vw | 1355.1 | 6.8 | 1355.1 | 4.6 | δ(C7H) |
| | | | | 1318.0 | 16.0 | 1316.6 | 17.1 | ν(CC)$_{ring}$; δ(OH) |
| 1315 | w | 1314 | w | 1313.7 | 1.7 | 1317.6 | 2.3 | γ(C7H) |
| **1294**/1287 | vs | **1291**/1284 | vs | 1289.8 | 36.5 | − | − | ν(CC)$_{ring}$; δ(CH)$_{ring}$ |
| 1278 | m | 1274 | m | − | − | 1278.9 | 46.2 | ν(CC)$_{ring}$; δ(CH)$_{ring}$ |
| 1269/1259 | vw | 1265 | vw | 1264.0 | 27.4 | 1260.5 | 8.1 | ν(CO) + ν(C5C10); ν(CC)$_{ring}$ |
| 1228 | vw | 1228 | vw | − | − | 1218.4 | 18.2 | ν(C2C7) |
| 1219 | s | 1216 | s | 1208.7 | 28.2 | − | − | ν(C2C7) |
| | | | | 1174.4 | 42.8 | 1175.0 | 38.3 | δ(OH); δ(C−H)$_{ring}$ |
| 1180 | s | 1179 | m | − | − | 1168.7 | 3.2 | ν(CO) − ν(C5C10); δ(CH)$_{ring}$ |
| 1166 | w | 1162 | m | 1160.8 | 16.7 | − | − | ν(CO) − ν(C5C10); δ(CH)$_{ring}$ |
| 1149 | vs | 1147/**1145** | s | 1147.7 | 56.1 | − | − | δ(OH); δ(CH)$_{ring}$; ν(CC)$_{ring}$ |
| 1123 | m | 1120 | m | − | − | 1123.1 | 111.8 | δ(ring); δ(OH); ν(CO) |
| 1116 | m | 1115 | m | 1108.7 | 24.8 | − | − | ν(C7C8) − ν(C7C9) |
| 1089 | m | 1087 | m | 1082.6 | 22.1 | − | − | δ(ring) |
| **1065**/1059 | w | 1063 | w | 1060.7 | 8.3 | − | − | ρ(CH$_3$)$_{iso}$ (+) |
| 1051 | vw | 1049 | vw | − | − | 1047.4 | 22.5 | ρ(CH$_3$)$_{iso}$ (+) |
| 1037 | vw | 1034 | vw | 1036.7 | 3.2 | 1037.2 | 2.7 | ρ(CH$_3$) |
| 1007/996 | vw | 1008 | vw | − | − | − | − | |
| 959 | vw | 956 | vw | 948.3 | 2.7 | − | − | ρ(CH$_3$)$_{iso}$ (−) |
| **948**/940 | m | **945**/937 | m | 939.6 | 16.0 | 939.7 | 17.5 | ν(CO) − ν(C5C10); δ(ring) |
| 918 | vw | 918 | vw | 912.4 | 0.9 | 914.2 | 0.8 | ρ(CH$_3$)$_{iso}$ (−); γ(C7H) |
| 886 | vw | n.o. | | 870.8 | 1.0 | 872.8 | 0.5 | ν(C7C8) + ν(C7C9) |
| 849/**846** | vw | 850/**844** | vw | 831.6 | 7.0 | 832.4 | 6.8 | γ(C6H) |
| 814/**809**/807 | s | 805 | vs | 805.0 | 34.5 | 805.1 | 33.8 | γ(C3H) + γ(C4H) |
| 741 | vw | 738 | vw | 736.3 | 4.3 | 730.5 | 4.1 | δ(ring) |
| 700 | vw | n.o. | | 681.3 | 1.3 | 684.0 | 4.9 | ν(C2C7) − ν(C5C10); δ(ring) |
| 594 | w | 593 | w | 594.0 | 7.8 | 594.6 | 8.5 | γ(C5) |
| 577 | w | 577 | w | 571.5 | 12.9 | 576.9 | 6.4 | δ(ring) |

[a] Absorptions falling in the 3100–2800 cm$^{-1}$ range, corresponding to the CH and CH$_3$ stretching vibrations, were not analyzed. Experimental intensities (Int.) are expressed qualitatively: vs = very strong; s = strong; m = medium; w = weak; vw = very weak. The most intense component of each multiplet band is highlighted in bold. [b] Calculated harmonic wavenumbers are scaled by 0.980 and 0.950, for the regions below 2000 cm$^{-1}$ and above 3200 cm$^{-1}$, respectively. Some weak calculated absorptions with no correspondence in the experimental spectrum are not included (a full list of the vibrations calculated for the two conformers is provided in Table S3). [c] Based on the results provided by the "vibAnalysis" software [24] supported by ChemCraft animation of the vibrations. Atom numbering is presented in Scheme 1. Abbreviations: ν, stretching; δ, in-plane deformation; γ, out-of-plane deformation; τ, torsion; ρ, rocking; as, antisymmetric; s, symmetric; iso, isopropyl group, sh, shoulder. Signs "+" and "−" designate combinations of vibrations in the same phase and in the opposite phase, respectively.

The intense band observed in the experimental spectra at wavenumbers above 3600 cm$^{-1}$ is assigned to the stretching vibration of the OH group (νOH). In the spectrum of thymol recorded in Ar matrix, this band appears as a relatively broad and split feature with two main components at 3643 and 3638 cm$^{-1}$ and a shoulder at ~3635 cm$^{-1}$. OH stretching bands at practically the same positions have been identified in the IR spectra of phenol [48] (also having two main components) and of some phenol derivatives such as tyramine [49] and carvacrol [15], isolated in solid Ar. In the spectrum of thymol trapped in Xe, the νOH band is sharper and appears as a doublet at 3617/3611 cm$^{-1}$, shifted to lower frequencies, as compared with thymol in Ar. Such a shift of the OH-stretching frequency is typical for matrix-isolated compounds [32,33]. In this range, the spectrum of thymol in a xenon matrix exhibits a very similar profile to that of the population-weighted theoretical spectrum. This similarity suggests that the higher and lower frequency band components in the experimental spectrum are assigned, respectively, to the **gt** and **tt** conformers, which agrees with the order of wavenumbers predicted for the νOH absorptions of the two rotamers: 3645.5 cm$^{-1}$ ($A^{th}$ = 56.6 km mol$^{-1}$) for **gt** and 3641.5 cm$^{-1}$ ($A^{th}$ = 52.8 km mol$^{-1}$) for **tt**.

In the fingerprint region, a detailed comparison of the experimental and theoretical spectra allows the identification of seven bands that can be assigned to the most populated **gt** conformer. These bands appear between 1300 and 1000 cm$^{-1}$ and are centered in the spectra recorded in Ar/Xe at 1294/1291, 1219/1216, 1166/1162, 1149/1145, 1116/1115, 1089/1087 and 1065/1063 cm$^{-1}$, while the corresponding theoretical absorptions are predicted at 1289.8, 1208.7, 1160.8, 1147.7, 1108.7, 1082.6 and 1060.7 cm$^{-1}$. Spectral markers of the **tt** conformer are located at 1278/1274, 1228/1228, 1123/1120 and 1051/1049 cm$^{-1}$ (see Table 2), with the respective calculated absorptions predicted at 1278.9, 1218.4, 1123.1 and 1047.4 cm$^{-1}$.

As we have mentioned in the previous section, the **gc** and **tc** conformers are not expected to be present in the as-deposited matrixes because of their fast depopulation (on the microsecond time scale) in favor of the **gt** and **tt** forms by quantum mechanical tunneling during deposition of the matrixes. This is experimentally supported by a thorough comparison between the experimental and calculated spectra shown in Figure S2, where intense absorptions calculated for the two higher-energy forms have no correspondence in the experimental spectra.

The experimental populations of the two matrix-populated conformers were estimated from the integrated intensities of selected pairs of band components (Ar: 1294/1278, 1219/1228, 1149/1123; Xe: 3617/3611, 1291/1274, 1216/1228, 1145/1120), normalized by the calculated absolute intensities of the corresponding vibration modes. The obtained values were 75 ± 6% for **gt** and 25 ± 6% for **tt**, which are in good concordance with the theoretical predictions. This gives an approximate **gt**:**tt** experimental ratio of 3:1, which will be discussed later on in this paper.

One attempt to promote conversion between the two rotamers in the matrix, in order to allow their unequivocal identification, consisted of performing annealing experiments. Taking into consideration the magnitude of the energy barrier corresponding to the conversion of the **tt** into the **gt** conformer (~24 kJ mol$^{-1}$), this conformational relaxation may only be observed in a matrix with stronger relaxant properties and with the possibility of heating to the highest possible temperature. Xe matrix was therefore the obvious choice for performing these experiments. The sample was annealed up to 75 K, which is the upper limit of thermal stability of solid xenon. However, no changes in the band intensities were observed that could be attributed to the conversion of the higher energy conformer into the most stable form. Apparently, the high energy barrier separating the two matrix-isolated conformers, combined with necessity of internal torsion of a bulky isopropyl group, precludes the occurrence of a **tt** → **gt** relaxation in the Xe matrix, even upon heating the sample at a temperature as high as 75 K. According to the correlation proposed by Barnes [50], a temperature above 80 K would be needed to overcome an energy barrier of ~24 kJ mol$^{-1}$. Hence, the non-observation of a thermally induced **tt** → **gt** relaxation is not surprising under the present experimental conditions.

Another attempt to induce changes in the population of the two matrix populated conformers consisted of the excitation of monomers of thymol isolated in solid Xe with narrowband or broadband IR radiation. The narrowband IR excitations were undertaken by tuning the optical parametric oscillator at the first overtone of the OH stretching vibration (2νOH). Before performing the irradiations, a near-IR spectrum was recorded to identify the position of the 2νOH band. This band appears as a doublet with a main component centered at 7061 cm$^{-1}$, and a shoulder at 7047 cm$^{-1}$ (see Figure S3), which is very close to its position observed in the near-IR spectrum of thymol dissolved in CCl$_4$ solution (7056 cm$^{-1}$) [14]. The shape of the 2νOH band is identical to that of the fundamental νOH feature observed in the mid-IR spectrum, meaning that the higher and lower frequency components should be ascribed, respectively, to the 2νOH absorptions of the **gt** and **tt** conformers. Absorption of a near-IR photon with wavenumber in the 7061–7047 cm$^{-1}$ range introduces in the molecule an energy of ~84 kJ mol$^{-1}$. This amount of excitation energy is 3.5 times larger than the barrier height computed for the interconversion between the conformers in both directions, which, in principle, is more than sufficient to overcome the barrier for conformational isomerization for a free monomer in the gas phase by internal rotation of the isopropyl group. However, exposure of thymol isolated in a Xe matrix to laser light tuned to 7061 or 7047 cm$^{-1}$ for ~30 min did not result in any spectral modifications that could indicate that a conformational transformation had occurred. We also attempted to promote changes in the population of the two matrix-isolated conformers by letting the matrix be exposed to broadband IR-radiation emitted from a kanthal wire electrically heated to a reddish-orange glow, a methodology that has been successfully applied by us to induce changes in the relative abundances of the two conformers of the structurally similar carvacrol isolated in an Ar matrix [15]. However, in the present case, we did not observe any spectral indications of the occurrence of conformational changes.

A possible explanation for the lack of infrared-induced conformational isomerization of the isopropyl group in thymol **tt** or **gt** is that the energy deposited in the 2νOH mode is more efficiently transferred to the lower energy OH-rotamerization coordinate (Figure 5), which will result in the isomerization of **tt** into **tc** or **gt** into **gc**. However, because **tc** and **gc** are high-energy cis-OH conformers that should decay very fast by H-tunneling back to the most stable **tt** and **gt** trans-OH-forms, respectively, no spectral changes would be observed in the course of the infrared irradiation experiments monitored using steady-state spectroscopy. Interestingly, as shown in Figure 5, in comparison with thymol, the carvacrol molecule has the C$_3$H$_7$ and OH competitive rotamerizations with an opposite barrier height trend. That feature is likely to provide for carvacrol a more efficient energy transfer to its lower energy C$_3$H$_7$ rotamerization coordinate, even when the energy is deposited in the 2ν(OH) or ν(OH) mode. This explains why the conformational isomerization of the isopropyl group in carvacrol was previously observed upon broadband IR irradiation [15].

### 3.3. UV-Induced Transformations

Monomers of thymol **1** isolated in a low temperature Ar matrix were subjected to irradiation with broadband UV-light at λ > 200 nm emitted directly from a Xe/Hg lamp. The experimental conditions for the UV-irradiations were established taking into consideration the UV-spectrum of thymol in methanol solution, which exhibits absorption maxima at λ = 221 and 280 nm [13]. The times of exposition of the sample to the UV-light varied from 2 to 120 min. After each irradiation, an IR spectrum was collected to monitor the transformations occurring in the matrix-isolated compound.

Spectral indication of occurrence of photoreactions started to be observed shortly after the first four minutes of irradiation. The most notorious spectral change was the appearance of a relatively broad feature within the 2125–2090 cm$^{-1}$ range, centered at ~2106 cm$^{-1}$ (see Figure 6a). This band is very characteristic of the antisymmetric stretching vibration of the ketene group [ν$_{as}$( C=C=O)] [21,48,51–54], thus revealing the photogeneration of compounds bearing this fragment. According to the results of previous photochemical studies carried out for matrix-isolated compounds containing the phenol [15,21,48,49,54,55] moiety, the photoreaction pathway leading to the formation of ketenes is initiated by detachment of the H-atom from the O–H bond in **1**. Then, recombination of the released H-atom with the highly reactive phenoxyl-type radical **2** at the *ortho-* or *para-* ring carbon atoms with respect to the CO group, gives rise to alkyl-substituted cyclohexadienones, which may exist in three isomeric forms, labeled as **3a**, **3b** and **3c** (see Scheme 2). The preference of the H-atom to reconnect at these ring positions is theoretically supported by the results of an NBO analysis carried out at the B3LYP/6-311++G(d,p) level for **2**. In fact, besides the oxygen atom, atoms C2, C4 and C6 are those exhibiting, by far, the largest natural spin densities (see Figure S4). The presence of **3** in the UV-irradiated thymol/Ar matrix is spectroscopically confirmed by the appearance of two bands at 1677 and 1659 cm$^{-1}$ (Figure 6a), which are typical positions of the stretching vibration of a C=O group in cyclohexadienones [15,21,48,56]. Finally, cleavage of the weaker C–C bond at α-position with respect to the C=O group in **3a** or **3b**, yields open-chain conjugated ketenes **5a** or **5b**, respectively (Scheme 2).

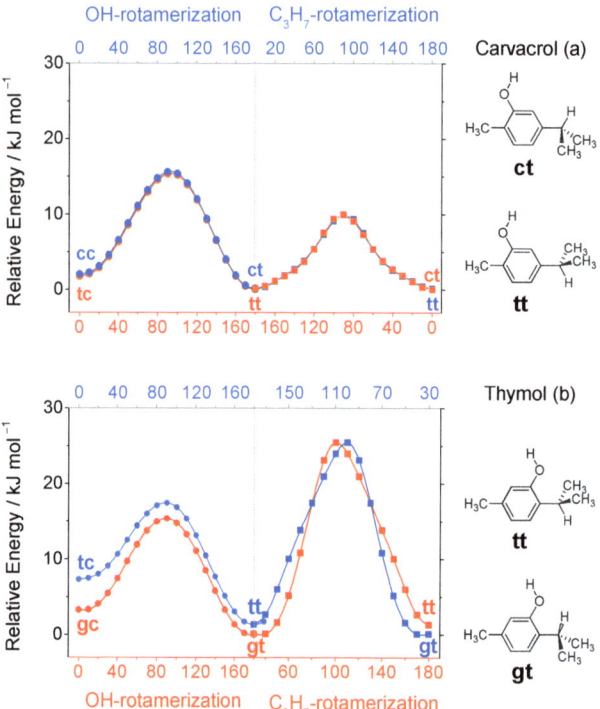

**Figure 5.** B3LYP/6-311++G(d,p) relaxed potential energy scans calculated for (**a**) carvacrol and (**b**) thymol regarding the rotamerization of the hydroxy (OH) and isopropyl (C$_3$H$_7$) groups. The bottom and top x-axis (red and blue) refer to the corresponding dihedral angle changes of the first and second most stable conformations, respectively, which are illustrated for each species at the right side of the panel. The conformers of carvacrol are designated according to the scheme adopted in this work for thymol: the first letter designates the orientation of the tertiary C–H bond of the isopropyl fragment relative to the C–O bond, while the second refers to the orientation of the O–H group relative to the vicinal alkyl group.

To theoretically support the proposed assignments, we have carried out full geometry optimizations and harmonic vibrational calculations for the relevant isomeric forms adopted by **3** and **5**, which are characterized in Tables S4–S6. For species **5**, only the structures adopting Z-configuration around the central C=C bond have been considered in the calculations. This assumption was based on the results of a recent study on the photochemistry of 1,3-cyclohexadiene isolated in solid parahydrogen, where only the Z-type isomers of the open-ring photoproducts were found to appear during the initial stages of the UV-irradiations [57]. The range of wavenumbers predicted for the $\nu_{as}$(C=C=O) mode of **5** was 2106–2084 cm$^{-1}$ (A$^{th}$ = 637–1485 km mol$^{-1}$) and for the $\nu$(C=O) mode of **3** was

1679–1654 cm$^{-1}$ (A$^{th}$ = 165–318 km mol$^{-1}$). These values are very close to the positions of the new observed absorptions of photoproducts, emerging upon UV-irradiations of thymol (Figure 6). We also tried to identify a spectral signature of **2**. However, the few strong infrared absorptions predicted for this species could not be observed experimentally, which is not surprising, due to its high reactivity.

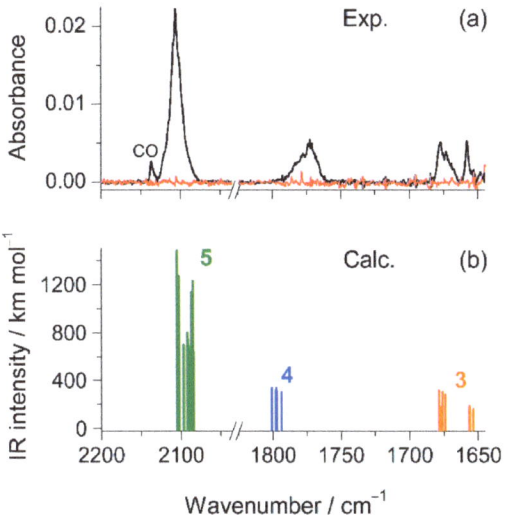

**Figure 6.** (**a**) Fragment of the IR spectrum of thymol **1** (Ar, 15 K) recorded after 4 min of UV-irradiation (black trace) and before any UV-irradiation (red trace). (**b**) B3LYP/6-311++G(d,p) calculated wavenumbers (scaled) and IR intensities corresponding to the antisymmetric C=C=O stretching modes of all surveyed isomeric forms of the open-ring ketenes **5** (green), C=O stretching modes of all Dewar valence isomers **4** (blue), and C=O stretching modes of all isomeric forms of the alkyl-substituted cyclohexadienones **3** (orange). For the structures of **3**, **4**, **5**, see Scheme 2.

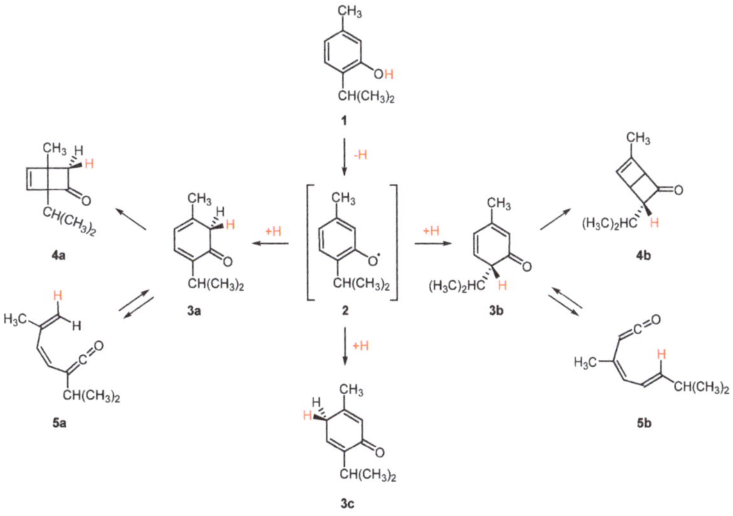

**Scheme 2.** UV-induced isomerizations experimentally observed for thymol **1** isolated in solid Ar at 15 K.

Alongside with the IR bands assigned to **3** and **5**, a new band centered at ~1773 cm$^{-1}$ was also observed in the infrared spectra recorded in the early stages of the UV-irradiations (Figure 6a). A band at nearly the same position has been previously detected in the photochemical studies performed for matrix-isolated phenol (1789 cm$^{-1}$) [48] and carvacrol (1775 cm$^{-1}$) [15] and has been ascribed to the stretching vibration of a C=O bond of a Dewar valence isomer. Therefore, an analogous species, labeled as **4** in Scheme 2, must be also formed during the UV-irradiations of **1**. It is most likely produced with the intermediacy of **3a** or **3b**, by analogy with the identical process observed for the electronically equivalent α-pyrone [52,58,59] and methyl-substituted α-pyrones [51,60]. The wavenumbers predicted for the possible structures adopted by **4** (see Table S7) fall between 1801 and 1794 cm$^{-1}$ ($A^{th}$ = 281–340 km mol$^{-1}$), see Figure 6b.

Increasing the time of exposition of the sample to the UV-radiation led to the occurrence of photodecomposition reactions, as revealed by the appearance of a narrow band at ~2137 cm$^{-1}$ which is a spectral indication of the formation of carbon monoxide (CO) (Figure 6a and Figure S5). This band was found to continuously grow up as the time of irradiation increases, whereas the opposite behavior was observed for the bands assigned to cyclohexadienones **3** (and the open-chain ketenes **5**), see Figure S5. By analogy with our recent discussion on the photochemical behavior of matrix-isolated carvacrol [15], the most probable species resulting from the photodecomposition of thymol and its photoproducts are alkyl-substituted cyclopentadienes. These photoproducts are most likely formed by decarbonylation of the alkyl-substituted cyclohexadienones **3**, which exist in photoequilibrium with the open-chain ketenes **5** (see Scheme 2) and as long as isomers **3** are consumed, isomers **5** are also consumed (indirectly) [15], see Figure S5. These species are constituted exclusively by carbon and hydrogen atoms, thus having intrinsically low intensities in the IR spectra. Furthermore, such decarbonylated products share common groups (methyl, isopropyl) with their precursors and their new bands may overlap with already existing bands of the precursors. Therefore, it is difficult to experimentally confirm their presence in the UV-irradiated matrix.

## 4. Concluding Discussion

In this work, we investigated the conformational structure of monomeric thymol by a combined experimental and computational approach. Geometry optimizations at the B3LYP/6-311++G(d,p) and MP2/aug-cc-pVDZ levels of theory predicted that thymol may adopt four unique conformations defined by the mutual orientation of the vicinal isopropyl and hydroxyl groups. Two of these conformations, with the gauche-isopropyl group, are doubly degenerate and represent mirror-equivalent structures with broad and shallow double-minima on the potential energy surface of the isopropyl torsion. The two conformations with the trans-isopropyl group represent narrow and deep single minima. The two cis-OH-conformers are destabilized due to the internal repulsions between the hydrogen atoms of the hydroxyl and isopropyl groups. The two trans-OH-forms (**gt** and **tt**) are the most stable and are expected to be the most populated forms in the gas phase of thymol and (consequently) also for thymol isolated in the cryogenic matrix.

Infrared spectroscopy was used to characterize the conformational composition of thymol experimentally. In agreement with the computations, the infrared spectrum of the as-prepared matrix revealed that only the two most stable thymol conformers (trans-OH-forms **gt** and **tt**, with the OH group pointing away from the vicinal isopropyl group) co-exist in the matrix. The experimental integrated intensities of the marker bands due to the **gt** and **tt** forms, reduced by their respective computed absolute infrared intensities, suggest an approximate 3:1 ratio in favor of **gt**. The two higher energy cis-OH forms, which may have residual populations in the gas phase prior to deposition, were shown to convert into their more stable trans-OH-counterparts by fast quantum mechanical tunneling during deposition of the matrix.

By analogy with the previously studied carvacrol (differing from thymol only by position of the OH group), we attempted to promote a partial conformational isomerization between the **gt** and **tt** conformers. For this purpose, we irradiated the matrix-isolated thymol either with broadband IR light (provided by electrically heated kanthal wire) or with narrowband near-IR light (provided by an optical parametric oscillator tuned to the frequency of the first overtone of the OH-stretching mode). Despite all of these attempts, the initial populations of the **gt** and **tt** conformers did not change. The annealing of thymol isolated in a xenon matrix up to the temperatures over 75 K (i.e., up to the limit of the thermal stability of solid xenon), aiming at relaxation of the higher-energy **tt** form into the global energy minimum **gt** form, did not result in the conformational interconversion either. From these experiments, one could safely postulate that conformational populations of the **gt** and **tt** forms in a freshly deposited matrix, in the 3:1 ratio, correspond to the ratio of conformational populations of these forms in the gas phase (at room temperature) during the deposition of the sample. Taking into consideration the computed relative Gibbs free energies of the **gt** and **tt** forms at room temperature, such a conformational composition is only possible when the **gt** and **gc** forms contribute to the Boltzmann equilibrium with the weighting factors of two and the **tt** and **tc** forms with the weighting factors of one. This represents an experimental proof of the fact that the two mirror-equivalent $g^+t$ and $g^-t$ forms of thymol, separated by a torsional energy barrier not exceeding 0.7 kJ mol$^{-1}$, are indeed two individual minima, and not a single symmetry-averaged **ct** minimum.

The lack of conformational relaxation of matrix-isolated thymol was explained with the aid of computations of potential energy surfaces for the torsion of the hydroxyl and isopropyl groups in thymol and in the related molecules. Furthermore, the torsional barrier for isopropyl group in thymol is nearly 2.5 times higher than in its isomeric carvacrol. On the other hand, the torsional barrier for the hydroxyl group in carvacrol and thymol are approximately equal, but in thymol it is about a half as low compared to the torsional barrier of its isopropyl group. In our mechanistic interpretation, the energy provided by vibrational excitation (infrared or near-IR) of thymol is dissipated much more efficiently via the hydroxyl torsion, and therefore the isopropyl torsion does not occur.

The matrix-isolated thymol was subsequently irradiated with broadband UV-light ($\lambda > 200$ nm), and the resulting transformations were monitored by IR spectroscopy. As is typical of phenol derivatives, the photoexcitation of thymol resulted in production of corresponding alkyl-substituted cyclohexadienones. Mechanistically, such a photochemical behavior was explained in terms of the OH group cleavage (with generation of an alkyl-substituted phenoxyl radical), followed by recombination of the released H-atom (also a radical) at the *ortho-* or *para-* positions of the ring (with respect to the CO group). The natural spin densities of the phenoxyl-type radical, extracted from the NBO calculations, indicate that precisely the *ortho-* or *para-* carbon atoms of the ring are the most favorable positions for the above-mentioned recombination of the radical pair. The cyclohexadienone photoproducts were found to undergo further isomerizations, yielding Dewar isomeric species and open-chain conjugated ketenes. Decarbonylation of the photoproducts was also observed for longer irradiation times.

On the whole, the present investigation fills a gap in the knowledge about the structural, vibrational, and photochemical properties of monomeric thymol. This fundamental knowledge constitutes a basis for the understanding of the bioactivity of this and related compounds.

**Supplementary Materials:** The following supporting information can be downloaded at: https://www.mdpi.com/article/10.3390/photochem2020028/s1. Tables S1 and S2, Cartesian coordinates for the conformers of thymol optimized at B3LYP/6-311++G(d,p) and MP2/aug-cc-pVDZ levels of theory; Table S3, Scaled wavenumbers (cm$^{-1}$) and IR intensities (km mol$^{-1}$) calculated for the thymol conformers by means of B3LYP harmonic vibrational calculations; Tables S4–S7, Geometries and selected energetic and vibrational parameters calculated at the B3LYP level for different photoproducts resulting from the UV-isomerizations of matrix-isolated thymol; Figure S1, B3LYP potential energy scans for the intramolecular torsion of the isopropyl group computed for thymol, carvacrol, and *p*-cymene; Figure S2, Detailed comparison of the experimental spectra of thymol isolated in Ar and Xe matrices with the spectra calculated for the four conformers of thymol; Figure S3, Fragment

(7100–7020 cm$^{-1}$) of the near-IR spectrum of thymol isolated in Xe showing the band assigned to the first OH-stretching overtone; Figure S4, Natural spin densities computed for the heavy atoms of the two conformers of the phenoxyl-type radical of thymol by means of a Natural Bond Orbital (NBO) analysis carried out at the UB3LYP/6-311++G(d,p) level of theory; Figure S5, Evolution of the bands due to photoproducts emerging in the 2160–2080 cm$^{-1}$ and 1690–1650 cm$^{-1}$ regions, as function of the time of UV-irradiation.

**Author Contributions:** A.J.L.J., conceptualization, investigation, computations, writing—original draft preparation; C.M.N., writing—review and editing, formal analysis, project administration; I.R., laboratory work, computations, writing—review and editing, formal analysis. All authors have read and agreed to the published version of the manuscript.

**Funding:** This work was supported by Project POCI-01-0145-FEDER-028973 funded by FEDER, via Portugal 2020-POCI, and by National Funds via the Portuguese Foundation for Science and Technology (FCT). The Coimbra Chemistry Centre—Institute of Molecular Sciences (CQC-IMS) is supported by FCT through projects UIDB/00313/2020 and UIDP/00313/2020 co-funded by COMPETE and the IMS special complementary funds provided by FCT. The Chemical Process Engineering and Forest Products Research Centre (CIEPQPF) is supported by FCT through projects UIDB/EQU/00102/2020 and UIDP/EQU/00102/2020 (National Funds). C.M.N. acknowledges FCT for an Auxiliary Researcher grant. The authors also acknowledge LaserLab Coimbra for experimental facilities.

**Institutional Review Board Statement:** Not applicable.

**Informed Consent Statement:** Not applicable.

**Acknowledgments:** The authors acknowledge Katarzyna M. Marzec and José P. L. Roque for their participation in some of the matrix-isolation experiments.

**Conflicts of Interest:** The authors declare no conflict of interest.

## References

1. Nagoor Meeran, M.F.; Javed, H.; Al Taee, H.; Azimullah, S.; Ojha, S.K. Pharmacological properties and molecular mechanisms of thymol: Prospects for its therapeutic potential and pharmaceutical development. *Front. Pharmacol.* **2017**, *8*, 380. [CrossRef]
2. Sahoo, C.R.; Paidesetty, S.K.; Padhy, R.N. The recent development of thymol derivative as a promising pharmacological scaffold. *Drug Dev. Res.* **2021**, *82*, 1079–1095. [CrossRef]
3. Escobar, A.; Pérez, M.; Romanelli, G.; Blustein, G. Thymol bioactivity: A review focusing on practical applications. *Arab. J. Chem.* **2020**, *13*, 9243–9269. [CrossRef]
4. Jamali, T.; Kavoosi, G.; Jamali, Y.; Mortezazadeh, S.; Ardestani, S.K. In-vitro, in-vivo, and in-silico assessment of radical scavenging and cytotoxic activities of *Oliveria decumbens* essential oil and its main components. *Sci. Rep.* **2021**, *11*, 14281. [CrossRef]
5. Bautista-Hernández, I.; Aguilar, C.N.; Martínez-Ávila, G.C.G.; Torres-León, C.; Ilina, A.; Flores-Gallegos, A.C.; Kumar Verma, D.; Chávez-González, M.L. Mexican oregano (Lippia graveolens Kunth) as source of bioactive compounds: A review. *Molecules* **2021**, *26*, 5156. [CrossRef]
6. Nieto, G. A review on applications and uses of thymus in the food industry. *Plants* **2020**, *9*, 961. [CrossRef]
7. Pandey, S.K.; Upadhyay, S.; Tripathi, A.K. Insecticidal and repellent activities of thymol from the essential oil of *Trachyspermum ammi* (Linn) Sprague seeds against Anopheles stephensi. *Parasitol. Res.* **2009**, *105*, 507–512. [CrossRef]
8. Scoralik, M.G.; Daemon, E.; de Oliveira Monteiro, C.M.; Maturano, R. Enhancing the acaricide effect of thymol on larvae of the cattle tick Rhipicephalus microplus (Acari: Ixodidae) by solubilization in ethanol. *Parasitol. Res.* **2012**, *110*, 645–648. [CrossRef] [PubMed]
9. Thozet, A.; Perrin, M. Structure of 2-isopropyl-5-methylphenol (thymol). *Acta Crystallogr. Sect. B* **1980**, *36*, 1444–1447. [CrossRef]
10. Schmitz, D.; Shubert, V.A.; Giuliano, B.M.; Schnell, M. The broadband microwave spectra of the monoterpenoids thymol and carvacrol: Conformational landscape and internal dynamics. *J. Chem. Phys.* **2014**, *141*, 034304. [CrossRef]
11. Sin, K.-R.; Kim, C.-J.; Ko, S.-G.; Hwang, T.-M.; Han, Y.-N.; Pak, Y.-N. Inclusion of thymol into cucurbiturils: Density functional theory approach with dispersion correction and natural bond orbital analysis. *J. Inclusion Phenom. Macrocycl. Chem.* **2022**, *102*, 533–542. [CrossRef]
12. Saraiva, A.G.Q.; Saraiva, G.D.; Albuquerque, R.L.; Nogueira, C.E.S.; Teixeira, A.M.R.; Lima, L.B.; Cruz, B.G.; de Sousa, F.F. Chemical analysis and vibrational spectroscopy study of essential oils from Lippia sidoides and of its major constituent. *Vib. Spectrosc.* **2020**, *110*, 103111. [CrossRef]
13. Rajkumar, P.; Selvaraj, S.; Suganya, R.; Velmurugan, D.; Gunasekaran, S.; Kumaresan, S. Vibrational and electronic spectral analysis of thymol an isomer of carvacrol isolated from Trachyspermum ammi seed: A combined experimental and theoretical study. *Chem. Data Collect.* **2018**, *15–16*, 10–31. [CrossRef]

14. Beć, K.B.; Grabska, J.; Kirchler, C.G.; Huck, C.W. NIR spectra simulation of thymol for better understanding of the spectra forming factors, phase and concentration effects and PLS regression features. *J. Mol. Liq.* **2018**, *268*, 895–902. [CrossRef]
15. Lopes Jesus, A.J.; Fausto, R.; Reva, I. Conformational Space, IR-Induced, and UV-induced chemistry of carvacrol isolated in a low-temperature argon matrix. *J. Phys. Chem. A* **2021**, *125*, 8215–8229. [CrossRef] [PubMed]
16. Becke, A.D. Density-functional exchange-energy approximation with correct asymptotic behavior. *Phys. Rev. A* **1988**, *38*, 3098–3100. [CrossRef]
17. Lee, C.T.; Yang, W.T.; Parr, R.G. Development of the Colle-Salvetti correlation-energy formula into a functional of the electron-density. *Phys. Rev. B* **1988**, *37*, 785–789. [CrossRef] [PubMed]
18. Vosko, S.H.; Wilk, L.; Nusair, M. Accurate spin-dependent electron liquid correlation energies for local spin density calculations: A critical analysis. *Can. J. Phys.* **1980**, *58*, 1200–1211. [CrossRef]
19. Møller, C.; Plesset, M.S. Note on an approximation treatment for many-electron systems. *Phys. Rev.* **1934**, *46*, 618–622. [CrossRef]
20. Pople, J.A.; Head-Gordon, M.; Raghavachari, K. Quadratic configuration interaction. A general technique for determining electron correlation energies. *J. Chem. Phys.* **1987**, *87*, 5968–5975. [CrossRef]
21. Reva, I.; Lopes Jesus, A.J.; Nunes, C.M.; Roque, J.P.L.; Fausto, R. UV-Induced Photochemistry of 1,3-Benzoxazole, 2-Isocyanophenol, and 2-Cyanophenol Isolated in Low-Temperature Ar Matrixes. *J. Org. Chem.* **2021**, *86*, 6126–6137. [CrossRef] [PubMed]
22. Zhurko, G.A. Chemcraft—Graphical Program for Visualization of Quantum Chemistry Computations, Version 1.8. 2020. Available online: http://www.chemcraftprog.com (accessed on 18 May 2022).
23. Frisch, M.J.; Trucks, G.W.; Schlegel, H.B.; Scuseria, G.E.; Robb, M.A.; Cheeseman, J.R.; Scalmani, G.; Barone, V.; Mennucci, B.; Petersson, G.A.; et al. *GAUSSIAN 09, Revision D.01*; Gaussian, Inc.: Wallingford, CT, USA, 2013.
24. Teixeira, F.; Cordeiro, M.N.D.S. Improving vibrational mode interpretation using bayesian regression. *J. Chem. Theory Comput.* **2019**, *15*, 456–470. [CrossRef] [PubMed]
25. Lesarri, A.; Shipman, S.T.; Neill, J.L.; Brown, G.G.; Suenram, R.D.; Kang, L.; Caminati, W.; Pate, B.H. Interplay of phenol and isopropyl isomerism in propofol from broadband chirped-pulse microwave spectroscopy. *J. Am. Chem. Soc.* **2010**, *132*, 13417–13424. [CrossRef]
26. Zhao, Y.; Jin, Y.; Hao, J.; Yang, Y.; Wang, L.; Li, C.; Jia, S. Rotamers of p-isopropylphenol studied by hole-burning resonantly enhanced multiphoton ionization and mass analyzed threshold ionization spectroscopy. *Spectrochim. Acta Part A* **2019**, *207*, 328–336. [CrossRef] [PubMed]
27. Richardson, P.R.; Chapman, M.A.; Wilson, D.C.; Bates, S.P.; Jones, A.C. The nature of conformational preference in a number of p-alkyl phenols and p-alkyl benzenes. *Phys. Chem. Chem. Phys.* **2002**, *4*, 4910–4915. [CrossRef]
28. Schaefer, T.; Addison, B.M.; Sebastian, R.; Wildman, T.A. Orientations of the hydroxyl and isopropyl groups in the cis and trans conformers of 2-isopropylphenol and 2-isopropyl-6-methylphenol. *Can. J. Chem.* **1981**, *59*, 1656–1659. [CrossRef]
29. Rozenberg, M.; Fausto, R.; Reva, I. Variable temperature FTIR spectra of polycrystalline purine nucleobases and estimating strengths of individual hydrogen bonds. *Spectrochim. Acta Part A* **2021**, *251*, 119323. [CrossRef] [PubMed]
30. Lopes Jesus, A.J.; Rosado, M.T.S.; Leitão, M.L.P.; Redinha, J.S. Molecular structure of butanediol isomers in gas and liquid states: Combination of DFT calculations and infrared spectroscopy studies. *J. Phys. Chem. A* **2003**, *107*, 3891–3897. [CrossRef]
31. Lopes Jesus, A.J.; Rosado, M.T.S.; Reva, I.; Fausto, R.; Eusébio, M.E.S.; Redinha, J.S. Structure of isolated 1,4-Butanediol: Combination of MP2 calculations, NBO analysis, and matrix-isolation infrared spectroscopy. *J. Phys. Chem. A* **2008**, *112*, 4669–4678. [CrossRef]
32. Lopes Jesus, A.J.; Rosado, M.T.S.; Reva, I.; Fausto, R.; Eusébio, M.E.; Redinha, J.S. Conformational study of monomeric 2,3-Butanediols by matrix-isolation infrared spectroscopy and DFT calculations. *J. Phys. Chem. A* **2006**, *110*, 4169–4179. [CrossRef] [PubMed]
33. Rosado, M.T.S.; Lopes Jesus, A.J.; Reva, I.D.; Fausto, R.; Redinha, J.S. Conformational cooling dynamics in matrix-isolated 1,3-Butanediol. *J. Phys. Chem. A* **2009**, *113*, 7499–7507. [CrossRef]
34. Lopes Jesus, A.J.; Nunes, C.M.; Reva, I.; Pinto, S.M.V.; Fausto, R. Effects of entangled IR radiation and tunneling on the conformational interconversion of 2-Cyanophenol. *J. Phys. Chem. A* **2019**, *123*, 4396–4405. [CrossRef]
35. Lopes Jesus, A.J.; Reva, I.; Nunes, C.M.; Roque, J.P.L.; Pinto, S.M.V.; Fausto, R. Kinetically unstable 2-isocyanophenol isolated in cryogenic matrices: Vibrational excitation, conformational changes and spontaneous tunneling. *Chem. Phys. Lett.* **2020**, *742*, 137069. [CrossRef]
36. Akai, N.; Kudoh, S.; Takayanagi, M.; Nakata, M. Cis-Trans isomerization equilibrium in hydroquinone in low-temperature argon and xenon matrices studied by FTIR spectroscopy. *Chem. Phys. Lett.* **2002**, *356*, 133–139. [CrossRef]
37. Akai, N.; Kudoh, S.; Nakata, M. Photoisomerization and tunneling isomerization of tetrachlorohydroquinone in a low-temperature argon matrix. *J. Phys. Chem. A* **2003**, *107*, 3655–3659. [CrossRef]
38. Borden, W.T. Reactions that involve tunneling by carbon and the role that calculations have played in their study. *WIREs Comput. Mol. Sci.* **2016**, *6*, 20–46. [CrossRef]
39. Nunes, C.M.; Reva, I.; Fausto, R. Direct observation of tunnelling reactions by matrix isolation spectroscopy. In *Tunnelling in Molecules: Nuclear Quantum Effects from Bio to Physical Chemistry*; Kästner, J., Kozuch, S., Eds.; The Royal Society of Chemistry: Cambridge, UK, 2021; pp. 1–60. [CrossRef]
40. Pettersson, M.; Lundell, J.; Khriachtchev, L.; Räsänen, M. IR spectrum of the other rotamer of formic acid, cis-HCOOH. *J. Am. Chem. Soc.* **1997**, *119*, 11715–11716. [CrossRef]

41. Maçôas, E.M.S.; Khriachtchev, L.; Pettersson, M.; Fausto, R.; Räsänen, M. Rotational isomerism in acetic acid: The first experimental observation of the high-energy conformer. *J. Am. Chem. Soc.* **2003**, *125*, 16188–16189. [CrossRef]
42. Lapinski, L.; Reva, I.; Rostkowska, H.; Halasa, A.; Fausto, R.; Nowak, M.J. Conformational Transformation in Squaric Acid Induced by Near-IR Laser Light. *J. Phys. Chem. A* **2013**, *117*, 5251–5259. [CrossRef]
43. Kuş, N.; Fausto, R. Effects of the matrix and intramolecular interactions on the stability of the higher-energy conformers of 2-fluorobenzoic acid. *J. Chem. Phys.* **2017**, *146*, 124305. [CrossRef]
44. Gerbig, D.; Schreiner, P.R. Hydrogen-Tunneling in Biologically Relevant Small Molecules: The Rotamerizations of α-Ketocarboxylic Acids. *J. Phys. Chem. B* **2015**, *119*, 693–703. [CrossRef]
45. Bazsó, G.; Magyarfalvi, G.; Tarczay, G. Tunneling lifetime of the ttc/VIp conformer of glycine in low-temperature matrices. *J. Phys. Chem. A* **2012**, *116*, 10539–10547. [CrossRef] [PubMed]
46. Lopes Jesus, A.J.; Reva, I.; Araujo-Andrade, C.; Fausto, R. Conformational changes in matrix-isolated 6-methoxyindole: Effects of the thermal and infrared light excitations. *J. Chem. Phys.* **2016**, *144*, 124306. [CrossRef]
47. Marzec, K.M.; Reva, I.; Fausto, R.; Proniewicz, L.M. Comparative matrix isolation infrared spectroscopy study of 1,3- and 1,4-diene monoterpenes (α-Phellandrene and γ-Terpinene). *J. Phys. Chem. A* **2011**, *115*, 4342–4353. [CrossRef] [PubMed]
48. Giuliano, B.M.; Reva, I.; Lapinski, L.; Fausto, R. Infrared spectra and ultraviolet-tunable laser induced photochemistry of matrix-isolated phenol and phenol-$d_5$. *J. Chem. Phys.* **2012**, *136*, 024505. [CrossRef]
49. Giuliano, B.M.; Melandri, S.; Reva, I.; Fausto, R. Conformational space and photochemistry of tyramine isolated in argon and xenon cryomatrixes. *J. Phys. Chem. A* **2013**, *117*, 10248–10259. [CrossRef]
50. Barnes, A.J. Matrix isolation vibrational spectroscopy as a tool for studying conformational isomerism. *J. Mol. Struct.* **1984**, *113*, 161–174. [CrossRef]
51. Breda, S.; Lapinski, L.; Reva, I.; Fausto, R. 4,6-Dimethyl-α-pyrone: A matrix isolation study of the photochemical generation of conjugated ketene, Dewar valence isomer and 1,3-dimethyl-cyclobutadiene. *J. Photochem. Photobiol. A* **2004**, *162*, 139–151. [CrossRef]
52. Breda, S.; Reva, I.; Lapinski, L.; Fausto, R. Matrix isolation FTIR and theoretical study of α-pyrone photochemistry. *Phys. Chem. Chem. Phys.* **2004**, *6*, 929–937. [CrossRef]
53. Tidwell, T.T. Spectroscopy and physical properties of ketenes. In *Ketenes II*; John Wiley and Sons: Hoboken, NJ, USA, 2006; pp. 27–53.
54. Kuş, N.; Sagdinc, S.; Fausto, R. Infrared spectrum and UV-induced photochemistry of matrix-isolated 5-Hydroxyquinoline. *J. Phys. Chem. A* **2015**, *119*, 6296–6308. [CrossRef] [PubMed]
55. Krupa, J.; Olbert-Majkut, A.; Reva, I.; Fausto, R.; Wierzejewska, M. Ultraviolet-tunable laser induced phototransformations of matrix isolated isoeugenol and eugenol. *J. Phys. Chem. B* **2012**, *116*, 11148–11158. [CrossRef]
56. Samanta, A.K.; Pandey, P.; Bandyopadhyay, B.; Chakraborty, T. Keto–enol tautomers of 1,2-cyclohexanedione in solid, liquid, vapour and a cold inert gas matrix: Infrared spectroscopy and quantum chemistry calculation. *J. Mol. Struct.* **2010**, *963*, 234–239. [CrossRef]
57. Miyazaki, J.; Toh, S.Y.; Moore, B.; Djuricanin, P.; Momose, T. UV photochemistry of 1,3-cyclohexadiene isolated in solid parahydrogen. *J. Mol. Struct.* **2021**, *1224*, 128986. [CrossRef]
58. Chapman, O.L.; McIntosh, C.L.; Pacansky, J. Photochemistry of a-Pyrone in Argon at 8 °K. *J. Am. Chem. Soc.* **1973**, *95*, 244–246. [CrossRef]
59. Pong, R.G.S.; Shirk, J.S. Photochemistry of α-Pyrone in solid argon. *J. Am. Chem. Soc.* **1973**, *95*, 248–249. [CrossRef]
60. Breda, S.; Lapinski, L.; Fausto, R.; Nowak, M.J. Photoisomerization reactions of 4-methoxy- and 4-hydroxy-6-methyl-α-pyrones: An experimental matrix isolation and theoretical density functional theory study. *Phys. Chem. Chem. Phys.* **2003**, *5*, 4527–4532. [CrossRef]

*Communication*

# Demonstration of a Stereospecific Photochemical Meta Effect

Hoai Pham [1], Madelyn Hunsley [2], Chou-Hsun Yang [3], Haobin Wang [3] and Scott M. Reed [3,*]

[1] Department of Chemistry, University of Colorado Boulder, Boulder, CO 80309, USA; Hoai.Pham@colorado.edu
[2] Pfizer Boulder Research & Development, Boulder, CO 80301, USA; mlhunsley@gmail.com
[3] Department of Chemistry, University of Colorado Denver, Denver, CO 80204, USA; chou.yang@ucdenver.edu (C.-H.Y.); haobin.wang@ucdenver.edu (H.W.)
* Correspondence: scott.reed@ucdenver.edu; Tel.: +1-303-315-7634

**Abstract:** A fundamental goal of photochemistry is to understand how structural features of a chromophore can make specific bonds within a molecule prone to cleavage by light, or photolabile. The meta effect is an example of a regiochemical explanation for photolability, in which electron donating groups on an aromatic ring cause photolability selectively at the meta position. Here, we show, using a chromophore containing one ring with a meta-methoxy group and one ring with a para-methoxy group, that two stereoisomers of the same compounds can react with light differently, based simply on the three-dimensional positioning of a meta anisyl ring. The result is that the stereoisomers of the compound with the same configuration at both stereogenic centers are photolabile while the stereoisomers with opposite configuration do not react with light. Furthermore, time-dependent density functional theory (TD-DFT) calculations show distinct excitation pathways for each stereoisomer.

**Keywords:** diastereoselectivity; photochemistry; protecting groups; diastereodifferentiating

## 1. Introduction

The regioselective meta effect was first reported in the photochemical solvolysis of nitrophenyl phosphonate and sulfonate esters [1] and later in a series of *m*- and *p*-isomers of nitrophenyl and cyanophenyl trityl ethers [2]. The meta effect has been used in the design of photoreleasable protecting groups (PPGs) [3,4], including a broad range of chromophores used to protect carbonyl compounds [5,6]. Early calculations showed that excited singlet states of *m*-methoxy-substituted benzylic acetates led to heterolysis and that increasing meta substitution enhanced photolability [7,8]. More recently, we have reported indirect pathways to photocleavage that occur after an ultrafast back electron transfer to the ground electronic state for *m*-methoxy substituted aromatic compounds. For suitably substituted compounds, the electronic-nuclear couplings facilitate sufficient energy transfer to cause a dissociation reaction [9].

While the meta effect has been thought of as a regioselective phenomenon, in principle, a more subtle change in the chromophore, such as a stereochemical change, could also influence whether a bond is photolabile. While differential reactivity to light is well documented in diastereomers, such as is seen in stereoselective bond-forming photoreactions [10–12] and diastereoselective photoisomerization reactions [13], we are aware of no examples in which a change to the absolute configurations of stereocenters results in a bond becoming photolabile.

Here, we report on diastereomers that either do or do not photocleave based on the configuration of their stereogenic centers. This is in contrast to the few known cases of diastereomer differentiating photocleavage reactions where two different products result based on the stereochemistry of the input molecule [10,14,15]. In this diastereomer differentiating reaction, only the stereoisomer with a meta methoxy ring positioned anti to

a third chromophore is photolabile. Given the utility of photolabile bonds in the design of PPGs [16], for patterning surfaces [17], automating DNA synthesis [18], releasing biological substrates [19], and for the design of molecular logic gates and actuators [20,21], the on/off type of reaction reported here is likely to be a useful extension of the meta effect.

## 2. Synthesis

The synthesis (Scheme 1) utilizes a similar strategy to one reported for derivatizing salicylic acid to form a carbonyl PPG [3] but takes advantage of the fact that nitriles react with only a single equivalent of Grignard reagent, thereby allowing two different groups to be added sequentially to a nitrile and then the resultant ketone, generating an asymmetric center at the benzylic position. Specifically, benzyl protected 2-hydroxy benzonitrile was reacted with the Grignard of *m*-bromoanisole to form ketone **1**. Addition of the Grignard of *p*-bromoanisole to ketone **1** in a subsequent reaction produced alcohol **2**. Removal of the benzyl protecting group to produce diol **3** was accomplished with hydrogen over palladium on carbon. Finally, 3-phenylpropanal and pTSA were added to produce acetal **4** while introducing a second stereogenic center to produce a pair of racemates.

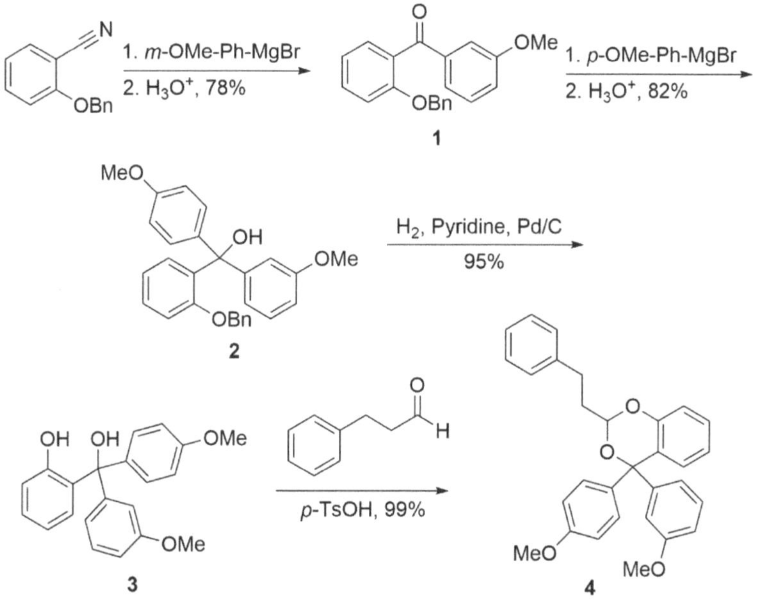

**Scheme 1.** Synthetic route to acetal stereoisomers.

## 3. Results

Crystals of **4** were obtained from ethyl acetate and hexanes and the structure determined by x-ray crystallography (see Supplementary Information). The unit cell contained four molecules of (2*S*,4*R*)-**4** and four molecules of (2*R*,4*S*)-**4** consistent with the absence of optical activity. The synthesis results in four stereoisomers, two with like stereogenic centers, (2*S*,4*S*)-**4** and (2*R*,4*R*)-**4**, and two with unlike stereogenic centers, (2*S*,4*R*)-**4** and (2*R*,4*S*)-**4**. The racemate with unlike [22] stereogenic centers (*u*) ((2*S*,4*R*)-**4** and (2*R*,4*S*)-**4**) was the predominant form isolated at 76%, with the racemate with like stereogenic centers (*l*) comprising the remaining 24%. The *u* racemate has an acetal $^{13}$C NMR peak at 92.45 ppm and was readily distinguishable from the *l* racemate at 92.87 ppm (Figure 1). Similarly, $^1$H NMR was also used to distinguish the two racemates (see Supplementary Information).

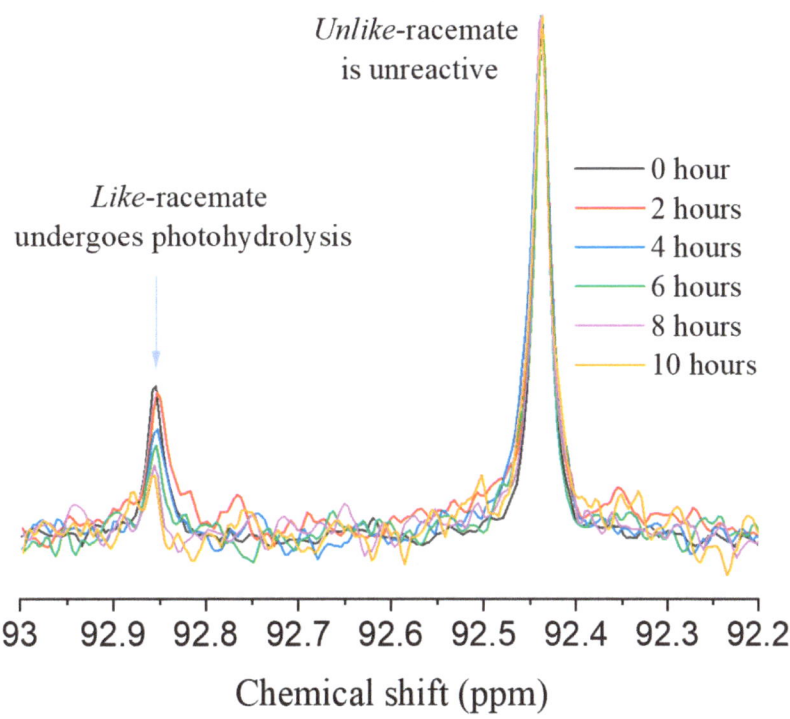

**Figure 1.** $^{13}$C NMR time series taken during irradiation of acetal 4 with UV light.

When a 0.02 M mixture of *l* and *u* racemates was irradiated with UV light from a 75 W Xenon lamp in a 90:10 MeCN:H$_2$O mixture at room temperature, 3-phenylpropanal and diol 3 were released (Scheme 2). Over 10 h, little photohydrolysis of the *u* racemate at 92.45 ppm was observed whereas 67% of the *l* racemate was converted to 3-phenylpropanal and diol 3 as measured by integrating the $^{13}$C acetal peak. A similar trend was observed in the $^1$H NMR confirming the decrease in the *l* racemate over 10 h although overlap of the two acetal peaks and the two methoxy signals made quantification of the $^1$H NMR difficult (see Supplementary Information, Figure S2).

Calculations were performed using the quantum chemical program packages Q-Chem [23] and Gaussian 16 [23,24]. For both racemates, the first excited state, $S_1$, has a stable structure not far away from the ground state Franck–Condon vertical excitation point (the stable structure of $S_0$). This is illustrated in Table 1, where the reorganization energies are listed for the four stereoisomers with respect to both states $S_0$ and $S_1$ (calculation details in Supplementary Information). The reorganizational energies are relatively small compared with the electronic excitation energies. Similarity of the $S_0$ and $S_1$ states is also demonstrated by the root-mean-square deviation in the atom locations between $S_0$ and $S_1$ being only 0.1507 Å. Generalized Mulliken-Hush (GMH) analysis [25,26] reveals that the electronic couplings are quite large (>50 kcal/mol). From our previous time-dependent density functional theory (TD-DFT) studies of similar systems [9] this suggests an ultrafast (sub-picosecond) internal conversion process upon excitation from $S_0$ to $S_1$. Here, the electronic transition associated with the internal conversion process is in the adiabatic regime, which is characterized by (damped) electronic coherence [9]. Thus, photoexcitation acts as an "energy pump": through absorption followed by internal conversion, the photo energy is converted to the nuclear energy that is used to drive the dissociation reaction at the ground state. The efficiency of this "energy pump" depends on the electronic-nuclear coupling, which is illustrated in Figure 2 by the nuclear force vectors at the Franck-Condon geometries. The more the force vector is aligned with the reaction coordinate for the dissociation reaction, the more likely the cleavage is to occur.

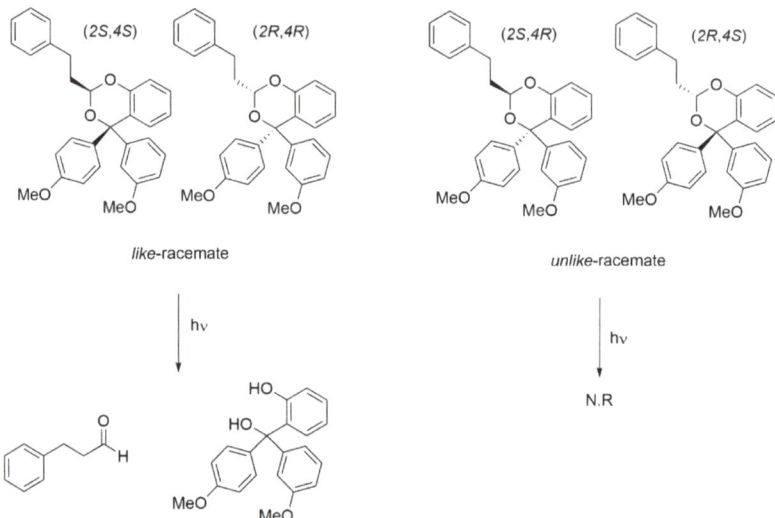

**Scheme 2.** Stereoselective photorelease.

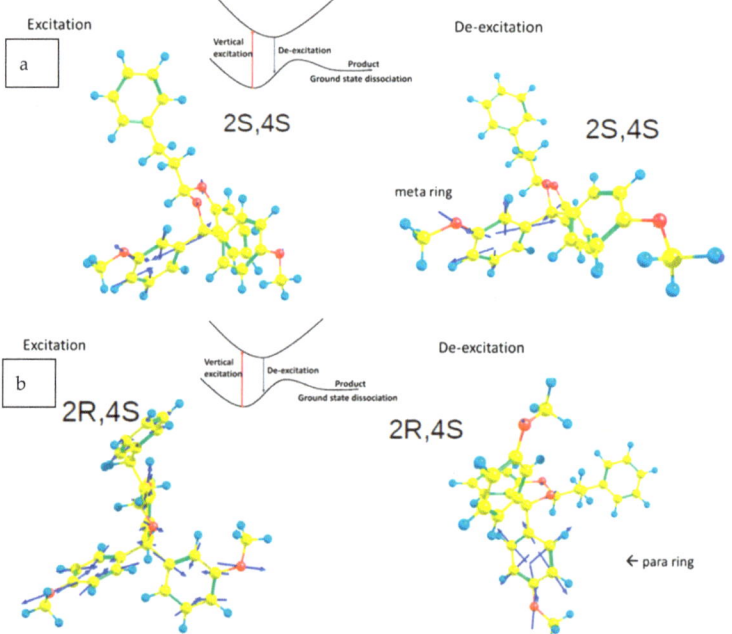

**Figure 2.** During photohydrolysis, all 4 stereoisomers are present but the lowest energy pathway available varies with the stereochemistry. The reaction pathway is represented from $S_0$ to $S_1$ (left) from $S_1$ to $S_0$ (right) for the *l* diastereomers (**a**) and the *u* diastereomers (**b**) of **4**. The blue arrows show the vector for the force on each atom during excitation (left) or de-excitation (right) in the lowest energy pathway.

Table 1. Excitation energy (Ex), de-excitation energy (DeEx) and reorganization energies ($\lambda$) (kcal/mol) in different electronic states among all stereoisomers. $\lambda$ ($S_1 \rightarrow S_0$) and $\lambda$ ($S_0 \rightarrow S_1$) were used for energy difference calculations.

| Stereoisomer | Ex ($S_1$) | $\lambda$ ($S_1 \rightarrow S_0$) | DeEx ($S_1$) | $\lambda$ ($S_0 \rightarrow S_1$) |
| --- | --- | --- | --- | --- |
| 2R,4R | 119.6 | 3.5 | 112.8 | 3.4 |
| 2S,4R | 120.3 | 3.1 | 114.3 | 2.9 |
| 2R,4S | 120.3 | 3.1 | 114.3 | 2.9 |
| 2S,4S | 119.6 | 3.5 | 112.8 | 3.4 |

When the molecule has two like stereogenic centers, ((2S,4S)-4 or (2R,4R)-4), the primary excitation pathway involves motion in the meta anisyl ring (Figure 2a). In this case, the meta methoxy ring is positioned anti to the aromatic ring of the phenylethyl group which increases its reactivity. Furthermore, the loss of energy also involves motion that is focused on the meta ring which is known to cause photolability. The overall result is that these structures are more prone to photocleavage. When the molecule has two unlike stereogenic centers, ((2R,4S)-4 or (2S,4R)-4) and the meta-methoxy ring is syn to the phenylethyl group, the excitation pathway is diffused, involving both the meta and para anisyl rings (Figure 2b). Additionally, here, the loss of energy occurs along with increased motion of atoms in the para substituted rings rather than the meta. In this case, the racemate is not prone to cleavage.

Although stereoselective bond-forming photoreactions [10–12] and distereoselective photoisomerization reactions [13] are somewhat common, analogous bond breaking reactions are not. One example we find of a diastereomer differentiating photocleavage reaction comes from -arylbutyrophenones that alternately undergo Yang cyclization or elimination for different diastereomers [14]. Another example are 3-(2-phthalimido-propionate)-yl PPGs [15] in which the the diastereomer undergoes an E2 elimination to produce a trans alkene in high yield, whereas the erythro compound is unable to populate an antiperiplanar conformation of carboxylate and acetate; therefore, instead, the two groups are removed sequentially in an E1cb mechanism producing a mixture of trans and cis alkene. In both examples there are two different reaction pathways for the different diastereomers. Additionally, the stereogenic centers in those cases are on adjacent carbons whereas here they are on a dioxane ring such that a spacer exists between the stereogenic centers in the 2 and 4 position. In the current system where there is not a second reaction pathway, the molecule either photoreleases its benzylic substituent or it does not. This system, therefore, will be well-suited for applications involving PPGs, surface modifiers, or in vitro or in vivo substrate release. Another category of prior photocleavage examples involve a Norrish type II photoelimination; however, the observed diastereoselectivity originates from a chiral auxiliary [27].

The unique pathway for one diastereomer over the other suggests that interactions between the different aromatic rings are important to the photocleavage mechanism. Either the energy pump is more effective in the $u$ racemate with the anti-phenyl ring or de-excitation force in the meta anisyl ring is more aligned with the dissociative reaction coordinate. Conversely for the syn-arranged rings, either the pump is less effective or de-excitation force in the para ring is less aligned with bond dissociation.

## 4. Discussion and Future Directions

It is likely that other ring systems and other separations between stereogenic centers could result in the same phenomena and that the two stereogenic centers could both be placed on the PPG as opposed to having one originating from the protected substrate as seen here. The inclusion of stereogenic centers on different ring positions could be a general method for creating distance between the stereogenic centers while maintaining communication between the rings. It should be possible to create a set of protecting groups that have identical chemical functionality but with one reactive to both light and acidity and one reactive to only acidity, creating a simple logic gate with actuator [21]. Other potential applications include monitoring racemizing conditions by the observation of a photoreleased substrate, and racemization could also be used to slowly introduce photolability to a substrate over time.

Alternatively, if both stereogenic centers were contained within the protected substrate, this approach could be used to isolate one diastereomer from the other. Here, achiral light causes diastereomers to react differently due to the arrangement of their chromophores and this suggests that chiral light could in turn affect an enantiospecific transformation using similar compounds which suggests this class of compounds could be used for chiral separations by photoderacemization [28], or that proteins or other chiral environments could be used to induce changes in the stereoselectivity of this reaction [29].

The design of new molecules as stereoselective PPGs can build off the structural lessons learned here as well as the computational techniques. In similar systems, attention to the positioning of aromatic rings near the PPG could again be used to enhance stereoselectivity. For novel compounds, TD-DFT can preview the likely initial steps in the photoreaction pathway. This will facilitate exploration of new chromophores for applications in stereoselective photochemistry.

The fact that two different positionings of the meta ring behaved differently also points out that only a single meta ring is needed to cause this PPG to release. This also demonstrates that communication between the rings, for example by energy transfer, is slow enough that the two positions act independently. This informs new strategies for PPG design and suggests new methods for the design of orthogonal protecting groups [6,30] Additionally, the computational methods described here could be used to predict which stereoisomers would be most likely to react with light.

**Supplementary Materials:** The following supporting information can be downloaded at: https://www.mdpi.com/article/10.3390/photochem2010006/s1, Figure S1. Crystal structure of (2R,4S)-4-(4-methoxyphenyl)-2-phenethyl-4-(3-methoxyphenyl)-4H-benzo[d][1,3]dioxine; Figure S2. $^1$H NMR time series taken during 10 h irradiation of a 76:24 (u:l) mixture of acetal 4 with UV; Figure S3. $^1$H NMR spectrum of (2-(benzyloxy)phenyl)(3-methoxyphenyl)methanone; Figure S4. $^{13}$C NMR spectrum of (2-(benzyloxy)phenyl)(3-methoxyphenyl)methanone; Figure S5. $^1$H NMR spectrum of 2-(benzyloxy)phenyl)(3-methoxyphenyl)(4-methoxyphenyl)methanol; Figure S6. $^{13}$C NMR spectrum of 2-(benzyloxy)phenyl)(3-methoxyphenyl)(4-methoxyphenyl)methanol; Figure S7. $^1$H NMR spectrum of (2-hydroxy(3-methoxyphenyl)(4-methoxy phenyl)methyl)phenol; Figure S8. $^{13}$C NMR spectrum of (2-hydroxy(3-methoxyphenyl)(4-methoxy phenyl)methyl)phenol; Figure S9. $^1$H NMR spectrum of acetal 4, (±)-4-(4-methoxyphenyl)-2-phenethyl-4-(3-methoxyphenyl)-4H-benzo[d][1,3]dioxine; Figure S10. $^{13}$C NMR spectrum of 4, (±)-4-(4-methoxyphenyl)-2-phenethyl-4-(3-methoxyphenyl)-4H-benzo[d][1,3]dioxine; Figure S11. IR spectrum of (2-(benzyloxy)phenyl)(3-methoxyphenyl)methanone; Figure S12. IR spectrum of 2-(benzyloxy)phenyl)(3-methoxyphenyl)(4-methoxyphenyl)methanol; Figure S13. IR spectrum of (2-hydroxy(3-methoxyphenyl)(4-methoxy phenyl)methyl)phenol; Figure S14. IR spectrum of (±)-4-(4-methoxyphenyl)-2-phenethyl-4-(3-methoxyphenyl)-4H-benzo[d][1,3]dioxine..

**Author Contributions:** Conceptualization, S.M.R. and H.W.; synthesis, H.P. and M.H.; methodology and calculations, C.-H.Y.; writing—original draft preparation, H.P.; writing—review and editing, S.M.R., C.-H.Y. and H.W. All authors have read and agreed to the published version of the manuscript.

**Funding:** NMR instrumentation was supported by NSF grant 1726947. H.P. acknowledges support from the CU Denver EUReCA Program and Undergraduate Research Opportunity Program and SMR acknowledges support from the CU Denver Office of Research Services. H.W. acknowledges the support from the National Science Foundation CHE-1954639. This work used the Extreme Science and Engineering Discovery Environment (XSEDE), which is supported by NSF grant number ACI-1548562, and resources of the National Energy Research Scientific Computing Center (NERSC), which is supported by the Office of Science of the U.S. Department of Energy under Contract No. DE-AC02-05CH11231.

**Data Availability Statement:** NMR data is provided in the Supplementary Information. X-ray data has been submitted to the Cambridge crystallographic data centre.

**Acknowledgments:** We thank Brian Newell (CSU) for X-ray crystallographic services.

**Conflicts of Interest:** The authors declare no conflict of interest.

## References

1. Havinga, E.; de Jongh, R.O.; Dorst, W. Photochemical acceleration of the hydrolysis of nitrophenyl phosphates and nitrophenyl sulphates. *Recl. Trav. Chim. Pays-Bas* **1956**, *75*, 378–383. [CrossRef]
2. Zimmerman, H.E.; Somasekhara, S. Mechanistic Organic Photochemistry. III. Excited State Solvolyses. *J. Am. Chem. Soc.* **1963**, *85*, 922–927.
3. Wang, P.; Hu, H.; Wang, Y. Novel photolabile protecting group for carbonyl compounds. *Org. Lett.* **2007**, *9*, 1533–1535. [CrossRef]
4. Yang, H.; Zhang, X.; Zhou, L.; Wang, P. Development of a photolabile carbonyl-protecting group toolbox. *J. Org. Chem.* **2011**, *76*, 2040–2048. [CrossRef]
5. Wang, P.; Hu, H.; Wang, Y. Application of the excited state meta effect in photolabile protecting group design. *Org. Lett.* **2007**, *9*, 2831–2833. [CrossRef]
6. Wang, P.; Wang, Y.; Hu, H.; Spencer, C.; Liang, X.; Pan, L. Sequential removal of photolabile protecting groups for carbonyls with controlled wavelength. *J. Org. Chem.* **2008**, *73*, 6152–6157. [CrossRef]
7. Zimmerman, H.E. The Meta Effect in Organic-Photochemistry-Mechanistic and Exploratory Organic-Photochemistry. *J. Am. Chem. Soc.* **1995**, *117*, 8988–8991.
8. Zimmerman, H.E. The meta-ortho effect in organic photochemistry; Mechanistic and exploratory organic photochemistry. *J. Phys. Chem. A* **1998**, *102*, 5616–5621. [CrossRef]
9. Yang, C.-H.; Denne, J.; Reed, S.; Wang, H. Computational study on the removal of photolabile protecting groups by photochemical reactions. *Comput. Theor. Chem.* **2019**, *1151*, 1–11. [CrossRef]
10. Wrobel, M.N.; Margaretha, P. Diastereomer-differentiating photoisomerization of 5-(cyclopent-2-en-1-yl)-2,5-dihydro-1H-pyrrol-2-ones. *Chem. Commun.* **1998**, *5*, 541–542. [CrossRef]
11. Bach, T.; Hehn, J.P. Photochemical Reactions as Key Steps in Natural Product Synthesis. *Angew. Chem. Int. Ed. Engl.* **2011**, *50*, 1000–1045. [CrossRef]
12. Inoue, Y. Asymmetric photochemical reactions in solution. *Chem. Rev.* **1992**, *92*, 741–770.
13. Cheung, E.; Chong, K.C.; Jayaraman, S.; Ramamurthy, V.; Scheffer, J.R.; Trotter, J. Enantio- and diastereodifferentiating cis,trans-photoisomerization of 2beta,3beta-diphenylcyclopropane-1alpha-carboxylic acid derivatives in organized media. *Org. Lett.* **2000**, *2*, 2801–2804. [CrossRef]
14. Singhal, N.; Koner, A.L.; Mal, P.; Venugopalan, P.; Nau, W.M.; Moorthy, J.N. Diastereomer-differentiating photochemistry of beta-arylbutyrophenones: Yang cyclization versus type II elimination. *J. Am. Chem. Soc.* **2005**, *127*, 14375–14382. [CrossRef]
15. Soldevilla, A.; Griesbeck, A.G. Chiral photocages based on phthalimide photochemistry. *J. Am. Chem. Soc.* **2006**, *128*, 16472–16473. [CrossRef]
16. Klán, P.; Šolomek, T.; Bochet, C.G.; Blanc, A.; Givens, R.; Rubina, M.; Popik, V.; Kostikov, A.; Wirz, J. Photoremovable protecting groups in chemistry and biology: Reaction mechanisms and efficacy. *Chem. Rev.* **2013**, *113*, 119–191. [CrossRef]
17. Romano, A.; Roppolo, I.; Rossegger, E.; Schlögl, S.; Sangermano, M. Recent Trends in Applying Ortho-Nitrobenzyl Esters for the Design of Photo-Responsive Polymer Networks. *Molecules* **2020**, *13*, 2777. [CrossRef]
18. Pirrung, M.C.; Wang, L.; Montague-Smith, M.P. 3′-nitrophenylpropyloxycarbonyl (NPPOC) protecting groups for high-fidelity automated 5′ –> 3′ photochemical DNA synthesis. *Org. Lett.* **2001**, *3*, 1105–1108. [CrossRef]
19. Stensrud, K.; Noh, J.; Kandler, K.; Wirz, J.; Heger, D.; Givens, R.S. Competing pathways in the photo-Favorskii rearrangement and release of esters: Studies on fluorinated p-hydroxyphenacyl-caged GABA and glutamate phototriggers. *J. Org. Chem.* **2009**, *74*, 5219–5227. [CrossRef] [PubMed]
20. Upendar, R.G.; Axthelm, J.; Hoffmann, P.; Taye, N.; Glaser, S.; Gorls, H.; Hopkins, S.L.; Plass, W.; Neugebauer, U.; Bonnet, S.; et al. Co-Registered Molecular Logic Gate with a CO-Releasing Molecule Triggered by Light and Peroxide. *J. Am. Chem. Soc.* **2017**, *139*, 4991–4994.
21. Balzani, V.; Credi, A.; Venturi, M. Molecular logic circuits. *Chemphyschem* **2003**, *4*, 49–59. [CrossRef]
22. Seebach, D.; Prelog, V. The Unambiguous Specification of the Steric Course of Asymmetric Syntheses. *Angew. Chem. Int. Ed. Engl.* **1982**, *9*, 654–660. [CrossRef]
23. Shao, Y.; Gan, Z.; Epifanovsky, E.; Gilbert, A.T.B.; Wormit, M.; Kussmann, J.; Lange, A.W.; Behn, A.; Deng, J.; Feng, X.; et al. Advances in molecular quantum chemistry contained in the Q-Chem 4 program package. *Mol. Phys.* **2015**, *113*, 184–215. [CrossRef]
24. Frisch, M.J.; Trucks, G.W.; Schlegel, H.B.; Scuseria, G.E.; Robb, M.A.; Cheeseman, J.R.; Scalmani, G.; Barone, V.; Petersson, G.A.; Nakatsuji, H.; et al. *Gaussian 16 Rev. C.01*; Gaussian, Inc.: Wallingford, CT, USA, 2016.
25. Cave, R.J.; Newton, M.D. Generalization of the Mulliken-Hush treatment for the calculation of electron transfer matrix elements. *Chem. Phys. Lett.* **1996**, *249*, 15–19. [CrossRef]
26. Cave, R.J.; Newton, M.D. Calculation of electronic coupling matrix elements for ground and excited state electron transfer reactions: Comparison of the generalized Mulliken–Hush and block diagonalization methods. *J. Chem. Phys.* **1997**, *106*, 9213–9226. [CrossRef]

27. Henin, F.; Muzart, J.; Pete, J.-P.; M'boungou-M'passi, A.; Rau, H. Enantioselective Protonation of a Simple Enol: Aminoalcohol-Catalyzed Ketonization of a Photochemically Produced 2-Methylinden-3-ol. *Angew. Chem. Int. Ed. Engl.* **1991**, *30*, 416–418. [CrossRef]
28. Shin, N.Y.; Ryss, J.M.; Zhang, X.; Miller, S.J.; Knowles, R.R. Light-driven deracemization enabled by excited-state electron transfer. *Science* **2019**, *366*, 364–369. [CrossRef] [PubMed]
29. Levi-Minzi, N.; Zandomeneghi, M. Photochemistry in Biological Matrices: Activation of Racemic Mixtures and Interconversion of Enantiomers. *J. Am. Chem. Soc.* **1992**, *114*, 9300–9304. [CrossRef]
30. Bochet, C.G. Orthogonal Photolysis of Protecting Groups. *Angew. Chem. Int. Ed. Engl.* **2001**, *40*, 2071–2073. [CrossRef]

Article

# Spectroscopic and DFT Study of Alizarin Red S Complexes of Ga(III) in Semi-Aqueous Solution

Licínia L. G. Justino *, Sofia Braz and M. Luísa Ramos

CQC-IMS, Department of Chemistry, University of Coimbra, Rua Larga, 3004-535 Coimbra, Portugal
* Correspondence: liciniaj@ci.uc.pt

**Abstract:** A combined spectroscopic and computational approach has been used to study in detail the complexation between Ga(III) and ARS in solution. The NMR results revealed the formation of four Ga(III)/ARS complexes, at pH 4, differing in their metal:ligand stoichiometries or configuration, and point to a coordination mode through the ligand positions C-1 and C-9. For equimolar metal:ligand solutions, a 1:1 $[Ga(ARS)(H_2O)_4]^+$ complex was formed, while for 1:2 molar ratio solutions, a $[Ga(ARS)_2(H_2O)_2]^-$ complex, in which the two ligands are magnetically equivalent, is proposed. Based on DFT calculations, it was determined that this is a centrosymmetric structure with the ligands in an *anti* configuration. For solutions with a 1:3 molar ratio, two isomeric $[Ga(ARS)_3]^{3-}$ complexes were detected by NMR, in which the ligands have a *mer* and a *fac* configuration around the metal centre. The DFT calculations provided structural details on the complexes and support the proposal of a 1,9 coordination mode. The infrared spectroscopy results, together with the calculation of the infrared spectra for the theoretically proposed structures, give further support to the conclusions above. Changes in the UV/vis absorption and fluorescence spectra of the ligand upon complexation revealed that ARS is a highly sensitive fluorescent probe for the detection of Ga(III).

**Keywords:** $Ga^{3+}$; alizarin sulfonate; ARS; NMR; ATR-FTIR; DFT; UV/vis; fluorescent probe

Citation: Justino, L.L.G.; Braz, S.; Ramos, M.L. Spectroscopic and DFT Study of Alizarin Red S Complexes of Ga(III) in Semi-Aqueous Solution. *Photochem* **2023**, *3*, 61–81. https://doi.org/10.3390/photochem3010005

Academic Editor: Stefanie Tschierlei

Received: 16 December 2022
Revised: 25 January 2023
Accepted: 26 January 2023
Published: 31 January 2023

**Copyright:** © 2023 by the authors. Licensee MDPI, Basel, Switzerland. This article is an open access article distributed under the terms and conditions of the Creative Commons Attribution (CC BY) license (https://creativecommons.org/licenses/by/4.0/).

## 1. Introduction

Gallium is widely distributed in the earth's crust and has a variety of applications in industry, such as in semiconductor materials for red, orange, and yellow light-emitting diodes (LEDs) [1], and in the production of low melting point alloys [2]. The antineoplastic potential of compounds of gallium, in particular of some simple salts such as Ga(III) nitrate, was also recognized soon after the discovery of the antitumor properties of cisplatin. Ga(III) shows coordination properties similar to other group IIIa metal ions, such as Al(III) and In(III), and it also shares some properties with Fe(III) in terms of ligand affinity, coordination geometry, ionic radius, electronegativity, etc. This is thought to enable its ability to interfere with the cellular iron metabolism and seems to be crucial for its antineoplastic effects [3–5]. Although gallium has gained relevance in the biomedical area by presenting anti-proliferative activity in some malignant tumours in humans, its presence in water for human consumption can cause immune system diseases and reduced blood leukocyte counts [6]. Its use in industry may lead to contamination of groundwater and its possible impact on human health and the environment makes it important to develop methods capable of detecting and quantifying the distribution of gallium in view of its remediation.

The use of molecular chemosensors that show fluorescence responses modulated by the interaction with metal ions has numerous advantages, such as high sensitivity and specificity, low cost, simplicity of operation, as well as the possibility of monitoring biological samples in real time with fast responses [7], and has application in many diverse areas, from analytical chemistry, medicinal chemistry, and biochemistry, to clinical and environmental sciences.

Anthraquinones and their derivatives are organic compounds that are highly relevant due to their enormous versatility in many diverse applications, such as in industry as dyes [8–10], in medicine as pharmaceutical drugs [11,12], and in chemistry as analytical reagents [13,14]. Hydroxyanthraquinones are particularly important for applications based on chromatic properties, since their lowest excited singlet state, with $\pi,\pi^*$ character, enables them to absorb visible light and have bright colours, contrary to anthraquinone which has a $n,\pi^*$ lowest excited singlet state [15,16]. The optical properties of the substituted anthraquinones depend on factors such as the nature and position of the substituents and the establishment of hydrogen bonds and other intermolecular interactions [15]. Alizarin, (1,2-di-hydroxyanthraquinone, 1,2-HAQ) is one of the most stable natural pigments and the main colouring component (together with purpurin, 1,2,4-HAQ) of the natural pigment madder, extracted from *Rubia tinctorum* L., and is widely used to dye textiles. Nowadays, alizarin is usually obtained from synthesis. Alizarin has also been investigated as a photosensitizer in dye-sensitized solar cells, showing high incident photon-to-current conversion efficiencies [17–19]. By introducing a sulfonate group into the alizarin structure, a derivative with greater water solubility is obtained (1,2-dihydroxyanthraquinone-3-sulfonate, or alizarin red S (ARS)), whose photophysics and photochemistry has been extensively studied due to its chromatic and fluorescence properties. ARS also shows interesting electrochemical behaviour and its performance as a negolyte in redox flow batteries has been investigated recently [20]. Its complexation behaviour with various metal ions has also been the subject of intense investigation, since its phenolic and carbonyl groups (Scheme 1) offer the possibility for metal–ligand complexation.

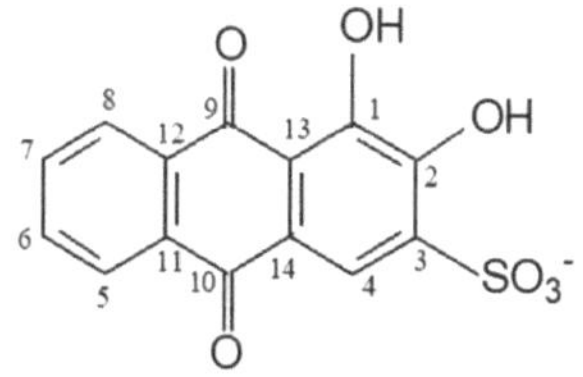

**Scheme 1.** Monoanionic form of alizarin sulfonate (ARS) and numbering scheme used in the discussion of results.

For a bidentate chelation with ARS, two coordination modes are possible: (i) via the (deprotonated) hydroxyl groups in positions 1 and 2 or (ii) via the (deprotonated) hydroxyl group in position 1 and the adjacent carbonyl group in position 9. Both modes of coordination have been proposed, depending on the metal ion, and in some cases related with the solution's pH. Coordination through positions 1,9 has been proposed to be favoured in aqueous acidic solution, while coordination through positions 1,2 has been associated with alkaline media. With Al(III), in acidic solution, it was suggested that ARS forms 1:1 and 1:2 (Al(III):ARS) complexes [21–23] coordinating to the metal through positions 1 and 2 [22,23] or through positions 1 and 9 [24]; a 1:3 complex, with 1,9-chelation, has also been suggested for the Al(III)/alizarin system in alkaline water suspensions [25]. Fewer studies are found concerning the In(III) and Ga(III) alizarinate complexes. The kinetic and thermodynamic parameters were studied for the 1:1 In(III):ARS complex in highly acidic aqueous solution, and a 1,2-chelation mode was proposed [26]. An early spectrophotometric study [27] showed that Ga(III) can bind to ARS forming a reddish chelate in the pH range of 3.0–5.0 in water. A 1:2 stoichiometry was proposed for the Ga(III)/ARS complex, and a tentative suggestion that ARS may bind Ga(III) in a 1,2 coordination mode was made.

Taking into account the importance of understanding the coordination chemistry of ARS, from both an analytical and fundamental point of view, and the relevance of studying Ga(III) compounds, we have carried out a speciation and structural characterization study of the complexes formed in the Ga(III)/ARS system in a water:methanol (1:1, $v/v$) solution. We have used a variety of spectroscopic and computational methods that can bring additional insight on the number, type, and structure of the complexes formed in this system. The structural characterization studies were carried out both in solution, using nuclear magnetic resonance spectroscopy (NMR), and in the solid state, using infrared spectroscopy. These studies were complemented with density functional theory (DFT) and time-dependent DFT quantum chemical calculations which allowed a detailed structural characterization of the ligand and complexes and the interpretation of their spectroscopic properties. Additionally, the absorption and fluorescence properties of the system were also studied, with the purpose of evaluating the potential of ARS as a sensor for Ga(III) ions.

## 2. Materials and Methods

### 2.1. Starting Materials and Preparation of Samples

Analytical grade gallium nitrate and commercially available 1,2-Dihydroxy-9,10-anthraquinone-3-sulfonate (Alizarin Red S, ARS) were used as received. For the NMR experiments, as the ligand and the mixtures of the ligand with metal have low solubility in $H_2O$ and consequently in $D_2O$, the solutions were tentatively prepared in other solvents as well as in mixtures of solvents. Although the mixture $D_2O/CD_3OD$ (50%:50%) was found to be the most appropriate for this study, additional experiments were also carried out in DMSO. For solutions in $D_2O/CD_3OD$, the pH was adjusted by the addition of DCl and NaOD; the pH* values quoted are the direct pH meter readings (at room temperature) after standardization with aqueous ($H_2O$) buffers. The solvents from the samples used in the NMR experiments were subsequently evaporated at room temperature and the resulting red powder was used in the ATR-FTIR experiments. For the UV/vis experiments, a sodium alizarin sulfonate solution of concentration $4.67 \times 10^{-5}$ mol/dm$^3$ and a gallium(III) nitrate solution ($2.46 \times 10^{-5}$ mol/dm$^3$) were prepared in $CH_3OH:H_2O$ 1:1 ($v/v$). The pH of the solutions was adjusted to pH 4 by the addition of $HClO_4$ and NaOH. The samples were stored in the dark until used.

### 2.2. Instrumentation

The $^1H$ and $^{13}C$ NMR spectra were obtained on a Bruker Avance 500 NMR spectrometer. The $^{13}C$ spectra were recorded using proton decoupling techniques (Waltz-16) taking advantage of the nuclear Overhauser effect. The methyl signal of tert-butyl alcohol was used as the internal reference for $^1H$ ($\delta$ 1.20) and $^{13}C$ ($\delta$ 31.20) for the spectra of solutions in $D_2O/CD_3OD$, and the residual signals of the solvent were used for $^1H$ ($\delta$ 2.50) and $^{13}C$ ($\delta$ 39.51), relative to TMS, for the spectra of solutions in DMSO. The ATR-FTIR spectra were obtained using a Thermoscientific Fourier Transform Infrared Spectrometer—Nicolet iS5 iD7 ATR (resolution 1 cm$^{-1}$), with the aid of the OMNIC program for spectral visualization. The vibrational modes were assigned with the help of visualization with the animation module of the GaussView program. The ultraviolet/visible absorption spectra were obtained using a Shimadzu spectrometer UV-2100 and the fluorescence spectra were obtained using a Horiba-Jobin-Yvon Fluorolog 3.2.2 spectrometer. The fluorescence emission spectra were recorded with excitation at 447 nm.

### 2.3. Computational Details

The geometries of the conformers and tautomers of ARS were optimized at the density functional (DFT) level of theory, using the B3LYP [28–31] hybrid exchange and correlation functional and the extended split-valence triple-$\zeta$ plus polarization 6-311++G(2d,2p) basis set. The geometries considered for the complexes were optimized with the same functional and also with the B3PW91 [29,32–36] hybrid functional, the CAM-B3LYP [37] long-range corrected hybrid functional, and the wB97X-D [38] long-range corrected hybrid functional which includes dispersion corrections, and the split-valence triple-$\zeta$ plus polarization 6-311G(d,p) basis set. All the structures were optimized considering the bulk solvent effects of water through the IEFPCM ("integral equation formalism variant of the polarizable continuum model") [39,40]. The vibrational frequencies were calculated for all the optimized geometries to verify the nature of the stationary points and ensure that they are energy minima. The structures proposed for the complexes were then reoptimized at the B3LYP/6-311++G(d,p) level and their harmonic vibrational frequencies were calculated and scaled with a factor of 0.978. This is the standard scale factor used for this calculation method, and aims to correct limitations introduced by the incomplete basis set used, the incomplete treatment of electronic correlation, and vibrational anharmonicity [41]. The theoretical infrared spectra presented were simulated using Lorentzian functions with a full-width-at-half-maximum (FWHM) of 6 or 4 cm$^{-1}$, centred at the scaled calculated frequencies. The B3LYP/GIAO ("gauge-including atomic orbital") method was used for the calculation of the $^1$H and $^{13}$C nuclear magnetic shielding constants ($\sigma$) of the lowest energy conformer of the ligand, using the 6-311++G(d,p) basis set. The nuclear magnetic shielding constants of tetramethylsilane (TMS) were calculated at the same theoretical level and the NMR chemical shifts were obtained relative to TMS from the equation $\delta = \sigma_{TMS} - \sigma$. Time-dependent DFT (TD-DFT) was used to calculate the UV/vis absorption spectra of the ligand and complexes using the CAM-B3LYP functional [37]. This long-range corrected hybrid functional, in which the amount of Hartree–Fock exchange interaction increases with the interelectronic distance, provides values of the vertical excitation energies which are in good agreement with the experimental ones for a wide variety of organic molecules [42]. The 6-311++G(d,p) basis set was used in these calculations. DFT and TD-DFT calculations were carried out with the Gaussian16 [43] quantum chemistry program and the GaussView 6.0 program was used to visualize the structures and molecular orbitals.

## 3. Results and Discussion

### 3.1. Structure and Energetics of Alizarin Red S

1,2-dihydroxyanthraquinone-3-sulfonate (ARS) is a polycyclic aromatic compound composed of three fused six-member rings, with two carbonyl groups in the central ring at the C-9 and C-10 positions. Additionally, one of outer rings bears two hydroxyl groups at the C-1 and C-2 positions and a sulfonate group at position C-3 (Scheme 1). Taking into account the pKa values reported for the two phenolic protons (pK$_{a2}$ = 5.8 and pK$_{a3}$ = 10.8) [44] and that the sulfonate group ionizes at very low pH, the monoanionic form of ARS (Scheme 1) is the dominant form in the pH region explored in this work (pH~4). Its structure, in addition to presenting several conformers formed by rotation of the CC-OH coordinates, may also present tautomerism involving the transfer of hydrogen atoms between the carbonyl and hydroxyl groups (keto–enol tautomerism). The structures and relative zero-point-corrected electronic energies of the different forms of ARS in water were previously studied by Joó et al. using DFT calculations (6-311+G(d,p) and TZVP basis sets and several different functionals) [45]. In our work, we have carried out additional studies on the structures and energetics of these species to determine their relative Gibbs energies (which include thermal corrections) which enable us to calculate their populations in water solutions at room temperature. These are essential for our theoretical studies of the spectroscopic properties of the ligand and Ga(III) complexes of ARS. Additionally, analysis of the structural details of the intramolecular hydrogen bonding in the most stable forms was also carried out.

In order to determine the most stable forms of ARS in aqueous solution, we have optimized the structures of the possible conformers and tautomers of the monoanionic form by DFT at the B3LYP/6-311++G(2d,2p) level of theory, considering the bulk solvent effects of water. The optimized structures of the most stable forms of ARS in water, and a summary of their relative electronic energies, zero-point corrected electronic energies, Gibbs energies, symmetries, and equilibrium populations at 298.15 K are given in Figure 1 and Table 1, respectively. The equilibrium populations were estimated from the Boltzmann equation and calculated relative Gibbs energies.

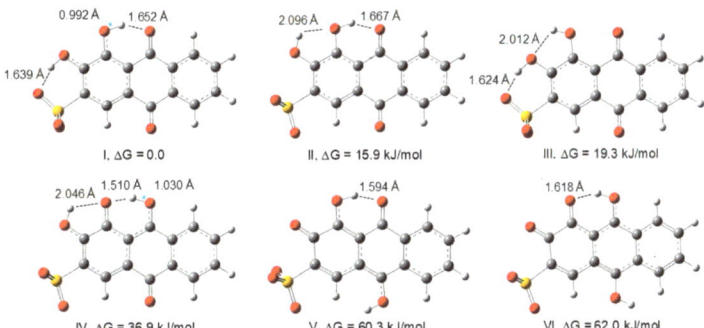

**Figure 1.** DFT/B3LYP/6-311++G(2d,2p) optimized geometries of the most stable conformers and tautomers of the monoanionic structure of ARS, considering the bulk effects of the water solvent.

**Table 1.** Relative electronic energies ($\Delta E_{el}$) (kJ mol$^{-1}$), zero-point-corrected electronic energies ($\Delta E_{Total}$), Gibbs energies at 298.15 K ($\Delta G_{298K}$), and equilibrium populations (%) estimated from the relative Gibbs energies ($P_{298}$), calculated for the lowest energies conformers and tautomers of the monoanionic structure of ARS (B3LYP/6-311++G(2d,2p)), considering the bulk effects of the H$_2$O solvent.

| Structure | I | II | III | IV | V | VI |
|---|---|---|---|---|---|---|
| Symmetry | $C_1$ | $C_1$ | $C_s$ | $C_1$ | $C_s$ | $C_s$ |
| $\Delta E_{el}$ | 0.0 | 18.8 | 26.0 | 40.7 | 71.8 | 73.6 |
| $\Delta(E_{Total})$ | 0.0 | 18.1 | 24.8 | 38.1 | 70.4 | 71.0 |
| $\Delta G_{298K}$ | 0.0 | 15.9 | 19.3 | 36.9 | 60.3 | 62.0 |
| $P(\%)_{298}$ | 99.8 | 0.2 | 0.0 | 0.0 | 0.0 | 0.0 |

Besides conformers **I**, **II** and **III**, an additional conformer, in which the two hydroxyl hydrogen atoms point to each other, was also optimized, but it converged to structure **III**. The additional tautomers calculated, with higher energies, are presented in the supplementary material (Figure S1). Structure **I** is stabilized by two strong intramolecular hydrogen bonds, with O–H$\cdots$O distances of 1.639 and 1.652 Å. The strongest hydrogen bond involves the hydroxyl group in position 2 as the H-donor and the oxygen atom of the sulfonate group lying in the plane of the rings as the acceptor, and the second hydrogen bond is established between the oxygen atom of the carbonyl group in position 9 and the hydrogen atom of the hydroxyl group in position 1. The structure is very close to planar, however the corresponding structure with $C_s$ symmetry is not a minimum. Therefore, the lowest energy conformer belongs to the $C_1$ point group. Structures **II**, **III**, and **IV** also present two intramolecular hydrogen bonds; however, in these conformers (**II**, **III**) and tautomer (**IV**), one of the bonds is relatively weaker, with OH$\cdots$O bond lengths close to 2 Å, making these structures significantly less stable, with $\Delta G$ values of 15.9, 19.3, and 36.9 kJ/mol, respectively, relative to conformer **I**. Tautomers **V** and **VI** present only one intramolecular hydrogen bond and, consequently, have very high energies. These results allow the conclusion that the dominant conformer of ARS in solution, with an estimated population of 99.8% at 298.15 K, is conformer **I**. The ordering of stabilities of the different structures in Figure 1 is in agreement with the results obtained by Joó et al. [45] at the B3LYP/TZVP level of theory. The structure of conformer **I** will be considered for the calculation of the spectroscopic properties of the free ligand in solution presented in the following sections in this work.

## 3.2. Complexation between Ga(III) and 1,2-Dihydroxy-9,10-anthraquinone-3-sulfonate (ARS)

### 3.2.1. NMR Studies on the Ga(III)/ARS System

Clear indications of metal–ligand binding are seen from broadening or coordination-induced shifts of the $^1$H and $^{13}$C signals of the ligand in the presence of the metal ion, compared with those of the free ligand, as well as, in favourable cases of slow exchange rate, from conformational changes after ligand complexation as indicated from the proton–proton coupling constants ($J_{H-H}$). This, together with the metal ion NMR when NMR active metallic isotopes are present, can provide valuable structural information, including the type of metal centre present in the complexes, as widely exemplified in our previous work on the complexation of metal ions, such as aluminium, gallium, and metal oxoions of vanadate, molybdate, and tungstate with relevant ligands [46–56]. Furthermore, the valuable structural information obtained from NMR and other spectroscopic techniques in this study, combined with the structural details accessible through DFT calculations, allow the complete structural understanding of the complexation [57,58].

For a complete structural characterization of the interaction of Ga(III) ions with ARS in aqueous solution, we have obtained $^1$H NMR spectra for solutions having an ARS concentration of 5 mmol dm$^{-3}$ and Ga(III) concentrations ranging from 0 to 5 mmol dm$^{-3}$, giving metal:ligand molar ratios from 1:1 to 1:3 at pH* 4. As the $^1$H signals observed appear to be very broadened, probably due to exchange processes dependent on the temperature, the NMR spectra were obtained both at room and low temperature. Although it is more difficult to have a reasonable signal/noise ratio due to its lower abundance, we have also obtained $^{13}$C spectra for the same solutions at 298.15 K. The results are shown in Figures 2 and 3 for the $^1$H and $^{13}$C NMR spectra of ARS alone and in the presence of gallium(III) nitrate, respectively. The $^1$H and $^{13}$C NMR experimental and calculated (DFT) spectral parameters for ARS are shown in Table S1.

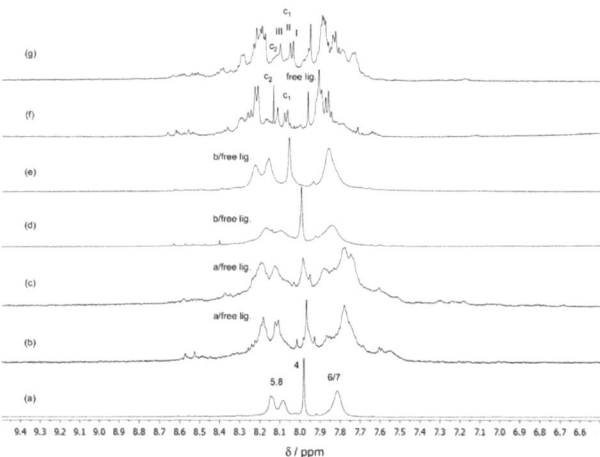

**Figure 2.** $^1$H NMR spectra of the solutions (50% D$_2$O/50% CD$_3$OD) of (**a**) ARS 5 mmol dm$^{-3}$, pH* = 3.96, temp. 298.15 K; (**b**) Ga/ARS 5:5 mmol dm$^{-3}$, pH* = 3.95, temp. 280.15 K; (**c**) Ga/ARS 5:5 mmol dm$^{-3}$, pH* = 3.95, temp. 298.15 K; (**d**) Ga/ARS 2.5:5 mmol dm$^{-3}$, pH* = 4.06, temp. 280.15 K; (**e**) Ga/ARS 2.5:5 mmol dm$^{-3}$, pH* = 4.06, temp. 298.15 K; (**f**) Ga/ARS 1.67:5.0 mmol dm$^{-3}$, pH* = 3.94, temp. 280.15 K; (**g**) Ga/ARS 1.67:5.0 mmol dm$^{-3}$, pH* = 3.94, temp. 298.15 K.

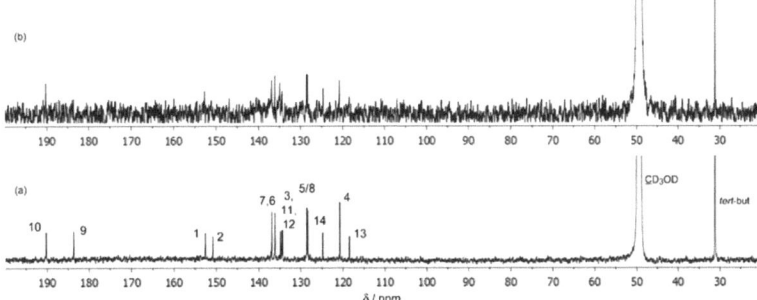

**Figure 3.** $^{13}C$ NMR spectra of the solutions (50% $D_2O$ e 50% $CD_3OD$) of (**a**) ARS 5 mmol dm$^{-3}$, pH* = 3.96; (**b**) Ga/ARS 5:5 mmol dm$^{-3}$, pH* = 3.95, temp. 298.15 K.

The $^1H$ NMR spectrum of the free ligand (ARS) in $D_2O/CD_3OD$ (1:1, $v/v$) at pH 3.96 showed a set of four signals (Figure 2a). The singlet at δ = 7.98 ppm was assigned to the H-4 proton. The multiplet centred at 7.82 ppm was assigned to protons 6 and 7, and the doublets (with low resolution) centred at 8.08 and 8.15 ppm were assigned to protons 5 and 8. The assignment of the $^{13}C$ signals is shown in Figure 3a. The assignment of the $^1H$ and $^{13}C$ NMR signals of ARS was completed with the help of the calculation of the nuclear magnetic shielding constants for the lowest energy conformer of ARS (conformer I, Figure 1) in aqueous solution that were subsequently converted into chemical shifts by subtraction from those calculated at the same theoretical level for TMS (Table S1).

As seen in Figure 2, upon addition of the Ga(III) metal ion, very broadened $^1H$ NMR signals were observed, particularly in 1:1 and 1:2 solutions, at both temperatures, suggesting the formation of 1:1 and/or 1:2 complexes strongly affected by the exchange processes (on the NMR time scale) between the free and complexed ligands. The corresponding spectra of the 1:3 molar ratio solutions showed, on the other hand, individual and distinct signals, even though they appeared to be slightly broadened (suggesting also some exchange on the NMR time scale). Although the signals in the 1:1 and 1:2 solutions became narrower with the decreasing temperature, it was not possible to detect distinct signals for the free and complexed ligands, probably due to the low concentration of the complexes and exchange, on the NMR time scale, between them and/or with the free ligand (more evident in solutions with 1:2 metal:ligand molar ratio). The corresponding spectra of the 1:3 (metal:ligand) molar ratio solutions showed, at 298.15 K, (besides the single signal at δ = 7.95, corresponding to proton H-4 of the free ligand) three additional single signals with similar intensities at δ values of 8.04, 8.05, and 8.10 ppm, suggesting the formation of one complex of 1:3 stoichiometry (complex $c_1$), in which the three ligands present a *mer* configuration (Figure 2g). It is proposed that this complex is in equilibrium with another isomer possessing *fac* configuration (complex $c_2$), according with the additional single signal detected (δ value of 8.14 ppm). The $^1H$ NMR spectra obtained at the lower temperature (280.15 K) showed narrower signals, the effect being visible in the spectrum of Figure 2f, suggesting a decrease of the exchange rate between the species present, allowing the detection of more separated individual signals. This observation gives further consistency to the suggestion that in solutions having a 1:3 (metal:ligand) molar ratio composition, two dominant 1:3 complexes ($c_1$ and $c_2$) having 1:3 stoichiometry were detected, in which the three coordinated ligands display *mer* and *fac* configurations around the metal centre, respectively. Complex $c_1$ (*mer*) was shown to be more stable than $c_2$ (*fac*) in the mixture of 50% $D_2O$/50% $CD_3OD$ solvents, and in the temperature range from 280.15 K to 298.15 K, although the concentrations of $c_1$ and $c_2$ decreased and increased, respectively, with increasing temperature, as indicated by the intensities of the $^1H$ NMR signals of the H-4 protons of the spectra obtained at 280.15 K and 298.15 K (Table 2).

**Table 2.** $^1$H NMR chemical shifts ($\delta$/ppm) of the H-4 proton for the free ligand ARS and isomers $c_1$ and $c_2$, and relative concentrations (in $D_2O/CD_3OD$, pH* 3,96) (the numbering is in accordance with Scheme 1).

|  | $\delta\ ^1$H RMN (exp.) [a] (H4) | Relative Concentration (%) |  | $\delta\ ^1$H RMN (exp.) [a] (H4) | Relative Concentration (%) |
| --- | --- | --- | --- | --- | --- |
|  | Temp. 280.15 K |  |  | Temp. 298.15 K |  |
| $c_1$ | 8.07 (I) | 57.5 | $c_1$ | 8.04 (I) | 51.4 |
|  | 8.08 (II) |  |  | 8.05 (II) |  |
|  | 8.11 (III) |  |  | 8.10 (III) |  |
| $c_2$ | 8.14 | 17.2 | $c_2$ | 8.14 (broad) | 24.0 |
| Free lig. | 7.96 | 25.3 | Free lig. | 7.96 | 24.6 |

[a] The numbering is based on the carbon atom to which the H atom is attached.

We have also obtained $^{13}$C spectra for the same solutions and the spectrum of the 1:1 solution is shown for illustration in Figure 3. It was not possible to achieve for this solution (1:1) a signal/noise ratio adequate to easily detect all the signals and increased difficulties were also found for the 1:2 and 1:3 solutions, due to the low abundance of the $^{13}$C isotope, the low concentration of the complexed species and the associated exchange processes between them. The $^{13}$C NMR spectrum of the solution of Ga(III)/ARS 5.0:5.0 mmol dm$^{-3}$ (1:1) (Figure 3b) showed that, relative to the free ligand (Figure 3a), intensity changes in the signals corresponding to the $^{13}$C nuclei of C-1, C-9, and C-13, suggesting the presence of one or more complexes in which the ligand was coordinated to the metal through the OH-1 and C(9)=O groups, with the intensity of C-13 also being affected as it is adjacent to both the C-1 and the C-9 atoms. Further support comes from the $^1$H spectra of Ga$^{3+}$/ARS 5:5 mmol dm$^{-3}$ solutions at temp. 298.15 K in DMSO, as shown in Figure 4, taking into account the intense broadening of the OH-1 signal in comparison with the slight broadening of OH-2, in the presence of the metal, suggesting that the coordination of the ligand through the OH-2 group will be unlikely at low pH, in accordance with the pK$_a$ values of the two phenolic protons (pK$_{a2}$ = 5.8 and pK$_{a3}$ = 10.8) [44].

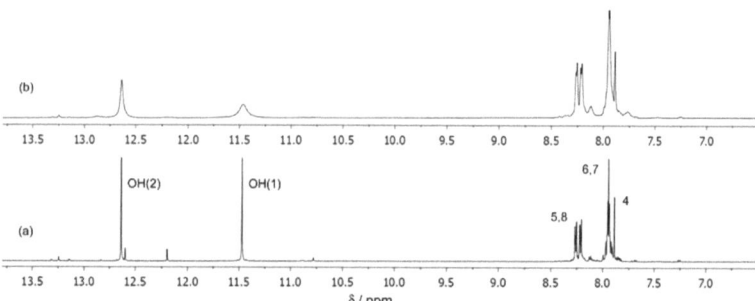

**Figure 4.** $^1$H NMR spectra of the solutions of (**a**) ARS 5 mmol dm$^{-3}$, pH* = 3.96; (**b**) Ga/ARS 5:5 mmol dm$^{-3}$ in DMSO, temp. 298.15 K.

Considering now the geometrical details, one, two, and three molecules of ARS coordinated one mononuclear Ga$^{3+}$ metal centre through the deprotonated OH-1 and carbonyl (C(9)=O) groups, forming six coordinated near-octahedral complexes with the minor species **a** and **b** and the dominant species, the isomers $c_1$ and $c_2$, respectively, as will be discussed in the DFT calculations section. One and two molecules of ARS coordinated the Ga$^{3+}$ metal ion in complexes **a** and **b**, respectively, with the remaining positions occupied by coordinated water molecules. For complex **a**, [Ga(ARS)(H$_2$O)$_4$]$^+$, a single set of $^1$H signals was observed as expected, taking into account that a single molecule of ARS was coordinated. Complex **b** is expected to be a 1:2 (metal:ligand) species, $[Ga(ARS)_2(H_2O)_2]^-$. The observation of only one set of $^1$H signals, although very broadened, means that the two complexed ligand moieties are magnetically equivalent (i.e., they have the same $^1$H and $^{13}$C chemical shifts), and indicates a symmetrical 1:2 complex. Several different structures can be proposed for complex **b** (different arrangements of *cis* or *trans* structures). However, it is not straightforward to definitively assign its structure from the NMR data. To determine which of the forms is the most stable, the structures of the 1:2 isomers were optimized at the DFT level, as will be discussed in detail below. With 1:3 (metal:ligand) complexes, although the two isomers detected, complexes $c_1$ and $c_2$, have a 1:3 (metal:ligand) stoichiometry $[Ga(ARS)_3]^{3-}$, the dominant isomer $c_1$ shows a *mer* geometry of the three coordinated ligands around the metal centre, while the *fac* geometry was observed for the minor complex $c_2$, as will be discussed in detail in the DFT calculations section. The coexistence of *fac* and *mer* isomers, with the *mer* isomer being more stable than the *fac*, was found in previous studies on the complexation of the Ga(III) metal ion with maltolate and hydroxythiopyrone [59,60]. For 1:3 (metal:ligand) Ga(III) metal ion complexes, the *mer* isomers were found to be generally more stable than their respective *fac* congeners and the relative stabilities have been found to be associated with the balance between bonding and steric factors. The *mer* isomer reduces intra-ligand repulsions, while the *fac* isomer is favoured where stronger covalent M–ligand bonds can be formed due to more extensive through-ligand conjugation mediated by metal *d* orbitals [61]. While previous studies on the complexation of $Ga^{3+}$ with 8-HQS [47] and 8-HQ [61,62] in solution are consistent with only *mer*-isomers being present in significant amounts at room temperature for 8-HQ and, for $Ga^{3+}$/8-HQS, over the whole temperature range up to 353.15 K [47], a mixture of *mer* and *fac* isomers has been identified, at room temperature, for the system Ga(III)/maltolate, both in solution and in the solid state (showing in the solid state the proportion of 0.67 for *mer* and 0.33 for *fac*) [59]. A mixture of *mer* and *fac* isomers has also been found for Ga(III)/hydroxythiopyrone, whose complexes can undergo fast isomerization in aqueous solution resulting in the coexistence of the *fac* and *mer* isomers, in spite of the *fac* geometry present in the solid state [60].

### 3.2.2. DFT Structural Characterization of the Ga(III)/ARS Complexes

Taking into account the structural information provided by the NMR results, which suggests the formation of complexes with 1:1, 1:2, and 1:3 Ga(III):ARS stoichiometries, DFT studies were carried out to obtain additional details on their geometries. To determine the lowest energy structures, several isomers were considered for each stoichiometry and, for the 1:2 complex, the requirement of a symmetrical structure in which the two ligands are magnetically equivalent was also taken into account. Additionally, since our NMR results suggest a 1,9 coordination mode, which is in contrast with the previous proposal [27] of coordination though positions 1 and 2, in order to bring additional insight on this question, the two potential modes of coordination for alizarinate complexes were investigated in these calculations: (i) chelation through the carbonyl in position C-9 and the adjacent hydroxyl group in position C-1 (labelled **hc**), or (ii) chelation through the hydroxyl groups in positions C-1 and C-2 (labelled **hh**). Given that the literature value for the p$K_{a3}$ of ARS is 10.8 [44], and considering the pH range in which our experiments were carried out, the hydroxyl group in position C-2 was set as protonated in these calculations. Whenever relevant, structures involving tautomerization of the ligand were also optimized. Several functionals were used in these calculations, in order to benchmark their ability to reproduce the experimental NMR results. The calculations were carried out with the B3LYP [28–31] and B3PW91 [29,32–36] hybrid functionals and with the long-range corrected CAM-B3LYP [37] and wB97X-D [38] functionals (wB97X-D includes additional dispersion corrections). The bulk solvent effects of water were taken into account in all cases; however, for the B3PW91 and wB97X-D calculations in water, the 1:3 **fac-hc** structures did not converge to a minimum in the potential energy surface, and therefore, for these two functionals, the relative Gibbs energies ($\Delta G$) given in Table 3 for the 1:3 structures were obtained from gas phase calculations. Table 3 summarizes the relative Gibbs energies and populations of the isomers at 298.15 K, and their point group symmetries. The comparison of the predicted populations of complexes with the NMR results revealed that, in general, the B3LYP functional allows a better reproduction of the experimental data, as it correctly predicted the formation of a dominant 1:1 complex, a dominant 1:2 complex, and significant amounts of both the fac and mer 1:3 isomers, as detected by NMR. In contrast, functionals CAM-B3LYP, B3PW91, and w-B97X-D predicted significant amounts of the **hh** 1:1 complex (not detected by NMR). Concerning the 1:2 stoichiometry, all functionals correctly predict the formation of one dominant complex, as found by NMR. For the 1:3 stoichiometry, and considering the calculations in water, both B3LYP and CAM-B3LYP overestimated the amount of the **fac-hc** isomer relatively to the **mer-hc** isomer; however, the error was smaller with B3LYP. The gas phase results obtained with functionals B3PW91 and w-B97X-D will be discussed and compared with those obtained with B3LYP later in this section. Taking into account these conclusions, the following discussion of the structures of complexes will be based on the B3LYP results. The optimized geometries obtained with the B3LYP functional are depicted in Figures 5–7.

**Table 3.** Relative Gibbs energies between isomers at 298.15 K ($\Delta G_{298}$/kJ mol$^{-1}$) and corresponding equilibrium populations ($P_{298}$/%) estimated from the relative Gibbs energies calculated using different functionals and the 6-311G(d,p) basis set (symmetries of structures are also indicated).

| M:L | Structure [a] | Sym | B3LYP | | CAM-B3LYP | | B3PW91 | | w-B97X-D | |
|---|---|---|---|---|---|---|---|---|---|---|
| | | | $\Delta G_{298}$ | $P_{298}$ | $\Delta G_{298}$ | $P_{298}$ | $\Delta G_{298}$ | $P_{298}$ | $\Delta G_{298}$ | $P_{298}$ |
| 1:1 | hc (**a**) | $C_1$ | 0.0 | 86.8 | 0.0 | 69.8 | 0.0 | 77.6 | 0.0 | 52.2 |
| | hh | $C_1$ | 4.7 | 13.1 | 2.1 | 30.2 | 3.1 | 22.4 | 0.2 | 47.8 |
| | hh-T1 | $C_1$ | 21.4 | 0.0 | 21.7 | 0.0 | 18.3 | 0.0 | 23.7 | 0.0 |
| | hh-T2 | $C_1$ | 51.2 | 0.0 | 58.8 | 0.0 | 51.0 | 0.0 | 58.1 | 0.0 |
| 1:2 | anti-hc (**b**) | $C_i$ | 0.0 | 98.8 | 0.0 | 96.7 | 0.0 | 97.5 | 0.0 | 87.3 |
| | syn-hc | $C_s$[b] | 10.9 | 1.2 | 8.3 | 3.3 | 9.1 | 2.5 | 4.8 | 12.7 |
| | anti-hh | $C_i$ | 26.0 | 0.0 | 27.1 | 0.0 | 24.3 | 0.0 | 23.0 | 0.0 |
| | syn-hh | $C_s$[b] | 32.3 | 0.0 | 30.5 | 0.0 | 28.9 | 0.0 | 36.9 | 0.0 |
| | anti-hh-T | $C_i$ | 44.0 | 0.0 | 53.5 | 0.0 | 40.2 | 0.0 | 51.8 | 0.0 |
| | syn-hh-T | $C_s$[b] | 48.4 | 0.0 | 58.1 | 0.0 | 44.1 | 0.0 | 58.0 | 0.0 |
| 1:3 | fac-hc (**c₂**) | $C_3$[b] | 0.0 | 61.4[d] | 0.0 | 76.7 | 12.2[c] | 0.7 | 29.8[c] | 0.0 |
| | mer-hc (**c₁**) | $C_1$ | 1.1 | 38.6[d] | 3.0 | 23.3 | 0.0[c] | 99.3 | 0.0[c] | 100 |
| | mer-hh | $C_1$ | 72.1 | 0.0 | 63.9 | 0.0 | 73.4[c] | 0.0 | 139.4[c] | 0.0 |
| | mer-hh-T | $C_1$ | 69.1 | 0.0 | 92.0 | 0.0 | 256.3[c] | 0.0 | 316.6[c] | 0.0 |
| | fac-hh | $C_3$[b] | 72.9 | 0.0 | 67.2 | 0.0 | 78.5[c] | 0.0 | 144.5[c] | 0.0 |
| | fac-hh-T | $C_3$[b] | 71.8 | 0.0 | 92.2 | 0.0 | 285.9[c] | 0.0 | 204.7[c] | 0.0 |

[a] The label "hc" indicates coordination through the carbonyl and the adjacent hydroxyl group; the label "hh" indicates coordination through both hydroxyl groups. [b] The symmetry in these structures is approximately $C_s$ and $C_3$, respectively. [c] For the B3PW91 and wB97X-D functionals, the 1:3 **fac-hc** structure did not converge to a minimum; therefore, for the 1:3 structures, the $\Delta G$ values presented were obtained from gas phase calculations. [d] The corresponding gas phase populations are 3.6% and 96.4% for the fac and mer isomers, respectively.

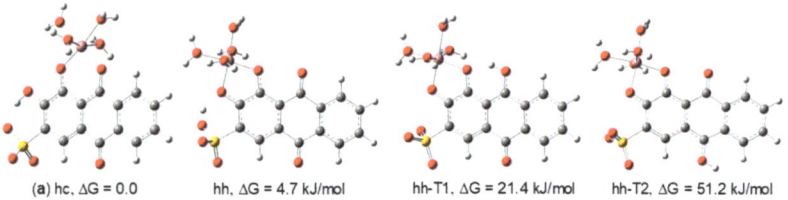

(a) hc, $\Delta G = 0.0$    hh, $\Delta G = 4.7$ kJ/mol    hh-T1, $\Delta G = 21.4$ kJ/mol    hh-T2, $\Delta G = 51.2$ kJ/mol

**Figure 5.** [Ga(ARS)(H$_2$O)$_4$]$^+$ isomeric structures considered to determine the structure of the 1:1 Ga(III)/ARS complex (optimized at the B3LYP/6-311G(d,p) in water).

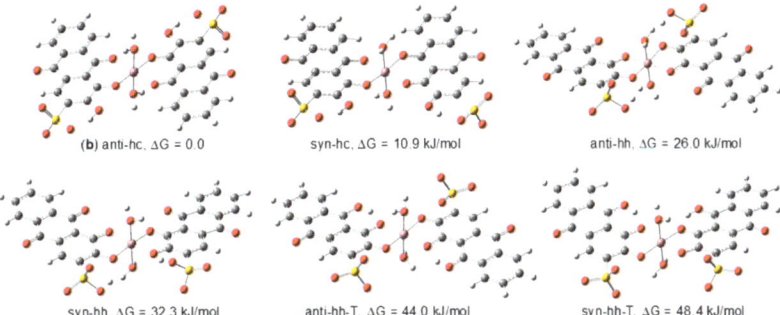

**Figure 6.** [Ga(ARS)$_2$(H$_2$O)$_2$]$^-$ isomeric structures considered to determine the structure of the 1:2 Ga(III)/ARS complex (optimized at the B3LYP/6-311G(d,p) in water).

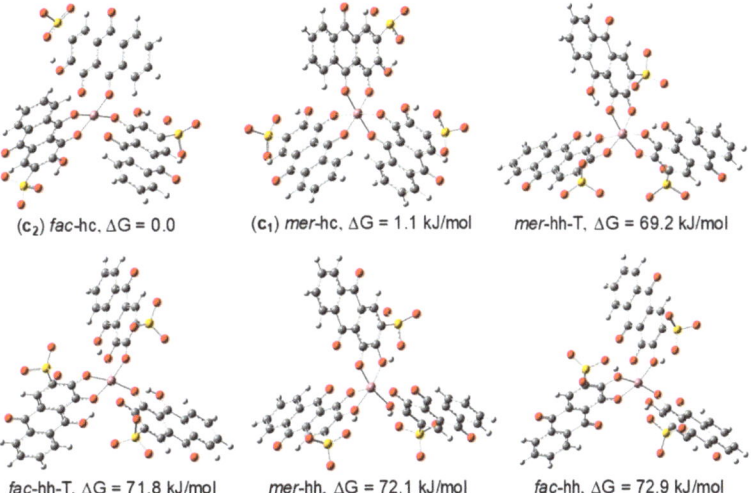

**Figure 7.** [Ga(ARS)$_3$]$^{3-}$ isomeric structures considered to determine the structure of the 1:3 Ga(III)/ARS complexes (optimized at the B3LYP/6-311G(d,p) in water).

The relative Gibbs energies of the structures considered for the 1:1 stoichiometry (Figure 5) indicate that, in this case, the structure involving coordination through the carbonyl in position C-9 and the adjacent hydroxyl group in position 1 (structure **a**, **hc**) is the most stable, with a population of 86.8% at the B3LYP/6-311G(d,p) level. Structure **hh**, in which there is coordination to the metal through both hydroxyl groups in positions 1 and 2 is 4.7 kJ/mol higher in energy and has a population of 13.1%. In structure **hh**, the hydrogen atom of the hydroxyl group in C-2 is involved in weak interactions with both the coordinated oxygen atom in C-2 and one of the oxygen atoms from the sulfonate group. Due to the lability of these interactions, we have also optimized the geometries of the tautomers **hh-T1** and **hh-T2**; however, these were found to be much higher in energy and not relevant from an experimental point of view. On the other hand, it is not expected that the oxygen atom in position C-2 remains protonated after binding to the metal. It is possible, therefore, that in acidic solution a weak interaction between H$^+$ and the O$^-$ in position C-2 persists in structure **hh**, as suggested by these calculations. Therefore, we propose structure **a** for the complex of 1:1 stoichiometry observed in the NMR spectra. According to its relative Gibbs energy, the **hh** structure may also be present in very low concentration. Its population is, however, too low to allow its detection in a mixture where complex **a** is dominant and there are exchange processes (on the NMR time scale) between the free and complexed ligand molecules.

To determine the geometry of the 1:2 (metal:ligand) complex we have considered six possible structures (Figure 6), with $C_i$ or $C_s$ symmetry, in accordance with the NMR data which points to a structure in which the two ligands are magnetically equivalent. Besides the two possible modes of coordination (**hc** or **hh**), the anti (ligands pointing in opposite directions) or syn (ligands pointing in the same direction) relative arrangements of the ligands were also analysed. The tautomers of the **hh** isomers were considered, as before for the 1:1 structures.

From the analysis of the relative energies of the isomers, we can conclude that the **anti-hc (b)** structure is the most stable one, with a population of 98.8% (B3LYP/6-311G(d,p). This suggests that in acidic solutions of 1:2 (metal:ligand) molar ratio, one dominant complex, coordinating to the metal through positions C-1 and C-9, should be present, in total accordance with the NMR findings for the 1:2 solutions at 298.15 and 280.15 K.

Finally, for the 1:3 stoichiometry complexes, six possible structures were optimized, in which mer ($C_1$ symmetry) and fac ($C_3$ symmetry) configuration structures with **hh** or **hc** coordination were considered. Additionally, tautomers of the **hh**-type structures were also calculated (Figure 7). The results obtained for the relative Gibbs energies and populations in water at 298.15 K indicate that it is possible that the **fac-hc** and **mer-hc** structures coexist in solution, with populations of 61.4 and 38.6% (B3LYP/6-311G(d,p), respectively. Again, the DFT results are consistent with the NMR data which shows that the fac and mer isomers of Ga(III)/ARS coexist in solutions of molar ratio 1:3. The calculated mer/fac ratio does not, however, reproduce the experimental ratio estimated from the NMR spectra, which favours the mer isomer instead. Nonetheless, the B3LYP calculations correctly predicted significant amounts of both complexes in equilibrium. This discrepancy is most likely due to the limitations of the solvation model used in the DFT calculations, which does not treat specific interactions (such as H-bonding) between the solvent and complexes and can easily change the relative thermodynamic stability between structures. Furthermore, also due to restrictions of the computational methods, we have considered water as the only solvent, not taking into account the less polar mixture of $CH_3OH/H_2O$ (1:1) used in the solution studies. Support for this conjecture comes from additional calculations at the B3LYP/6-311G(d,p) level in gas phase, for which the population of the mer isomer is 96,4% and that of the fac isomer is 3,6% (the B3PW91 and the w-B97X-D functionals gave similar results), reversing the stability predicted for water solutions, and suggesting that the mer isomer is favoured in less polar media. This seems to suggest that the solvent polarity is an important factor in the stabilization of these two isomers.

### 3.2.3. ATR-FTIR Studies on the Ga(III)/ARS System

The Ga(III)/ARS 1:1, 1:2, and 1:3 molar ratio samples used in the NMR experiments (solutions in $CD_3OD/D_2O$) were left at room temperature until the solvents were completely evaporated and, in all cases, a red powder was obtained. These solids were then studied using ATR-FTIR and the spectra of the solid reagents, ARS and $Ga(NO_3)_3$, were also obtained for comparison.

Figure 8 shows the ATR-FTIR spectrum in the 1800 to 400 $cm^{-1}$ region of the 1:2 molar ratio solid sample (top spectrum), in comparison with the spectra of solid ARS and solid gallium nitrate. The top spectrum shows evidence of complexation between Ga(III) and the ligand, as demonstrated by the presence of the new bands at 1527.3 (band 2), 1482.1 (band 3), and 796.2 $cm^{-1}$ (band 9) and the changes in the intensity or shape of the bands at 1660.6 (band 1), 1445.7 (band 4), 1019.9 (band 8), and 680.1 $cm^{-1}$ (band 10). Moreover, the shape and intensity of band 5 at 1348.8 $cm^{-1}$ is probably also an indication of complexation, since it is much stronger than the free ligand band at this wavelength and much narrower than the metal salt band with a maximum at 1327.9 $cm^{-1}$. Additionally, bands 6 (1250.5 $cm^{-1}$) and 7 (1175.7 $cm^{-1}$) are slightly shifted relative to the ligand bands (1241.2 and 1164.5 $cm^{-1}$). All these changes are a clear indication of the formation of one or more complexes between Ga(III) and ARS. The assignment of these bands based on DFT calculations will be discussed in the following sections. The spectra obtained for the powder samples of the solutions with 1:1 and 1:3 Ga(III)/ARS molar ratios (Figure S2) showed the same type of features described above for the 1:2 powder sample, with slight changes relative to the latter. The band profile was similar in all these samples (Figure S2), with three bands in the region between 1700 and 1500 cm$^{-1}$, followed by a region of intense bands between 1500 and 1000 cm$^{-1}$, and less intense bands in the region between 900 and 500 cm$^{-1}$. The spectra of the ligand and complexes in the 2500–3700 cm$^{-1}$ region (Figure S3) were dominated by wide and intense bands that are due to the symmetrical and anti-symmetrical stretching vibrations of the OH groups of ARS, of coordinated $H_2O$ molecules in complexes, and of the hygroscopic $H_2O$ present in the samples. The CH vibrations were observed close to 3000 cm$^{-1}$. To provide a detailed analysis of the FTIR-ATR spectra of the samples, the vibrational frequencies and intensities were calculated by DFT at the B3LYP/6-311++G(d,p) theory level for the ARS ligand (conformer I) and for the optimized structures of the complexes. The calculated spectra for the ligand and for the 1:3 mer complex are shown in Figure 9 (the full spectrum can be seen in Figure S4) in comparison with the experimental spectra. The spectra calculated for the remaining complexes are given in Figure S5.

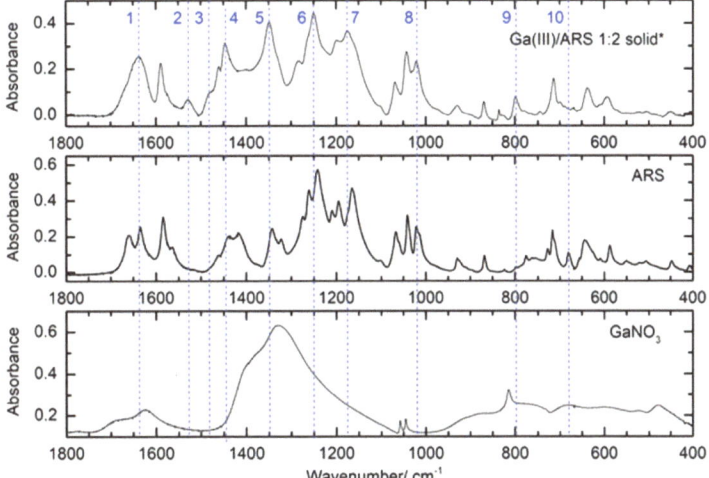

**Figure 8.** ATR-FTIR spectra (1800–400 cm$^{-1}$) of solid Ga(NO$_3$)$_3$ (bottom), solid ARS (middle), and the solid sample obtained from a 2.5:5 mmol dm$^{-3}$ Ga(III):ARS solution at pH = 4 (top); (*solid obtained after evaporation of the solvents from the Ga(III):ARS sample).

The vibrational spectrum calculated for the most stable conformer of ARS, conformer I, reproduced the main bands of the experimental spectrum of the solid ARS reasonably well, particularly in the spectral region above 1200 cm$^{-1}$. In the 1700–1500 cm$^{-1}$ spectral range, three bands were observed in the experimental spectrum at 1660, 1635, and 1584 cm$^{-1}$, which were predicted at 1653, 1621, and 1582 cm$^{-1}$, respectively, at the B3LYP/6-311++G(d,p) level of theory. The good agreement between the calculated and the experimental frequencies allowed the assignment of these bands with a good degree of confidence. According to the calculations, the band observed at 1660 cm$^{-1}$ in the experimental spectrum (predicted at 1653 cm$^{-1}$) has a predominant contribution from the $\upsilon$C10=O stretching mode, while the band observed at 1635 cm$^{-1}$ (predicted at 1621 cm$^{-1}$) is due to the $\upsilon$C9=O stretching mode combined with the bending of the C–OH groups. The lower frequency observed for the $\upsilon$C9=O stretching (relative to the $\upsilon$C10=O stretching) is reasoned by the fact that the C9=O group is involved in an intramolecular H-bond. The band predicted at 1582 cm$^{-1}$ and observed at 1584 cm$^{-1}$ has a major contribution from the bending of the C–OH groups. Regarding the changes observed in the experimental spectrum of the solid Ga(III):ARS 1:3 sample in comparison with the experimental spectrum for ARS (Figure 9), the most significant differences are the presence of the two new bands observed at 1533 cm$^{-1}$ (band 2) and 1480 cm$^{-1}$ (band 3) in the spectrum of the solid Ga(III):ARS 1:3 sample, which are absent from the ARS experimental spectrum. Accordingly, these bands, being absent from the theoretical spectrum of ARS, are predicted for the complex at 1543 cm$^{-1}$ and 1486 cm$^{-1}$, respectively, and correspond to the most important differences between the theoretical spectra of the complex and ligand. Band 2, predicted at 1543 cm$^{-1}$, is due to the $\nu$C9=O and $\nu$C1–O stretching modes, combined with ring $\nu$CC stretching. Band 3, predicted at 1486 cm$^{-1}$, involves contributions from these three modes, together with an additional contribution from the $\nu$C2–O stretching mode. These results are particularly relevant since they confirm the involvement of the C9=O group in the coordination to the metal, in complete agreement with the NMR and DFT findings. The shift of the $\nu$C9=O stretching mode to a lower wavelength upon complexation is expected, taking into account the weakening of the C9=O bond occurring with the ligand–metal electron donation and has also been reported for related complexes, such as the In(III)/ARS complex [26].

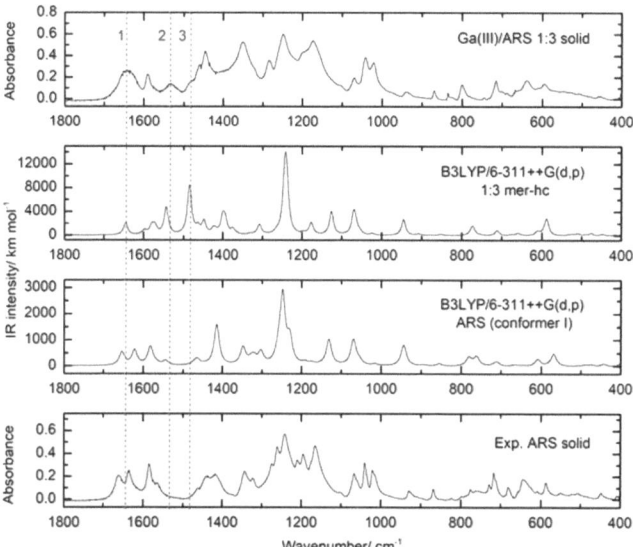

**Figure 9.** ATR-FTIR spectra (1800–400 cm$^{-1}$) of the solid powder samples of ARS and of the 1:3 Ga(III)/ARS sample (bottom and top spectra, respectively), compared with the B3LYP/6-311++G(d,p) calculated spectra for conformer **I** of ARS (middle bottom) and for the 1:3 mer structure (middle top). The calculated spectra were simulated using Lorentzian functions with a full-width-at-half-maximum (FWHM) of 6 cm$^{-1}$, centred at the scaled calculated frequencies.

The bands in the region below 1200 cm$^{-1}$ in the spectra of the ligand and complex were somewhat more difficult to assign, due to some differences between the theoretical and the experimental spectra. These differences are ascribable to intermolecular interactions, such as H-bonding, involving the polar groups of the ligand and also, possibly, hygroscopic water. Nonetheless, a tentative assignment of the bands of the complex can be made based on the calculations. Thus, the experimental broad bands with maxima at 1445 cm$^{-1}$ and 1249 cm$^{-1}$ correspond most likely to the bands predicted at 1401 cm$^{-1}$ and 1244 cm$^{-1}$, respectively, which have major contributions from ring $\nu$CC stretching vibrations. Additionally, the bands observed at 1070, 1042, and 1019 cm$^{-1}$ were assigned to the vibrational modes of the sulfonate group based on the calculated wavelengths for the anti-symmetrical stretching (1126 cm$^{-1}$), anti-symmetrical stretching combined with ring vibrations (1070 cm$^{-1}$), and symmetrical stretching combined with ring vibrations (945 cm$^{-1}$).

### 3.2.4. UV-Visible Absorption and Fluorescence Studies on the Ga(III)/ARS System

An early spectrophotometric study of the Ga(III)/ARS system in water [27] has suggested the formation of a 1:2 complex between the metal and ARS, with $\lambda_{max}$ at 490 nm, for pH values between 3 and 5. In the present work, UV-vis absorption spectra were obtained for a solution of ARS in water:methanol (1:1, $v/v$) at pH 4, with increasing concentrations of Ga(III) (Figure 10a). The absorption spectrum of ARS at pH 4 showed a maximum in the visible region at 424 nm. With the addition of Ga(III), a decrease in the intensity of the 424 nm band and the appearance of a new band with maximum at 482 nm were observed (Figure 10b), suggesting the formation of one or more Ga(III)/ARS complexes. The red shift observed upon complexation is probably due to the decrease of the frontier orbitals energy gap involved in the ligand-based transition (this will be analysed in detail in the following section based on TD-DFT calculations). A relatively well-defined isosbestic point can be identified at 447 nm, suggesting that under the conditions of our study, there is a dominant complex in equilibrium with the remaining species and the free ligand. Looking at the metal:ligand molar ratio range covered in our UV/vis study, the mer and fac complexes will be favoured at the beginning of the titration, when the ligand is in excess relatively to the metal; the 1:1 complex may be favoured at latter stages in conditions of excess of metal, and the dominant species will probably be the 1:2 complex, previously proposed to form in water [27].

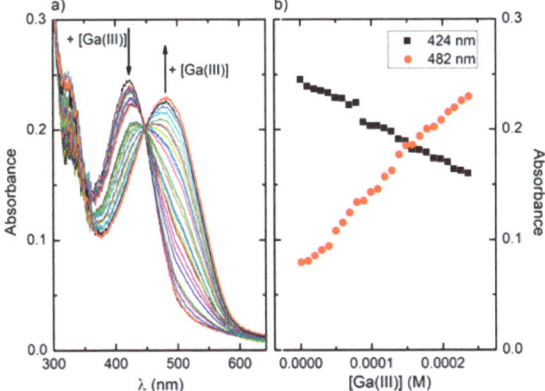

**Figure 10.** (a) Absorption spectra of ARS (4.67 × 10$^{-5}$ mol/dm$^3$) in H$_2$O:CH$_3$OH 1:1 ($v/v$) in the presence of increasing concentrations of Ga(III) (0–2.36 × 10$^{-4}$ mol/dm$^3$; steps of 9.84 × 10$^{-6}$ mol/dm$^3$) at pH 4; (b) absorbance intensity at 424 nm (black squares) and 482 nm (red circles) as a function of Ga(III) concentration at pH 4.

The fluorescence behaviour of ARS upon complexation with Ga(III) was studied with excitation at the absorption isosbestic point, 447 nm. Upon complexation with Ga(III), a decrease in the vibrational structure of the characteristic ARS emission band and an increase in the fluorescence at 625 nm were observed (Figure 11a). With the addition of an amount of Ga(III) as small as 9.84 × 10$^{-6}$ mol/dm$^3$, an increase of the fluorescence could already be observed (Figure 11b), indicating that ARS has high sensitivity for the detection of Ga(III) ions.

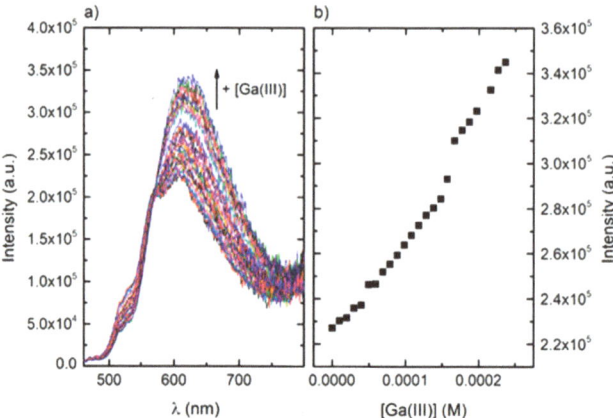

**Figure 11.** (a) Fluorescence emission spectra of ARS ($4.67 \times 10^{-5}$ mol/dm$^3$) in H$_2$O:CH$_3$OH 1:1 ($v/v$) at pH 4, collected with $\lambda_{exc}$ = 447 nm in the presence of increasing concentrations of Ga(III) (0–2.36 $\times$ $10^{-4}$ mol/dm$^3$; steps of $9.84 \times 10^{-6}$ mol/dm$^3$); (b) fluorescence intensity at 625 nm as a function of Ga(III) concentration at pH 4.

### 3.2.5. TD-DFT Studies on the Ga(III)/ARS System

The UV-visible absorption and fluorescence results can be analysed using TD-DFT calculations, which can help understand the underlying mechanism leading to the observed increase in fluorescence upon complexation. Calculations have been carried out for ARS (conformer **I**) and for the structures **a, b, c$_1$**, and **c$_2$** (Figures 5–7) and will be discussed in detail for the free ligand and the 1:2 **anti-hc** complex. Table 4 presents the vertical excitation energies, oscillator strengths, wavelengths, and major contributions to the first excited state of the free ligand and to some of the lowest energy excited states of the complexes. The calculations predicted an absorption band at 377 nm for the ligand, which corresponds to the experimental band observed at 424 nm. This band is due to excitation from the highest occupied molecular orbital, HOMO (H), to the lowest unoccupied molecular orbital, LUMO (L), of the ligand and has $\pi \rightarrow \pi^*$ character (Figure 12). On the other hand, the S$_1$ excited state of the centrosymmetric 1:2 **anti-hc** complex involves parity-forbidden transitions (the H-1, L+1, H and L orbitals can be seen in Figure 12, in the description of the S$_2$ excited state), and has an oscillator strength of zero. For this complex, an intense band was predicted at 426 nm (S$_2$ excited state), which corresponds to the experimental band observed at 482 nm, and involves electronic excitation from the H and H-1 orbitals, with $\pi$ character, to the L+1 and L orbitals, with $\pi^*$ character (Figure 12). These calculations predict a red shift of approximately 49 nm of the absorption maximum upon complexation, which is in good agreement with the experimental red shift of 58 nm observed when comparing the experimental absorption maxima for the ligand and complexes. It is also significant that the absorption maxima calculated for the 1:2 **anti-hc** and the fac and mer complexes have very similar values (a strong band at 426/427 nm, with a smaller contribution at 435/436 nm) indicating that it is possible that a mixture of these complexes in solution give rise to similar spectra in this region.

Table 4. Vertical excitation energies, oscillator strengths (*f*), wavelengths (λ), and major contributions calculated for ARS and the Ga(III)/ARS complexes (TD-DFT CAM-B3LYP/6-311++G(d,p), IEFPCM ($H_2O$)).

| Excited State | Energy (eV) | $\lambda_{calc.}$ (nm) | $\lambda_{exp.}$ (nm) | *f* | Major Contributions (%) | Character |
|---|---|---|---|---|---|---|
| | | | | | ARS | |
| $S_1$ | 3.29 | 377 | 424 | 0.2267 | H→L (100%) | $\pi \to \pi^*$ |
| | | | | | 1:1 hc complex (a) | |
| $S_1$ | 2.78 | 446 | 482 | 0.1901 | H→L (100%) | $\pi \to \pi^*$ |
| | | | | | 1:2 anti-hc complex (b) | |
| $S_1$ | 2.84 | 436 | 482 | 0.0000 | H-1→L+1 (41%) + H→L (59%) | $\pi \to \pi^*$ |
| $S_2$ | 2.91 | 426 | | 0.4307 | H-1→L (52%) + H→L+1 (48%) | $\pi \to \pi^*$ |
| | | | | | 1:3 mer ($c_1$) | |
| $S_1$ | 2.85 | 436 | 482 | 0.2675 | H-1→L (26%) + H-1→L+2 (20%) | $\pi \to \pi^*$ |
| $S_3$ | 2.90 | 427 | | 0.4348 | H→L (24%) + H-1→L (21%) | $\pi \to \pi^*$ |
| | | | | | 1:3 fac ($c_2$) | |
| $S_1$ | 2.85 | 435 | | 0.1532 | H-2→L (34%) + H→L+1 (22%) | $\pi \to \pi^*$ |
| $S_2$ | 2.85 | 435 | 482 | 0.1548 | H-1→L (33%) + H→L+2 (23%) | $\pi \to \pi^*$ |
| $S_3$ | 2.91 | 426 | | 0.4244 | H→L (46%) + H-1→L+2 (28%) | $\pi \to \pi^*$ |

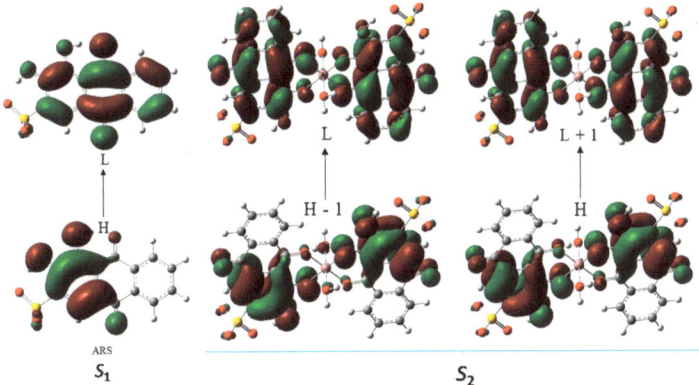

**Figure 12.** Dominant contributions to the $S_1$ excited state of the ligand (experimental band observed at 424 nm) and the $S_2$ excited state of the 1:2 **anti-hc** complex (experimental band observed at 482 nm) (TD-DFT CAM-B3LYP/6-311++G(d,p), IEFPCM, $H_2O$).

The $\pi \to \pi^*$ character of the transitions involved in the $S_1$ excited state of the 1:2 **anti-hc** complex (and in the 1:1 and 1:3 complexes) also reveals that the observed increase in fluorescence with complexation to the metal is possibly related, among other factors, with the decrease in non-radiative relaxation processes associated with the increased flexibility of the uncomplexed ligand.

## 4. Conclusions

A detailed study of the complexation between Ga(III) and alizarine red S was carried out using both spectroscopic and computational methods. The behaviour of the system was analysed at pH 4 in a water:methanol (1:1, v/v) solution for a variety of ligand:metal molar ratios, as well as in the solid state. This allowed to us to understand that, besides the previously 1:2 (metal:ligand) complex [27] reported for pH values between 3 and 5 in water, an additional 1:1 and two 1:3 complexes were formed in the conditions of our study. The structures of all the complexes were characterized in detail. The 1:1 complex was proposed to be a near-octahedral species with the formula [Ga(ARS)($H_2O$)$_4$]$^+$. The 1:2 complex is a centrosymmetric species with two magnetically equivalent ARS ligands in an anti relative arrangement. The two 1:3 complexes have, respectively, mer and fac configurations of the ligands around the metal centres. A 1,9 coordination mode between alizarine red S and Ga(III) was proposed for all the complexes. This conclusion obtains support from the different techniques used, namely from the NMR results, which showed significant effects on the intensity of the C-9 and C-1 carbon atoms and the C-1 hydroxyl proton; from the DFT calculations, which indicate that, irrespective of the stoichiometry, the structures with a 1,9 coordination mode were more stable than the analogous structures with a 1,2 coordination mode; and from the comparison of the infrared spectra calculated for the proposed structures with a 1,9 coordination mode with the experimental spectra with a good agreement between them. Our proposal is, however, in contrast with the previous tentative suggestion of a 1,2 coordination mode for the 1:2 Ga(III)/ARS complex in water [27]. Although our experimental studies have been carried out in water:methanol, the DFT relative energies calculated for the 1,2 and 1,9 structures in water (assuming a monoanionic ARS ligand in acidic medium) point to an increased stability of the Ga(III)/ARS complexes with a 1,9 coordination mode. Therefore, we believe this is the also most probable mode of coordination in these systems in water. Marked changes were observed in the UV/vis absorption spectra of ARS upon the addition of Ga(III), as well as in its fluorescence emission spectra, which shows a strong increase in the fluorescence intensity. These results support the use of ARS as a potential sensor for the detection of Ga(III).

**Supplementary Materials:** The following supporting information can be downloaded at: https://www.mdpi.com/article/10.3390/photochem3010005/s1, Figure S1: B3LYP/6-311++G(2d,2p)-optimized geometries of additional tautomers of ARS; Figure S2: ATR-FTIR spectra (1800–400 cm$^{-1}$) of the solid powder samples obtained from the Ga(III):ARS 1:3, 1:2, and 1:1 molar ratio solutions; Figure S3: ATR-FTIR spectra (4000–400 cm$^{-1}$) of the solid powder samples obtained from the Ga(III):ARS 1:3, 1:2, and 1:1 molar ratio solutions; Figure S4: ATR-FTIR spectra (4400–400 cm$^{-1}$) of the solid powder samples of ARS and of the 1:3 Ga(III)/ARS sample compared with the B3LYP/6-311++G(d,p) calculated spectra for the same compounds; Figure S5: ATR-FTIR spectra (4400–400 cm$^{-1}$) calculated at the B3LYP/6-311++G(d,p) level of theory for the 1:3 fac-hc, 1:2-hc, 1:1-hc, and 1:1-hh structures; Table S1: Experimental (D$_2$O/CD$_3$OD, pH* 3,96) and calculated (B3LYP/GIAO/6-311++G(d,p) level in water) $^1$H and $^{13}$C NMR chemical shifts ($\delta$/ppm) for the free ligand ARS.

**Author Contributions:** Conceptualization, L.L.G.J. and M.L.R.; methodology, L.L.G.J. and M.L.R.; validation, L.L.G.J., S.B. and M.L.R.; formal analysis, L.L.G.J., S.B. and M.L.R.; investigation, L.L.G.J., S.B. and M.L.R.; resources, L.L.G.J. and M.L.R.; data curation, L.L.G.J., S.B. and M.L.R.; writing—original draft preparation, L.L.G.J., S.B. and M.L.R.; writing—review and editing, L.L.G.J. and M.L.R.; visualization, L.L.G.J., S.B. and M.L.R.; supervision, L.L.G.J. and M.L.R.; project administration, L.L.G.J. and M.L.R.; funding acquisition, L.L.G.J. and M.L.R. All authors have read and agreed to the published version of the manuscript.

**Funding:** The CQC-IMS is supported by Fundação para a Ciência e a Tecnologia (FCT) through projects UI0313B/QUI/2020, UI0313P/QUI/2020, and LA/P/0056/2020.

**Acknowledgments:** The authors thank the Laboratory for Advanced Computing at the University of Coimbra for providing computing resources that have contributed to the reported research results and the UC-NMR facility which is supported in part by FEDER through the COMPETE Programme and by National Funds from FCT through grants REEQ/481/QUI/2006, RECI/QEQ-QFI/0168/2012, CENTRO-07-CT62-FEDER-002012, and Rede Nacional de Ressonância Magnética Nuclear (RNRMN).

**Conflicts of Interest:** The authors declare no conflict of interest.

## References

1. Sato, T.; Imai, M. Characteristics of nitrogen-doped GaAsP light-emitting diodes. *Jpn. J. Appl. Phys.* **2002**, *41*, 5995–5998. [CrossRef]
2. Borra, E.F.; Tremblay, G.; Huot, Y.; Gauvin, J. Gallium liquid mirrors: Basic technology, optical-shop tests and observations. *Astron. Soc. Pac.* **1997**, *109*, 319–325. [CrossRef]
3. Chua, M.-S.; Bernstein, L.R.; Li, R.; So, S.K.S. Gallium maltolate is a promising chemotherapeutic agent for the treatment of hepatocellular carcinoma. *Anticancer Res.* **2006**, *26*, 1739–1744. [PubMed]

4. Jakupec, M.A.; Galanski, M.; Arion, V.B.; Hartingerand, C.C.; Keppler, B.K. Antitumour metal compounds: More than theme and variations. *Dalton Trans.* **2008**, *2*, 183–194. [CrossRef]
5. Baran, E.J. La nueva farmacoterapia inorgánica XIX. Compuestos de galio. *Latin Am. J. Pharm.* **2008**, *27*, 776–779.
6. Chen, H.-W. Gallium, indium, and arsenic pollution of groundwater from a semiconductor manufacturing area of taiwan. *Bull. Environ. Contam. Toxicol.* **2006**, *77*, 289–296. [CrossRef]
7. Powe, A.M.; Das, S.; Lowry, M.; El-Zahab, B.; Fakayode, S.O.; Geng, M.L.; Baker, G.A.; Wang, L.; McCarroll, M.E.; Patonay, G.; et al. Molecular fluorescence, phosphorescence, and chemiluminescence spectrometry. *Anal. Chem.* **2010**, *82*, 4865–4894. [CrossRef]
8. Novotná, P.; Pacáková, V.; Bosáková, Z.; Štulík, K. High-performance liquid chromatographic determination of some anthraquinone and naphthoquinone dyes occurring in historical textiles. *J. Chromatogr. A* **1999**, *863*, 235–241. [CrossRef]
9. Orska-Gawryś, J.; Surowiec, I.; Kehl, J.; Rejniak, H.; Urbaniak-Walczak, K.; Trojanowicz, M. Identification of natural dyes in archeological Coptic textiles by liquid chromatography with diode array detection. *J. Chromatogr. A* **2003**, *989*, 239–248. [CrossRef]
10. Szostek, B.; Orska-Gawrys, J.; Surowiec, I.; Trojanowicz, M. Investigation of natural dyes occurring in historical Coptic textiles by high-performance liquid chromatography with UV–Vis and mass spectrometric detection. *J. Chromatogr. A* **2003**, *1012*, 179–192. [CrossRef]
11. Alves, D.S.; Perez-Fons, L.; Estepa, A.; Micol, V. Membrane-related effects underlying the biological activity of the anthraquinones emodin and barbaloin. *Biochem. Pharmacol.* **2004**, *68*, 549–561. [CrossRef]
12. Norton, S.A. Useful plants of dermatology. IV. Alizarin red and madder. *J. Am. Acad. Dermatol.* **1998**, *39*, 484–485. [CrossRef]
13. Diaz, A.N. Analytical applications of 1,10-anthraquinones: A review. *Talanta* **1991**, *38*, 571–588. [CrossRef]
14. Ghosh, A.; Jose, D.A.; Kaushik, R. Anthraquinones as versatile colorimetric reagent for anions. *Sens. Actuators B Chem.* **2016**, *229*, 545–560. [CrossRef]
15. Diaz, A.N. Absorption and emission spectroscopy and photochemistry of 1,10-anthraquinone derivatives: A review. *J. Photochem. Photobiol. A Chem.* **1990**, *53*, 141–167. [CrossRef]
16. Miliani, C.; Romani, A.; Favaro, G. Acidichromic effects in 1,2-di- and 1,2,4-trihydroxyanthraquinones. A spectrophotometric and Fuorimetric study. *J. Phys. Org. Chem.* **2000**, *13*, 141–150. [CrossRef]
17. Duncan, W.R.; Prezhdo, O.V. Temperature independence of the photoinduced electron injection in dye-sensitized $TiO_2$ rationalized by ab initio time-domain density functional theory. *J. Am. Chem. Soc.* **2008**, *130*, 9756–9762. [CrossRef]
18. Kaniyankandy, S.; Verma, S.; Mondal, J.A.; Palit, D.K.; Ghosh, H.N. Evidence of multiple electron injection and slow back electron transfer in alizarin-sensitized ultrasmall $TiO_2$ particles. *J. Phys. Chem. C* **2009**, *113*, 3593–3599. [CrossRef]
19. Nawrocka, A.; Krawczyk, S. Electronic excited state of alizarin dye adsorbed on $TiO_2$ nanoparticles: A study by electroabsorption (stark effect) spectroscopy. *J. Phys. Chem. C* **2008**, *112*, 10233–10241. [CrossRef]
20. Lima, A.R.F.; Pereira, R.C.; Azevedo, J.; Mendes, A.; de Melo, J.S.S. On the path to aqueous organic redox flow batteries: Alizarin red S alkaline negolyte. Performance evaluation and photochemical studies. *J. Mol. Liq.* **2021**, *336*, 116364. [CrossRef]
21. Safavi, A.; Abdollahi, H.; Mirzajani, R. Simultaneous spectrophotometric determination of Fe(III), Al(III) and Cu(II) by partial least-squares calibration method. *Spectrochim. Acta Part A* **2006**, *63*, 196–199. [CrossRef]
22. Sathish, R.S.; Kumar, M.R.; Rao, G.N.; Kumar, K.A.; Janardhana, C. A water-soluble fluorescent fluoride ion probe based on Alizarin Red S–Al(III) complex. *Spectrochim. Acta Part A* **2007**, *66*, 457–461. [CrossRef] [PubMed]
23. Wang, Y.; Xiong, L.; Geng, F.; Zhang, F.; Xu, M. Design of a dual-signaling sensing system for fluorescent ratiometric detection of $Al^{3+}$ ion based on the inner-filter effect. *Analyst* **2011**, *136*, 4809. [CrossRef] [PubMed]
24. Supian, S.M.; Ling, T.L.; Heng, L.Y.; Chong, K.F. Quantitative determination of Al(III) ion by using alizarin red S including its microspheres optical sensing Material. *Anal. Methods* **2013**, *5*, 2602–2609. [CrossRef]
25. Epstein, M.; Yariv, S. Visible-spectroscopy study of the adsorption of alizarinate by Al-montmorillonite in aqueous suspensions and in solid state. *J. Colloid Interface Sci.* **2003**, *263*, 377–385. [CrossRef]
26. Biver, T.; Kraiem, M.; Secco, F.; Venturini, M. On the mechanism of indium(III) complex formation with metallochromic indicators. *Polyhedron* **2018**, *156*, 6–13. [CrossRef]
27. Dwivedi, C.D.; Munshi, K.N.; Dey, A.K. Chelate formation of trivalent gallium with 1,2-dihydroxy-3-anthraquinone sulfonic acid. *Microchem. J.* **1965**, *9*, 218–226. [CrossRef]
28. Becke, A.D. Density-functional exchange-energy approximation with correct asymptotic behavior. *Phys. Rev. A* **1988**, *38*, 3098–3100. [CrossRef]
29. Becke, A.D. Density-functional thermochemistry. III. The role of exact exchange. *Chem. Phys.* **1993**, *98*, 5648–5652. [CrossRef]
30. Lee, C.T.; Yang, W.; Parr, R.G. Development of the Colle-Salvetti correlation-energy formula into a functional of the electron density. *Phys. Rev. B* **1988**, *37*, 785–789. [CrossRef]
31. Vosko, S.H.; Wilk, L.; Nusair, M. Accurate spin-dependent electron liquid correlation energies for local spin density calculations: A critical analysis. *Can. J. Phys.* **1980**, *58*, 1200–1211. [CrossRef]
32. Perdew, J.P. *Electronic Structure of Solids '91*; Ziesche, P., Eschrig, H., Eds.; Akademie Verlag: Berlin, Germany, 1991; Volume 11.
33. Perdew, J.P.; Chevary, J.A.; Vosko, S.H.; Jackson, K.A.; Pederson, M.R.; Singh, D.J.; Fiolhais, C. Atoms, molecules, solids, and surfaces: Applications of the generalized gradient approximation for exchange and correlation. *Phys. Rev. B* **1992**, *46*, 6671–6687. [CrossRef]

34. Perdew, J.P.; Chevary, J.A.; Vosko, S.H.; Jackson, K.A.; Pederson, M.R.; Singh, D.J.; Fiolhais, C. Erratum: Atoms, molecules, solids, and surfaces–Applications of the generalized gradient approximation for exchange and correlation. *Phys. Rev. B* **1993**, *48*, 4978. [CrossRef]
35. Perdew, J.P.; Burke, K.; Wang, Y. Generalized gradient approximation for the exchange-correlation hole of a many-electron system. *Phys. Rev. B* **1996**, *54*, 16533–16539. [CrossRef]
36. Burke, K.; Perdew, J.P.; Wang, Y. Derivation of a Generalized Gradient Approximation: The PW91 Density Functional. In *Electronic Density Functional Theory: Recent Progress and New Directions*; Dobson, J.F., Vignale, G., Das, M.P., Eds.; Plenum Press: New York, NY, USA, 1998; pp. 81–111.
37. Yanai, T.; Tew, D.P.; Handy, N.C. A new hybrid exchange–correlation functional using the Coulomb-attenuating method (CAM-B3LYP). *Chem. Phys. Lett.* **2004**, *393*, 51–57. [CrossRef]
38. Chai, J.-D.; Head-Gordon, M. Long-range corrected hybrid density functionals with damped atom–atom dispersion corrections. *Phys. Chem. Chem. Phys.* **2008**, *10*, 6615–6620. [CrossRef]
39. Miertus, S.; Scrocco, E.; Tomasi, J. Electrostatic interaction of a solute with a continuum. A direct utilization of ab initio molecular potentials for the prevision of solvent effects. *J. Chem. Phys.* **1981**, *55*, 117–129.
40. Tomasi, J.; Mennucci, B.; Cammi, R. Quantum mechanical continuum solvation models. *Chem. Rev.* **2005**, *105*, 2999–3093. [CrossRef]
41. Justino, L.L.G.; Reva, I.; Fausto, R. Thermally and vibrationally induced conformational isomerizations, infrared spectra, and photochemistry of gallic acid in low-temperature matrices. *J. Chem. Phys.* **2016**, *145*, 014304. [CrossRef]
42. Jaquemin, D.; Perpète, E.A.; Scuseria, G.E.; Ciofini, I.; Adamo, C. TD-DFT Performance for the Visible Absorption Spectra of Organic Dyes: Conventional versus Long-Range Hybrids. *J. Chem. Theory Comput.* **2008**, *4*, 123–135. [CrossRef]
43. Frisch, M.J.; Trucks, G.; Schlegel, H.B.; Scuseria, G.E.; Robb, M.A.; Cheeseman, J.R.; Scalmani, G.; Barone, V.; Petersson, G.; Nakatsuji, H.; et al. *Gaussian 16*, Revision B.01; Gaussian, Inc.: Wallingford, CT, USA, 2016.
44. Shalaby, A.A.; Mohamed, A.A. Determination of acid dissociation constants of alizarin red S, methyl orange, bromothymol blue and bromophenol blue using a digital camera. *RSC Adv.* **2020**, *10*, 11311–11316. [CrossRef] [PubMed]
45. Fehér, P.P.; Purgel, M.; Joó, F. Performance of exchange–correlation functionals on describing ground state geometries and excitations of alizarin red S: Effect of complexation and degree of deprotonation. *Comput. Theor. Chem.* **2014**, *1045*, 113–122. [CrossRef]
46. Ramos, M.L.; Justino, L.L.G.; Salvador, A.I.N.; de Sousa, A.R.E.; Abreu, P.E.; Fonseca, S.M.; Burrows, H.D. NMR, DFT and luminescence studies of the complexation of Al(III) with 8-hydroxyquinoline-5-sulfonate. *Dalton Trans.* **2012**, *41*, 12478–12489. [CrossRef] [PubMed]
47. Ramos, M.L.; Justino, L.L.G.; de Sousa, A.R.E.; Fonseca, S.M.; Geraldes, C.F.G.C.; Burrows, H.D. Structural and photophysical studies on gallium(III) 8-hydroxyquinoline-5-sulfonates. Does excited state decay involve ligand photolabilisation? *Dalton Trans.* **2013**, *42*, 3682–3694. [CrossRef] [PubMed]
48. Ramos, M.L.; Justino, L.L.G.; Fonseca, S.M.; Burrows, H.D. NMR, DFT and luminescence studies of the complexation of V(V) oxoions in solution with 8-hydroxyquinoline-5-sulfonate. *New J. Chem.* **2015**, *39*, 1488–1497. [CrossRef]
49. Ramos, M.L.; Justino, L.L.G.; Abreu, P.E.; Fonseca, S.M.; Burrows, H.D. Oxocomplexes of Mo(VI) and W(VI) with 8-hydroxyquinoline-5-sulfonate in solution: Structural studies and the effect of the metal ion on the photophysical behavior. *Dalton Trans.* **2015**, *44*, 19076–19089. [CrossRef]
50. Caldeira, M.M.; Ramos, M.L.; Cavaleiro, A.M.; Gil, V.M.S. Multinuclear NMR study of vanadium(V) complexation with tartaric and citric acids. *J. Mol. Struct.* **1988**, *174*, 461–466. [CrossRef]
51. Ramos, M.L.; Caldeira, M.M.; Gil, V.M.S. NMR study of the complexation of D-galactonic acid with tungsten(VI) and molybdenum(VI). *Carbohydr. Res.* **1997**, *297*, 191–200. [CrossRef]
52. Ramos, M.L.; Caldeira, M.M.; Gil, V.M.S. Multinuclear NMR study of complexation of D-galactaric and D-mannaric acids with tungsten(VI) oxoions. *J. Coord. Chem.* **1994**, *33*, 319–329. [CrossRef]
53. Ramos, M.L.; Caldeira, M.M.; Gil, V.M.S. NMR spectroscopy study of the complexation of L-mannonic acid with tungsten(VI) and molybdenum(VI). *Carbohydr. Res.* **1997**, *299*, 209–220. [CrossRef]
54. Ramos, M.L.; Caldeira, M.M.; Gil, V.M.S. NMR spectroscopy study of the complexation of D-gluconic acid with tungsten(VI) and molybdenum(VI). *Carbohydr. Res.* **1997**, *304*, 97–109. [CrossRef]
55. Ramos, M.L.; Pereira, M.M.; Beja, A.M.; Silva, M.R.; Paixão, J.A.; Gil, V.M.S. NMR and X-ray diffraction studies of the complexation of D-(−)quinic acid with tungsten(VI) and molybdenum(VI). *J. Chem. Soc. Dalton Trans.* **2002**, *10*, 2126–2131. [CrossRef]
56. Ramos, M.L.; Caldeira, M.M.; Gil, V.M.S. Multinuclear NMR study of the complexation of D-glucaric acid with molybdenum(VI) and tungsten(VI). *Inorg. Chim. Acta* **1991**, *180*, 219–224. [CrossRef]
57. Justino, L.L.G.; Ramos, M.L.; Kaupp, M.; Burrows, H.D.; Fiolhais, C.; Gil, V.M.S. Density functional theory study of the oxoperoxo vanadium(V) complexes of glycolic acid. Structural correlations with NMR chemical shifts. *Dalton Trans.* **2009**, *44*, 9735–9745. [CrossRef]
58. Justino, L.L.G.; Ramos, M.L.; Nogueira, F.; Sobral, A.J.F.N.; Geraldes, C.F.G.C.; Kaupp, M.; Burrows, H.D.; Fiolhais, C.; Gil, V.M.S. Oxoperoxo vanadium(V) complexes of L-lactic acid: Density functional theory study of structure and NMR chemical shifts. *Inorg. Chem.* **2008**, *47*, 7317–7326. [CrossRef]

59. Bernstein, L.R.; Tanner, T.; Godfrey, C.; Noll, B. Chemistry and pharmacokinetics of gallium maltolate, a compound with high oral gallium bioavailability. *Met.-Based Drugs* **2000**, *7*, 33–47. [CrossRef]
60. Enyedy, É.A.; Dömötör, O.; Varga, E.; Kiss, T.; Trondl, R.; Hartinger, C.G.; Keppler, B.K. Comparative solution equilibrium studies of anticancer gallium(III) complexes of 8-hydroxyquinoline and hydroxy(thio)pyrone ligands. *J. Inorg. Biochem.* **2012**, *117*, 189–197. [CrossRef]
61. Lima, C.F.R.A.C.; Taveira, R.J.S.; Costa, J.C.S.; Fernandes, A.M.; Melo, A.; Silva, A.M.S.; Santos, L.M.N.B.F. Understanding M–ligand bonding and mer-/fac isomerism in tris(8-hydroxyquinolinate) metallic complexes. *Phys. Chem. Chem. Phys.* **2016**, *18*, 16555–16565. [CrossRef]
62. Costa, J.C.S.; Lima, C.F.R.A.C.; Santos, L.M.N.B.F. Electron transport materials for organic light-emitting diodes: Understanding the crystal and molecular stability of the tris(8-hydroxyquinolines) of Al, Ga, and In. *J. Phys. Chem. C* **2014**, *118*, 21762–21769. [CrossRef]

**Disclaimer/Publisher's Note:** The statements, opinions and data contained in all publications are solely those of the individual author(s) and contributor(s) and not of MDPI and/or the editor(s). MDPI and/or the editor(s) disclaim responsibility for any injury to people or property resulting from any ideas, methods, instructions or products referred to in the content.

*Article*

# Theoretical Modeling of Absorption and Fluorescent Characteristics of Cyanine Dyes

Sonia Ilieva, Meglena Kandinska, Aleksey Vasilev and Diana Cheshmedzhieva *

Faculty of Chemistry and Pharmacy, Sofia University "St. Kliment Ohridski", 1164 Sofia, Bulgaria; silieva@chem.uni-sofia.bg (S.I.); ohmk@chem.uni-sofia.bg (M.K.); ohtavv@chem.uni-sofia.bg (A.V.)
* Correspondence: dvalentinova@chem.uni-sofia.bg; Tel.: +359-28161354

**Abstract:** The rational design of cyanine dyes for the fine-tuning of their photophysical properties undoubtedly requires theoretical considerations for understanding and predicting their absorption and fluorescence characteristics. The present study aims to assess the applicability and accuracy of several DFT functionals for calculating the absorption and fluorescence maxima of monomethine cyanine dyes. Ten DFT functionals and different basis sets were examined to select the proper theoretical model for calculating the electronic transitions of eight representative molecules from this class of compounds. The self-aggregation of the dyes was also considered. The pure exchange functionals (M06L, HFS, HFB, B97D) combined with the triple-zeta basis set 6-311+G(2d,p) showed the best performance during the theoretical estimation of the absorption and fluorescent characteristics of cyanine dyes.

**Keywords:** cyanine dyes; DFT; TDDFT; UV-VIS spectroscopy; fluorescence; aggregation; dimers; DNA

**Citation:** Ilieva, S.; Kandinska, M.; Vasilev, A.; Cheshmedzhieva, D. Theoretical Modeling of Absorption and Fluorescent Characteristics of Cyanine Dyes. *Photochem* 2022, 2, 202–216. https://doi.org/10.3390/photochem2010015

**Academic Editors:** Gulce Ogruc Ildiz and Licinia L.G. Justino

Received: 20 December 2021
Accepted: 1 March 2022
Published: 4 March 2022

**Publisher's Note:** MDPI stays neutral with regard to jurisdictional claims in published maps and institutional affiliations.

**Copyright:** © 2022 by the authors. Licensee MDPI, Basel, Switzerland. This article is an open access article distributed under the terms and conditions of the Creative Commons Attribution (CC BY) license (https://creativecommons.org/licenses/by/4.0/).

## 1. Introduction

Fluorescent dyes are widely used for the detection and quantification of nucleic acids (NA) and proteins and are applied in real-time PCR, gel electrophoresis, flow cytometry, microscopy, etc. [1–3]. This is due to the ability of specific fluorescent dyes to bind to various target biomolecules in a mostly noncovalent mode, leading to changes in the fluorescent properties of the respective dye. Any significant changes in the photophysical properties of the dye would be useful, but the most used one is the increase in the emission intensity of the dye upon binding. Cyanine dyes are a wide class of cationic compounds that have been proven to be efficient probes for nucleic acid detection [4] due to the fact that they have very low fluorescence intensity before binding, and this intensity increases significantly after binding to NA. Two types of nucleic acid binding have been demonstrated for these compounds: intercalation and minor groove binding [5–9]. Cyanine dyes are known to extend over the visible and near infrared spectrum due to changes in the length of the central polymethine bridge or the heterocycles [4,10]. The dye spectrum can be fine-tuned by introducing substituents into the aromatic heterocycles [11]. Because of their importance as fluorogenic acid probes, the cyanines have been the subject of versatile research in the last decade [7,11–18]. The development of new fluorescent probes for dsRNA is even more important nowadays [18,19]. The broad application of cyanine dyes in medicine and diagnostics [20–23] fuels the interest in them. The rational design of new functional materials requires a deep understanding of the driving forces behind the changes in the photophysical properties of the synthesized dyes as well as the binding mode toward the biomolecular targets. The rational approach to that problem requires a body of knowledge about the properties of the dye itself and the changes that these properties undergo upon binding. Undoubtedly, this knowledge includes experimental synthetic and spectroscopic studies as well as theoretical considerations of the dyes' properties for resolving the factors governing the binding and selectivity towards nucleic acids. Quantum chemical computations provide information that allow us to have a deeper understanding of the binding to the DNA mechanism, structure of the ligands and complexes, and photochemical and spectral characteristics of the dyes [11,14,24–28].

The present study aims to assess the applicability and accuracy of several DFT functionals for calculating the absorption and fluorescent maxima of monomethine cyanine dyes. It is generally accepted that DFT provides adequate description of the geometry and physico-chemical properties of organic compounds in their ground state [29]. Time-Dependent Density Functional Theory (TDDFT) formalism is considered to be an adequate and robust tool for the computation of the electronic structure and geometry in the excited state for various organic compounds [14,30,31]. In some earlier studies, the TDDFT approach was considered to have poor performance when studying the photophysical properties of cyanines due to the overstimulation of the electron transition energies for this class of compounds [11,26,31–33]. It has been discussed that TDDFT's poor performance in cyanines is due to the multi-reference nature of the electronic states of the dyes, especially when compared to the results of the CASPT2 method [34,35]. The applicability of the TDDFT approach for characterizing cyanine dyes has recently been reconsidered [14,24]. A comparison between the performance and accuracy of several Minnesota and PBE functionals with respect to Quantum Monte Carlo and CASPT2 calculations for cyanines is published from D. Truhlar and co-authors [14]. The work of Send et al. [24] also recognizes TDDFT by considering the electronic excitation of simple cyanine dyes compared to QMC, CASPT2, and the coupled cluster method (up to CC3). A conclusion was drawn that TDDFT is an adequate and reliable tool for the calculation of the excited-state electronic structures of organic molecules [32,36], including the class of cyanine dyes [14]. The combination of the functional and the basis set within the TDDFT method must be chosen for the specific fluorophore carefully. Thus, based on the already settled important issue that TDDFT does not show any shortcomings in the resolution of cyanine functionality, the aim of the present study concerns the TDDFT applications for a specific class of cyanines with more complex structure. The objective is to find functional-basis set combinations that are suitable for describing the specific chromophore for a group of monomethinecyanine dyes that are studied. We examined ten DFT functionals and different basis sets to select the proper theoretical model for calculating the electronic transitions of cyanine dyes. The theoretical results were validated against experimental data for the dyes studied.

## 2. Computational Methods

Quantum chemical computations were applied to simulate the geometry, electronic structure, and spectral properties for a series of cyanine dyes.

The geometry optimization and photophysical properties of the monomers of thiazole orange (TO) and the seven analogues that were studied were computed using the G16 software package [37]. The minimum energy structures in the ground state were optimized at the DFT [38] theory level using the B3LYP hybrid function in conjunction with the 6-31G(d,p) [39] basis set, and for the iodide counterions, only the SDD basis set and effective core potential were used [40–43]. The plausible TO dimers were optimized at M062X/6-31G(d,p) (SDD basis set used for the iodide counterions).

The effect of the medium was taken into account at each step by means of PCM formalism [44,45]. All of the computations were performed in a water medium to reproduce the experimental conditions (TE buffer). In order to verify that each optimized structure is a minimum of the potential energy surface, an analysis of the harmonic vibrational frequencies was performed using the same method/basis, set and no imaginary frequency was found. To determine the absorption wavelengths, the lowest energy absorption transitions were evaluated by TDDFT calculations of the vertical excitations. Ten different functional basis sets: B3LYP [46], PBE0 [47,48], M062X, M06, M06L [49], BH and HLYP [50], CAM-B3LYP [51], HFS [52,53], HFB [54], and B97D [55] with 6-311+G(2d,p) [56], were used to assess the accuracy of the functionals in predicting the spectral properties of cyanine dyes. The procedure for vertical absorption and emission computations is described in our previous work [30]. The following steps were included in the computations: (i) geometry optimization of the ground-state structure at the DFT level and the computation of vibrational frequencies using the same method/basis set to verify the optimized structure. (ii) TDDFT calculations of vertical excitations to estimate the lowest energy absorption transition. Six excited singlet states were considered, and the lowest energy transition with non-zero oscillator strength was taken into account for each of the monomer dyes from the series. Comparisons were made for two of the cyanine dyes using calculations that considered 12, 20, and 24 excited states. As a result, no significant difference in the vertical excitation energy was obtained. Twenty excited singlet states were considered for the dimer absorption spectra computations. (iii) After the excited state of interest was identified, TDDFT geometry optimization with equilibrium linear response solvation was performed. The optimizaton of the excited state at the TDDFT level starting from the ground state geometry was determined. Frequency calculations and the absence of imaginary frequencies confirm the equilibrium of the excited-state geometry. (iv) Fluorescence electronic transitions–TDDFT calculations of the vertical de-excitations based on the optimized geometry of the excited state. Vertical excitation and de-excitation energies were calculated without state-specific correction. The computed absorption and emission transitions in solution were compared to the experimental spectral data.

## 3. Results

### 3.1. Geometry Optimization

A series of eight asymmetric cyanine dyes (Scheme 1) were modelled with the help of DFT and TDDFT calculations to gather deep insights into the geometry and electron density distribution in the ground and excited states and to achieve a better understanding of the electronic spectra of the dyes. The cyanine dyes that were used in the present study were previously synthesized by our group [18,57,58].

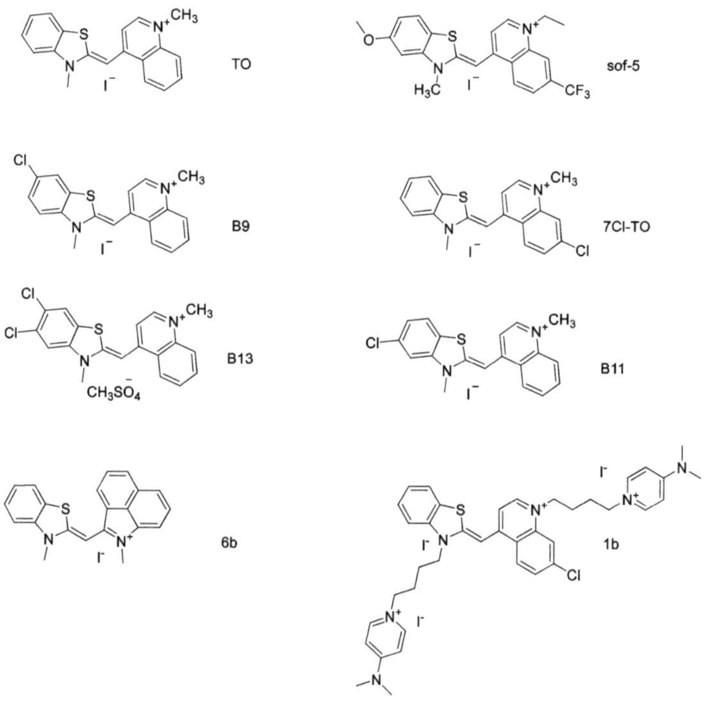

**Scheme 1.** Thiazole orange (TO) analogues. Dye labelling is the same as in the original papers.

The geometries of the *cis*- and *trans*- TO conformers optimized at the B3LYP/6-31G(d,p) theory levels are presented in Figure 1. For all of the studied cyanine dyes, the *trans* conformer is more stable than the *cis* conformer. Geometry parameters and some spectral characteristics of several of the conformational TO states are provided in Table 1. The energy difference between the *cis* and *trans* TO conformers is 5.4 kcal/mol in favour of the trans conformer. The main difference between the two conformational states, *cis* and *trans*, is the dihedral angle between the quinoline and benzothiazole heterocycles-$\tau(SC_2C_4C_6)$. The *cis* conformer is not planar, and the dihedral angle $\tau(S_1C_2C_4C_6)$ is 125.7°. Due to some steric hindrance between the two heterocycles, the *trans* conformer is not fully planar. In the *trans* conformer, the angle $\tau(S_1C_2C_4C_6)$ is 17.8°, thus leading to a conjugation between the two heterocycles through the methine bridge. These theoretical results are in agreement with the results of other conformational NMR studies [59].

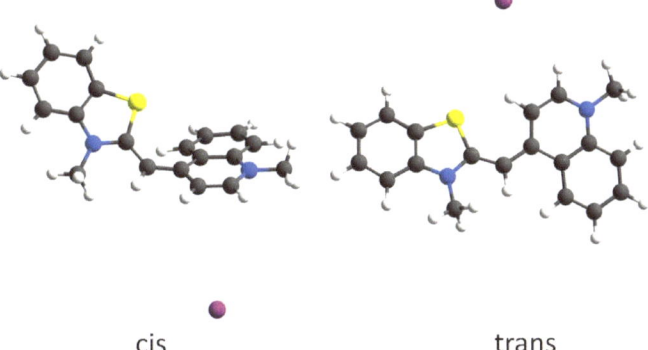

cis                                      trans

**Figure 1.** B3LYP/6-31G(d,p)-optimized geometry of *cis* and *trans* TO conformers. Color scheme: C–gray, S–yellow, H–white, N–blue, I–magenta.

**Table 1.** B3LYP/6-31G(d,p)-optimized geometry parameters, absorption maxima (nm), and oscillator strength (f) from PBE0/6-311+G(2d,p) computations for several conformational TO states.

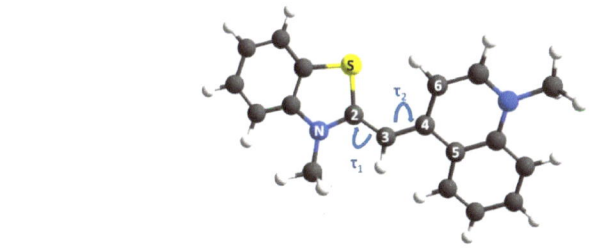

| Isomer | $\tau_1$ (NC$_2$C$_3$H) | $\tau_2$ (HC$_3$C$_4$C$_5$) | $\tau$ (C$_2$C$_3$C$_4$C$_6$) | $\Delta G$ (kcal/mol) | $\lambda_{abs}$ (nm) | f |
|---|---|---|---|---|---|---|
| cis | −14.8 | 157.7 | 145.6 | 5.4 | 476.8 | 0.6706 |
| trans | 6.4 | 10.0 | 12.7 | 0.0 | 447.2 | 0.9598 |
| trans 2 | 6.5 | 9.1 | 11.8 | 0.8 | 448.8 | 0.9000 |
| trans 3 | −6.5 | −10.6 | −13.2 | 1.5 | 447.2 | 0.9504 |

Furthermore, due to variations in the counterion position and the influence of the counterion position on some photophysical TO characteristics, different *trans* conformations were considered. The geometry parameters and spectral characteristics of three trans conformers with different counterion positions are provided in Table 1. The calculations show that the position of the counterion has little influence on the Gibbs free energy of the conformers and their absorption maxima.

## 3.2. Modelling Spectroscopic Properties

Computing the molecular properties of organic compounds in the ground state is more or less systematic, while the theoretical calculation of excited state properties, such as absorption and fluorescence maxima, is not so trivial. A careful calibration of the theoretical model applied to perform computational studies for a series of molecules with a particular chromophore is of critical importance.

The absorption maxima of the dyes that were studied were computed using the B3LYP/6-31G(d,p)-optimized geometry of the molecules. Vertical excitations were obtained from single point TDDFT computations on the optimized geometries using ten different functionals: B3LYP, PBE0, M062X, M06, M06L, BH and HLYP, CAM-B3LYP, HFS, HFB, and B97D, to find an accurate description of the electronic excitations as well as to predict the absorption and emission spectral characteristics. The vertical excitations/absorption maxima were computed in the ground-state geometry. All of the computations were made in a water medium to mimic the experimental conditions. The solvent effect was modeled using the SCRF formalism IEFPCM.

TDDFT allows transition energies as well as excited-state properties such as dipole moments and emitting geometries to be computed [60,61]. Despite its huge popularity, the reliability of TDDFT results depends significantly on the selected exchange–correlation (XC) functional. The accepted accuracy of TDDFT computations is 0.2–0.3 eV [62]. The chemical accuracy of the 0.1 eV difference between the calculated and measured absorption maxima has not yet been reached.

In the literature, one of the ways to overcome this is to use range-separated hybrids (RSH) [62–65]. These functionals incorporate a growing fraction of exact exchange with increasing inter-electronic distance and allow the charge-transfer phenomena to be modelled accurately. Range-separated functionals (RSF) are a subgroup of hybrid functionals. While conventional (global) hybrid functionals such as PBE0 or B3LYP use fixed Hartree–Fock and DFT exchange, RSFs mix the two contributions based on the spatial distance between two points.

The predicted absorption wavelengths of the lowest electronic energy transitions and respective oscillator strengths for the *cis* and *trans* TO conformers are listed in the last two columns of Table 1. Vertical excitation energies computed at the PBE0/6-311+G(2d,p) level of theory without any state-specific correction are reported in Table 1. It can be seen that the calculated absorption maximum of the *cis* conformer (477 nm) is batochromicaly shifted compared to the *trans* absorption maxima (447–449 nm). The transitions of the two conformers show different oscillator strength, which is lower for the *cis* conformer.

The molecular orbitals involved in the $S_0 \rightarrow S_1$ transitions for the *cis* and *trans* conformers were calculated. Figure 2 shows the ground-state orbital energy levels of the highest occupied molecular orbital (HOMO), the lowest unoccupied molecular orbital (LUMO), and the energy gap for the *cis* TO and the *trans* TO in water. The trends in electron density change can be illustrated through the molecular orbitals shape analysis. The HOMO has a higher electron density on the benzothiazole ring. As seen from Figure 2, the electron density is redistributed from the benzothiazole and moves toward the quinoline heterocycle of the *cis* and *trans* TOs. Strong charge transfer is observed from the benzothiazole ring toward the quinoline portion, accompanying the electronic transition in the *cis* TO conformer. The strong charge transfer observed for the *cis* conformer explains the red shift of the transition. For the other compounds being studied, only the most stable conformer was considered.

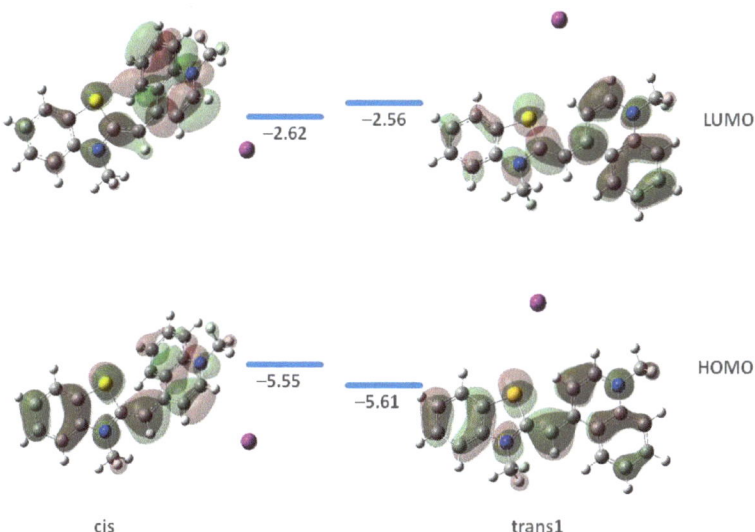

**Figure 2.** Molecular orbitals involved in the $S_0 \rightarrow S_1$ transition for *cis* and *trans* TO conformers computed at B3LYP/6-31G(d,p) in water medium.

The influence of the basis set on the absorption maxima calculations was considered. The theoretical vertical excitation energies were computed with six Pople-type basis sets. The results that were obtained are summarized in Table 2. As seen from Table 2, the addition of one diffuse function to the 6-31G(d,p) basis set instantly increases the value of the calculated absorption maximum, thus improving the results obtained with 6-31+G(d,p). The best results were obtained with the large 6-311+G(2d,p) basis set (2.71 eV(458 nm) the predicted maximum, 2.47 eV(502 nm)–experimental value). The difference between the theoretical estimation and the experimental value was 0.24 eV, which is an acceptable accuracy. Benchmark calculations on the absorption properties of various systems have demonstrated that the expected TDDFT accuracy is between 0.2 and 0.3 eV [30,62,66].

**Table 2.** Computed absorption maxima $\lambda_{abs}$ (nm) for TO computed with PBE0 functional and different basis sets.

| Basis Set | $\lambda_{abs}$ [nm] PCM | $\lambda_{abs}$ [nm] SMD |
|---|---|---|
| 6-31G(d,p) | 445 | 445 |
| 6-31+G(d,p) | 454 | 453 |
| 6-31++G(d,p) | 454 | 453 |
| 6-311G(d,p) | 451 | 450 |
| 6-311+G(d,p) | 452 | 455 |
| 6-311+G(2d,p) | 458 | 457 |
| Experiment $\lambda_{abs}$ (nm) | 502 [a] | |

[a] From Ref. [64].

Further calculations of vertical absorptions for a series of eight cyanine dyes were carried out with different functionals combined with the triple-zeta basis set 6-311+G(2d,p). Table 3 summarizes the vertical absorption energies calculated with ten different TDDFT functionals: B3LYP, PBE0, M062X, M06, M06L, BH and HLYP, CAM-B3LYP, HFS, HFB, and B97D. The applied functionals differ in the form of their exchange functional. This functional has various percentage HF exchange (0% to 54%). A comparison between the theoretical results obtained with the ten functionals combined with the triple-zeta basis set 6-311+G(2d,p) in a water medium and the experimental spectral data are presented in Table 3. The performance of the different functionals can be assessed by the mean absolute deviations (MAD) of the theoretical values from the experiment. The MAD values for each of the functionals used are given in the last row of Table 3.

**Table 3.** The deviation from experiment of the theoretical vertical excitation energies (in eV) computed with different functionals combined with the 6-311+G(2d,p) basis set in a water medium for a series of cyanine dyes.

| Dye | B3LYP | PBE0 | M062X | BH&HLYP | CAM B3LYP | M06 | M06L | HFS | HFB | B97D |
|---|---|---|---|---|---|---|---|---|---|---|
| TO | 0.24 | 0.30 | 0.39 | 0.52 | 0.42 | 0.27 | 0.20 | 0.01 | 0.02 | 0.06 |
| 1b | 0.25 | 0.31 | 0.40 | 0.54 | 0.44 | 0.29 | 0.21 | 0.01 | 0.04 | 0.06 |
| B9 | 0.23 | 0.29 | 0.39 | 0.52 | 0.43 | 0.26 | 0.16 | −0.03 | −0.01 | 0.03 |
| B11 | 0.23 | 0.29 | 0.38 | 0.52 | 0.42 | 0.26 | 0.16 | −0.03 | −0.01 | 0.03 |
| B13 | 0.22 | 0.28 | 0.38 | 0.52 | 0.42 | 0.25 | 0.16 | −0.06 | −0.03 | 0.01 |
| 6b | 0.11 | 0.17 | 0.33 | 0.41 | 0.35 | 0.15 | 0.07 | −0.11 | −0.09 | −0.07 |
| 7Cl−TO | 0.24 | 0.31 | 0.39 | 0.52 | 0.42 | 0.28 | 0.19 | 0.00 | 0.02 | 0.05 |
| sof-5 | 0.19 | 0.26 | 0.39 | 0.51 | 0.43 | 0.23 | 0.04 | −0.17 | −0.13 | −0.11 |
| MAD [a] | 0.21 | 0.28 | 0.38 | 0.51 | 0.42 | 0.25 | 0.15 | −0.05 | −0.02 | 0.01 |

[a] Mean absolute deviation from the experiment (MAD) $MAD = \frac{\sum |\lambda_{calc} - \lambda_{exp}|}{n}$, n–number of compounds.

All of the DFT functionals follow the experimental tendencies (Figure 3) and qualitatively describe the changes in the absorption maximum in the studied series. Such correlation between the experiment and theory indicates that the entire group of functionals is good enough for picturing the trends in the series.

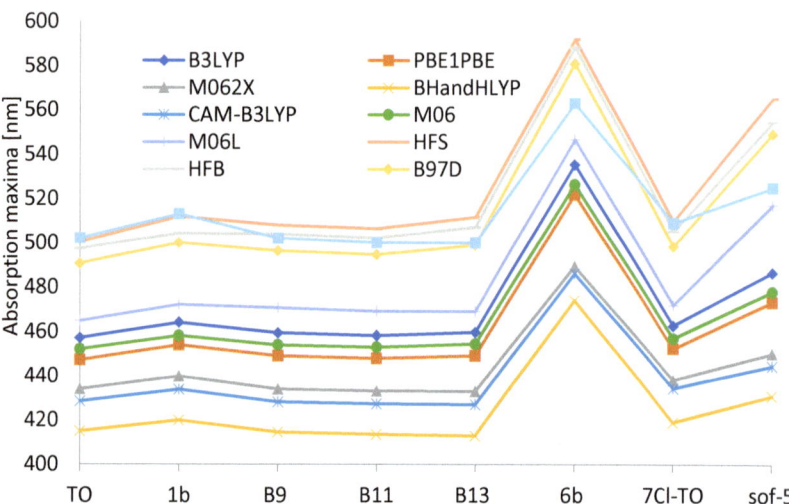

**Figure 3.** Trends in changes of the theoretical TDDFT vertical excitation energies (in nm) computed with different functionals combined with the 6-311+G(2d,p) basis set in a water medium and experimental absorption maxima from spectra in TE buffer for the series of cyanine dyes.

The deviation in the theoretically calculated absorption energies from the experimentally observed values for the electronic transitions (in eV) can be seen in Figure 4. It can be seen from Table 3 and Figure 4 that the hybrid functionals B3LYP, PBE0, and M06 (25–27% Hartree–Fock exchange) performed adequately for this class of dyes and can be used as a tool for calculating the electronic structures of monomethine cyanine dyes. Although they overestimate the transition energies, MAD is in the admissible range of 0.2–0.3 eV. The performance of the range-separated functional CAM-B3LYP is the worst in our case, although it has been recommended for the calculation of electronic charge transfer transitions [65,67,68].

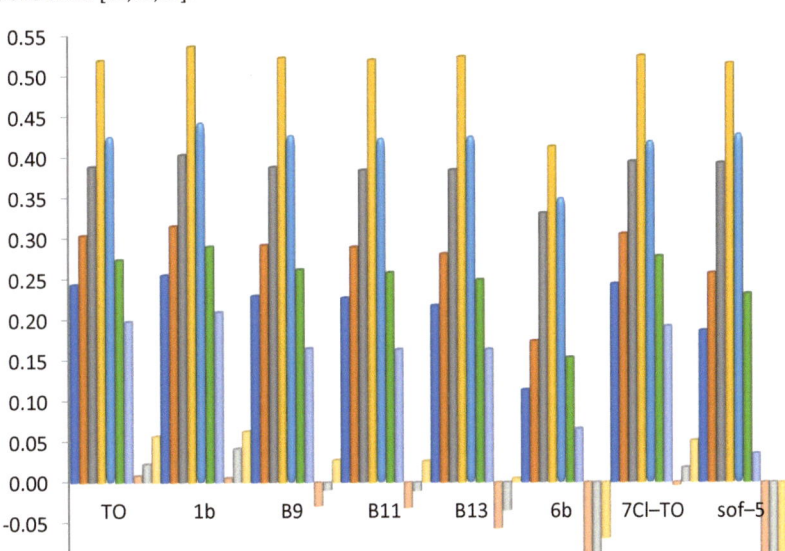

**Figure 4.** Deviations in the theoretical TDDFT vertical excitation energies (in eV) computed with different functionals combined with 6-311+G(2d,p) basis set in a water medium from the experimental data from spectra in TE buffer for the series of cyanine dyes.

In Gaussian 16, HFS stands for the Slater exchange. HFB is Becke's 1988 functional, which includes the Slater exchange with corrections involving the gradient of the density. The pure exchange functionals (M06L, HFS, HFB, B97D) show the best performance for the case study. All of these functionals have excellent performance and good predictability, the best being B97D with MAD 0.01 eV.

Based on these studies, we recommend that pure DFT functionals (M06L, HFS, HFB, B97D) be used to calculate the absorption properties of cyanine dyes.

### 3.3. Fluorescence

As mentioned earlier, the fluorescent response of cyanine dyes is sensitive to the environment. TO and its analogs have almost no fluorescence in organic solvents and exhibit a serious enhancement in the fluorescence intensity in viscous solutions (such as glycerin) and in DNA/RNA. The nature of fluorescence quenching in asymmetric cyanine dyes has been elucidated in several studies [66,69,70] and has been attributed to intramolecular torsion in the excited state. Easy rotation in solvent quenches the fluorescence. The fluorescence quantum yield increases when this rotation is obstructed. The geometry of the first TO excited state was optimized at B3LYP and PBE0 /6-311+G(2d,p) and is presented in Figure 5. The optimized ground-state geometry is planar, while the excited state is highly twisted. The vertical excitation leads to a locally excited state with a weak CT character (Figure 2) that has the same geometry as the ground state and has a planar structure. This local excited state has a transition energy of 2.51 eV (494nm). The optimization of the first excited state $S_1$ leads to a twisted geometry where the donor (quinolone moiety) and the acceptor (benzthiazole part) are perpendicular to each another (Figure 5). With the change in the dihedral angle between the two heterocycles ($\tau_2$—Figure 6) a fully twisted dark state is formed in accordance with previous findings [11–13]. This conformational changes in the structure of the fluorophore lead to the formation of a twisted intramolecular charge-transfer (TICT) excited state ($S_1$) [71]. The computed Stokes shift for TO is 1613 cm$^{-1}$, which agrees with the experimentally measured 1193 cm$^{-1}$. The fluorescence computations at B3LYP and PBE0 are in line with the experiment although the transition energies are slightly overestimated.

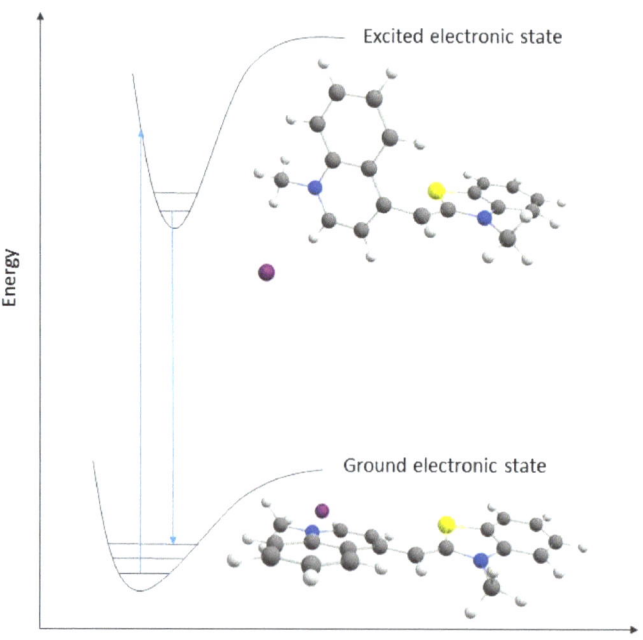

**Figure 5.** Schematic representation for ground ($S_0$) and excited state ($S_1$) geometries of TO.

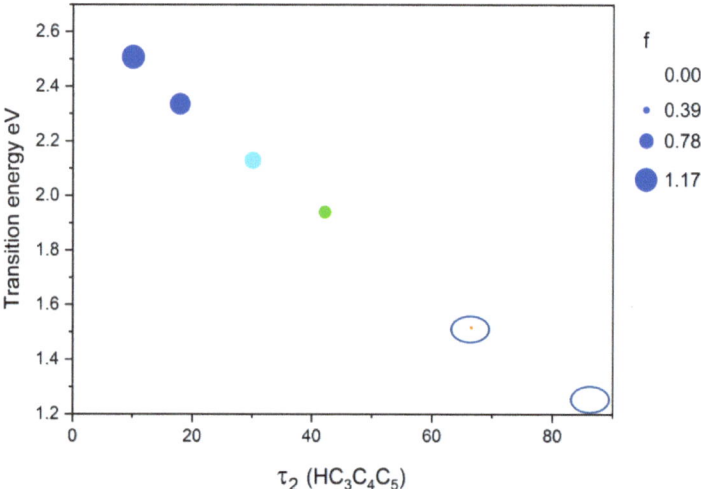

**Figure 6.** Excitation energies of TO as a function of the interplanar angle $\tau_2(HC_3C_4C_5)$. The radius of the circles is proportional to the predicted oscillator strength.

The conformational change in the excited state and the fact that the fluorescence intensity changes after binding of the dye to NA, thus leading to the fixation of the planar geometry, are the origin of the TO sensing mechanism and its analogs.

*3.4. Aggregation of the Dyes*

Monomethine cyanine dyes tend to aggregate in aqueous solution due to hydrophobic interactions. It has been suggested that the hydrophobicity and polarizability of dyes favor π-stacking interactions [72]. The self-aggregation of both TO and Benzthiazole Orange dyes has been found to occur but with deferent strength and other features depending on the dye itself [73]. An additional absorption band at 471 nm is observed in the absorption spectrum of TO alongside the dye's intrinsic absorption at 501 nm [73–75]. The band at 501 nm prevails at low concentrations, while at higher concentrations, the band at 471 predominates. This behavior is associated with the self-aggregation of the dye molecules and the formation of H dimers [74]. H-dimers are formed by interactions between the heterocyclic systems of two dye molecules located one above the other (the molecules are superimposed), while in the case of J dimers, molecule slipping relative to each other is observed. To address this behavior, we modeled four different π-stacked TO H-dimers, and their optimized geometry is shown in Figure 7.

During TO dimer optimization, the main challenge is maintaining π-stacking. The use of global hybrids such as B3LYP and PBE0 is not possible since the structure of the dimers falls apart during optimization. The dimer structures were optimized using the M062x functional. The presence of the iodide counterions was explicitly considered when modelling the structures. Mooi and Heyne [76] studied the effect of various counterions on aggregation. Counterions have been shown to play a significant role in terms of structural dimer organization and specific ionic effects.

The Gibbs free energy resulting from dimer formation was calculated at M062X/6-31G(d,p) from the following reaction: $2TO \rightarrow (TO)_2$. The optimized geometries of the H-dimers are shown in Figure 7, and the theoretical estimations are provided in Table 4. The most stable H-dimer is Dimer 2, where the donor part of the first molecule is above the acceptor part of the second TO molecule. The calculated Gibbs free energy resulting from the formation of the most stable dimer is $-6.10$ kcal/mol.

$K_D$ has been experimentally measured in two different TO dimerization studies, with the following results being achieved: $-3.1 \times 10^4$ M$^{-1}$ from [73] and $2.5 \times 10^4$ M$^{-1}$ [77]. The Gibbs free energy of formation of the dimers was computed from the experiment using the equation $\Delta G = -RT\ln K$. The respective values were $-6.12$ kcal/mol and $-6.00$ kcal/mol. Thus, a very good agreement between the experiment and theory ($-6.10$ kcal/mol) was obtained in the present study.

The predicted vertical absorption values for the dimers are blue shifted according to the monomer absorption. The absorption maxima of the lowest energy TO dimer (Dimer 2) were calculated with HFS and HFB functionals determined to be the most accurate for the monomer absorption prediction. The obtained theoretical values are provided in Table 5. As seen from the table, the absorption maxima computed with HFB and HFS functionals are in very good agreement with the experimental values [73,75]. For comparison, the PBE0-calculated values are also provided in the table.

Dimer formation can be used as a model for the aggregation characteristics of TO analogs as well as analysis of the change in the fluorescence of the dye after binding to DNA. It can be stated that aggregation-induced fluorescence is observed during the dye aggregation process and upon dye-binding to NA.

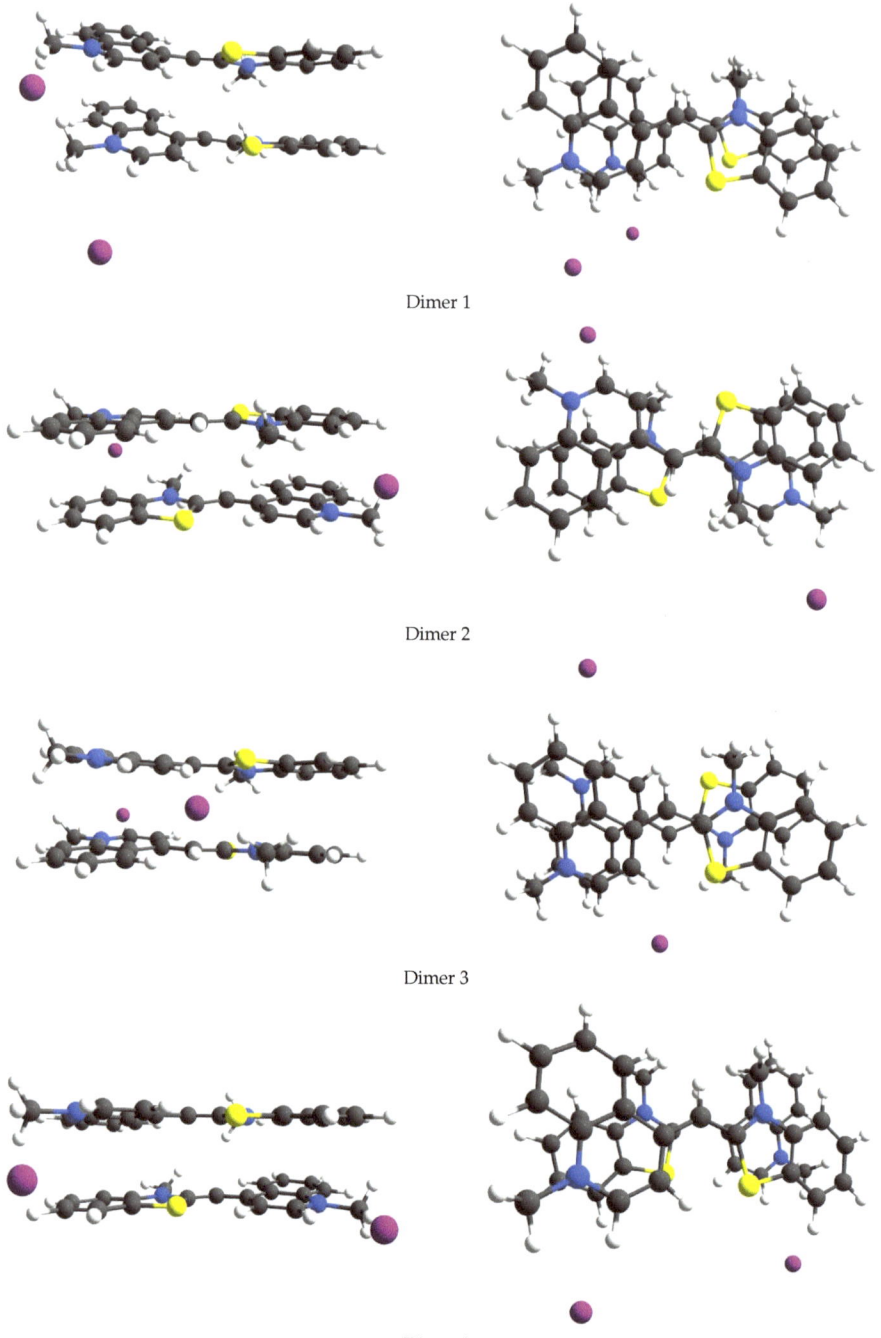

Dimer 1

Dimer 2

Dimer 3

Dimer 4

**Figure 7.** M062X/6-31G(d,p)-optimized molecular structures of TO H-dimers in water: left-top view, right-site view.

**Table 4.** M062X/6-31G(d,p) computed Gibbs free energy of formation (in kcal/mol) and absorption maxima (nm) of TO H-dimers.

| TO Dimer | ΔG (kcal/mol) | $\lambda_{abs}$ (nm) [a] |
|---|---|---|
| Dimer 1 | −1.4 | 415 |
| Dimer 2 | −6.1 | 423 |
| Dimer 3 | −5.9 | 417 |
| Dimer 4 | −4.6 | 426 |

[a] From PBE0/6-311+G(2d,p) computations.

**Table 5.** Calculated absorption maxima (nm) for TO monomer and dimer.

| Method/Basis Set | Monomer | Dimer |
|---|---|---|
| HFS/6-311+G(2d,p) | 500 | 484 |
| HFB/6-311+G(2d,p) | 497 | 477 |
| PBE0/6-311+G(2d,p) | 447 | 423 |
| experiment | 501 | 471 |

## 4. Conclusions

A series of eight asymmetric cyanine dyes was modelled using DFT and TDDFT calculations. The calibration of the theoretical model for performing computational studies for dye molecules containing a particular chromophore was performed by examining the accuracy of ten functionals and six Pople-type basis sets. Theoretical results were validated against the experimental data. It was shown that the addition of one diffuse function to the 6-31G(d,p) basis set instantly increased the value of the calculated absorption maximum, thus improving the results obtained with 6-31+G(d,p). The large triple-zeta basis set 6-311+G(2d,p) showed the best performance. The comparison between the theoretical results obtained with the ten functionals combined with the triple-zeta basis set in a water medium and the experimental spectral data evince that the pure exchange functionals M06L, HFS, HFB, and B97D had the best performance in the case study.

**Author Contributions:** Conceptualization, S.I. and D.C.; methodology, S.I. and D.C.; validation, S.I., A.V., M.K. and D.C.; formal analysis, S.I. and D.C.; investigation, D.C.; resources, S.I.; data curation, S.I. and D.C.; writing—original draft preparation, S.I. and D.C.; writing—review and editing, S.I. and D.C.; visualization, D.C.; supervision, S.I. and D.C.; project administration, S.I. and D.C.; funding acquisition, S.I. All authors have read and agreed to the published version of the manuscript.

**Funding:** This research was funded by the BULGARIAN NATIONAL SCIENCE FUND (BNSF), grant number KP-06-H39/11 from 09.12.2019–SOFIa.

**Acknowledgments:** This work was supported by the Bulgarian National Science Fund (BNSF) project SOFIa (KP-06-H39/11), granted on 9 December 2019.

**Conflicts of Interest:** The authors declare no conflict of interest.

## References

1. Pindur, U.; Jansen, M.; Lemster, T. Advances in DNA-Ligands with Groove Binding, Intercalating and/or Alkylating Activity: Chemistry, DNA-Binding and Biology. *Curr. Med. Chem.* **2005**, *12*, 2805–2847. [CrossRef] [PubMed]
2. Rask-Andersen, M.; Almén, M.S.; Schiöth, H.B. Trends in the exploitation of novel drug targets. *Nat. Rev. Drug Discov.* **2011**, *10*, 579–590. [CrossRef] [PubMed]
3. Saarnio, V.K.; Salorinne, K.; Ruokolainen, V.P.; Nilsson, J.R.; Tero, T.-R.; Oikarinen, S.; Wilhelmsson, L.M.; Lahtinen, T.M.; Marjomäki, V.S. Development of functionalized SYBR green II related cyanine dyes for viral RNA detection. *Dye. Pigment.* **2020**, *177*, 108282. [CrossRef]
4. Johnson, I.; Spence, M.T.Z. *The Molecular Probes Handbook—A Guide to Fluorescent Probes and Labeling Technologies*, 11th ed.; Life Technologies Corporation: Carlsbad, CA, USA, 2010.
5. Spielmann, H.P.; Wemmer, D.E.; Jacobsen, J.P. Solution Structure of a DNA Complex with the Fluorescent Bis-Intercalator TOTO Determined by NMR Spectroscopy. *Biochemistry* **1995**, *34*, 8542–8553. [CrossRef] [PubMed]
6. Yarmoluk, S.M.; Lukashov, S.S.; Ogul'Chansky, T.Y.; Losytskyy, M.Y.; Kornyushyna, O.S. Interaction of cyanine dyes with nucleic acids. XXI. Arguments for half-intercalation model of interaction. *Biopolymers* **2001**, *62*, 219–227. [CrossRef] [PubMed]

7. Zhytniakivska, O.; Girych, M.; Trusova, V.; Gorbenko, G.; Vasilev, A.; Kandinska, M.; Kurutos, A.; Baluschev, S.B. Spectroscopic and molecular docking studies of the interactions of monomeric unsymmetrical polycationic fluorochromes with DNA and RNA. *Dye. Pigment.* **2020**, *180*, 108446. [CrossRef]
8. Ziarani, G.M.; Moradi, R.; Lashgari, N.; Kruger, H.G. Cyanine Dyes. In *Metal-Free Synthetic Organic Dyes*; Elsevier: Amsterdam, The Netherlands, 2018; pp. 127–152. [CrossRef]
9. Armitage, B.A. Cyanine dye-DNA interactions: Intercalation, groove binding, and aggregation. In *DNA Binders and Related Subjects*; Waring, M.J., Chaires, J.B., Eds.; Springer: Berlin/Heidelberg, Germany, 2005; Volume 253, pp. 55–76. ISBN 3540228357.
10. Mishra, A.; Behera, R.K.; Behera, P.K.; Mishra, B.K.; Behera, G.B. Cyanines during the 1990s: A Review. *Chem. Rev.* **2000**, *100*, 1973–2012. [CrossRef]
11. Silva, G.L.; Ediz, V.; Yaron, D.; Armitage, B.A. Experimental and Computational Investigation of Unsymmetrical Cyanine Dyes: Understanding Torsionally Responsive Fluorogenic Dyes. *J. Am. Chem. Soc.* **2007**, *129*, 5710–5718. [CrossRef]
12. Rastede, E.E.; Tanha, M.; Yaron, D.; Watkins, S.C.; Waggoner, A.S.; Armitage, B.A. Spectral fine tuning of cyanine dyes: Electron donor–acceptor substituted analogues of thiazole orange. *Photochem. Photobiol. Sci.* **2015**, *14*, 1703–1712. [CrossRef]
13. Biancardi, A.; Biver, T.; Marini, A.; Mennucci, B.; Secco, F. Thiazole orange (TO) as a light-switch probe: A combined quantum-mechanical and spectroscopic study. *Phys. Chem. Chem. Phys.* **2011**, *13*, 12595–12602. [CrossRef]
14. Jacquemin, D.; Zhao, Y.; Valero, R.; Adamo, C.; Ciofini, I.; Truhlar, D.G. Verdict: Time-Dependent Density Functional Theory "Not Guilty" of Large Errors for Cyanines. *J. Chem. Theory Comput.* **2012**, *8*, 1255–1259. [CrossRef] [PubMed]
15. Vus, K.; Girych, M.; Trusova, V.; Gorbenko, G.; Kurutos, A.; Vasilev, A.; Gadjev, N.; Deligeorgiev, T. Cyanine dyes derived inhibition of insulin fibrillization. *J. Mol. Liq.* **2019**, *276*, 541–552. [CrossRef]
16. Gorbenko, G.; Trusova, V.; Kirilova, E.; Kirilov, G.; Kalnina, I.; Vasilev, A.; Kaloyanova, S.; Deligeorgiev, T. New fluorescent probes for detection and characterization of amyloid fibrils. *Chem. Phys. Lett.* **2010**, *495*, 275–279. [CrossRef]
17. Vasilev, A.; Deligeorgiev, T.; Kaloyanova, S.; Stoyanov, S.; Maximova, V.; Vaquero, J.J.; Alvarez-Builla, J. Synthesis of novel tetracationic asymmetric monomeric monomethine cyanine dyes—Highly fluorescent dsDNA probes. *Color. Technol.* **2011**, *127*, 69–74. [CrossRef]
18. Kandinska, M.; Cheshmedzhieva, D.; Kostadinov, A.; Rusinov, K.; Rangelov, M.; Todorova, N.; Ilieva, S.; Ivanov, D.; Videva, V.; Lozanov, V.; et al. Tricationic asymmetric monomeric monomethine cyanine dyes with chlorine and trifluoromethyl functionality—Fluorogenic nucleic acids probes. *J. Mol. Liq.* **2021**, *342*, 117501. [CrossRef]
19. Sato, Y. Design of Fluorescent Peptide Nucleic Acid Probes Carrying Cyanine Dyes for Targeting Double-Stranded RNAs for Analytical Applications. *Bull. Chem. Soc. Jpn.* **2020**, *93*, 406–413. [CrossRef]
20. Martineau, C.; Whyte, L.G.; Greer, C.W. Development of a SYBR safe™ technique for the sensitive detection of DNA in cesium chloride density gradients for stable isotope probing assays. *J. Microbiol. Methods* **2008**, *73*, 199–202. [CrossRef]
21. Gibson, J.F.; Kelso, S.; Skevington, J.H. Band-cutting no more: A method for the isolation and purification of target PCR bands from multiplex PCR products using new technology. *Mol. Phylogenetics Evol.* **2010**, *56*, 1126–1128. [CrossRef]
22. Clauß, M.; Springorum, A.C.; Hartung, J. Comparison of Different Fluorescence and Non-Fluorescence Staining Techniques for Rapid Detection of Airborne Micro-Organisms Collected on Room Temperature Vulcanizing (RTV) Silicones from Generated Aerosols and from Ambient Air. *Aerosol Sci. Technol.* **2012**, *46*, 818–827. [CrossRef]
23. Pérez-Cordero, J.-J.; Sánchez-Suárez, J.; Delgado, G. Use of a fluorescent stain for evaluating in vitro infection with Leishmania panamensis. *Exp. Parasitol.* **2011**, *129*, 31–35. [CrossRef]
24. Send, R.; Valsson, O.; Filippi, C. Electronic Excitations of Simple Cyanine Dyes: Reconciling Density Functional and Wave Function Methods. *J. Chem. Theory Comput.* **2011**, *7*, 444–455. [CrossRef] [PubMed]
25. Charaf-Eddin, A.; Le Guennic, B.; Jacquemin, D. Excited-states of BODIPY–cyanines: Ultimate TD-DFT challenges? *RSC Adv.* **2014**, *4*, 49449–49456. [CrossRef]
26. Grimme, S.; Neese, F. Double-hybrid density functional theory for excited electronic states of molecules. *J. Chem. Phys.* **2007**, *127*, 154116. [CrossRef] [PubMed]
27. Zhekova, H.; Krykunov, M.; Autschbach, J.; Ziegler, T. Applications of Time Dependent and Time Independent Density Functional Theory to the First π to π* Transition in Cyanine Dyes. *J. Chem. Theory Comput.* **2014**, *10*, 3299–3307. [CrossRef]
28. Le Guennic, B.; Jacquemin, D. Taking Up the Cyanine Challenge with Quantum Tools. *Acc. Chem. Res.* **2015**, *48*, 530–537. [CrossRef]
29. Koch, W.; Holthausen, M.C. *A Chemist's Guide to Density Functional Theory*; John Wiley & Sons: Hoboken, NJ, USA, 2001.
30. Cheshmedzhieva, D.; Ivanova, P.; Stoyanov, S.; Tasheva, D.; Dimitrova, M.; Ivanov, I.; Ilieva, S. Experimental and theoretical study on the absorption and fluorescence properties of substituted aryl hydrazones of 1,8-naphthalimide. *Phys. Chem. Chem. Phys.* **2011**, *13*, 18530–18538. [CrossRef]
31. Jacquemin, D.; Wathelet, V.; Perpète, E.A.; Adamo, C. Extensive TD-DFT Benchmark: Singlet-Excited States of Organic Molecules. *J. Chem. Theory Comput.* **2009**, *5*, 2420–2435. [CrossRef]
32. Jacquemin, D.; Perpète, E.A.; Ciofini, I.; Adamo, C.; Valero, R.; Zhao, Y.; Truhlar, D. On the Performances of the M06 Family of Density Functionals for Electronic Excitation Energies. *J. Chem. Theory Comput.* **2010**, *6*, 2071–2085. [CrossRef]
33. Jacquemin, D.; Perpète, E.A.; Scalmani, G.; Frisch, M.J.; Kobayashi, R.; Adamo, C. Assessment of the efficiency of long-range corrected functionals for some properties of large compounds. *J. Chem. Phys.* **2007**, *126*, 144105. [CrossRef]

34. Roos, B.O.; Fülscher, M.; Malmqvist, P.-Å.; Merchán, M.; Serrano-Andrés, L. Theoretical Studies of the Electronic Spectra of Organic Molecules. In *Quantum Mechanical Electronic Structure Calculations with Chemical Accuracy*; Springer: Dordrecht, The Netherlands, 1995; pp. 357–438.
35. Truhlar, D.G. Valence bond theory for chemical dynamics. *J. Comput. Chem.* **2007**, *28*, 73–86. [CrossRef]
36. Champagne, B.; Guillaume, M.; Zutterman, F. TDDFT investigation of the optical properties of cyanine dyes. *Chem. Phys. Lett.* **2006**, *425*, 105–109. [CrossRef]
37. Frisch, M.J.; Trucks, G.W.; Schlegel, H.B.; Scuseria, G.E.; Robb, M.A.; Cheeseman, J.R.; Scalmani, G.; Barone, V.; Petersson, G.A.; Nakatsuji, H.; et al. *Gaussian 16 Revision 16.A.03*; Gaussian, Inc.: Wallingford, CT, USA, 2016.
38. Labanowski, J.K.; Andzelm, J.W. (Eds.) *Density Functional Methods in Chemistry*; Springer: New York, NY, USA, 1991; ISBN 978-1-4612-7809-2.
39. Petersson, G.A.; Bennett, A.; Tensfeldt, T.G.; Al-Laham, M.A.; Shirley, W.A.; Mantzaris, J. A complete basis set model chemistry. I. The total energies of closed-shell atoms and hydrides of the first-row elements. *J. Chem. Phys.* **1988**, *89*, 2193–2218. [CrossRef]
40. Nazeeruddin, M.K.; De Angelis, F.; Fantacci, S.; Selloni, A.; Viscardi, G.; Liska, P.; Ito, S.; Takeru, B.; Grätzel, M. Combined Experimental and DFT-TDDFT Computational Study of Photoelectrochemical Cell Ruthenium Sensitizers. *J. Am. Chem. Soc.* **2005**, *127*, 16835–16847. [CrossRef] [PubMed]
41. Asaduzzaman, A.M.; Schreckenbach, G. Computational study of the ground state properties of iodine and polyiodide ions. *Theor. Chem. Acta* **2009**, *122*, 119–125. [CrossRef]
42. Lin, R.Y.-Y.; Chuang, T.-M.; Wu, F.-L.; Chen, P.-Y.; Chu, T.-C.; Ni, J.-S.; Fan, M.-S.; Lo, Y.-H.; Ho, K.-C.; Lin, J.T. Anthracene/Phenothiazine π-Conjugated Sensitizers for Dye-Sensitized Solar Cells using Redox Mediator in Organic and Water-based Solvents. *ChemSusChem* **2015**, *8*, 105–113. [CrossRef]
43. Sohrabi, M.; Amirnasr, M.; Farrokhpour, H.; Meghdadi, S. A single chemosensor with combined ionophore/fluorophore moieties acting as a fluorescent "Off-On" $Zn^{2+}$ sensor and a colorimetric sensor for $Cu^{2+}$: Experimental, logic gate behavior and TD-DFT calculations. *Sens. Actuators B Chem.* **2017**, *250*, 647–658. [CrossRef]
44. Cossi, M.; Barone, V.; Cammi, R.; Tomasi, J. Ab initio study of solvated molecules: A new implementation of the polarizable continuum model. *Chem. Phys. Lett.* **1996**, *255*, 327–335. [CrossRef]
45. Tomasi, J.; Mennucci, B.; Cammi, R. Quantum Mechanical Continuum Solvation Models. *Chem. Rev.* **2005**, *105*, 2999–3094. [CrossRef]
46. Becke, A.D. Density-functional thermochemistry. IV. A new dynamical correlation functional and implications for exact-exchange mixing. *J. Chem. Phys.* **1996**, *104*, 1040–1046. [CrossRef]
47. Perdew, J.P.; Ernzerhof, M.; Burke, K. Rationale for mixing exact exchange with density functional approximations. *J. Chem. Phys.* **1996**, *105*, 9982–9985. [CrossRef]
48. Adamo, C.; Barone, V. Toward reliable adiabatic connection models free from adjustable parameters. *Chem. Phys. Lett.* **1997**, *274*, 242–250. [CrossRef]
49. Zhao, Y.; Truhlar, D.G. The M06 suite of density functionals for main group thermochemistry, thermochemical kinetics, noncovalent interactions, excited states, and transition elements: Two new functionals and systematic testing of four M06-class functionals and 12 other functionals. *Theor. Chem. Acc.* **2008**, *120*, 215–241. [CrossRef]
50. Becke, A.D. A new mixing of Hartree-Fock and local density-functional theories. *J. Chem. Phys.* **1993**, *98*, 1372–1377. [CrossRef]
51. Yanai, T.; Tew, D.; Handy, N.C. A new hybrid exchange–correlation functional using the Coulomb-attenuating method (CAM-B3LYP). *Chem. Phys. Lett.* **2004**, *393*, 51–57. [CrossRef]
52. Hohenberg, P.; Kohn, W. Inhomogeneous Electron Gas. *Phys. Rev.* **1964**, *136*, B864–B871. [CrossRef]
53. Kohn, W.; Sham, L.J. Self-consistent equations including exchange and correlation effects. *Phys. Rev.* **1965**, *140*, A1133–A1138. [CrossRef]
54. Becke, A.D. Density-functional exchange-energy approximation with correct asymptotic behavior. *Phys. Rev. A* **1988**, *38*, 3098–3100. [CrossRef]
55. Grimme, S. Semiempirical GGA-type density functional constructed with a long-range dispersion correction. *J. Comput. Chem.* **2006**, *27*, 1787–1799. [CrossRef]
56. Krishnan, R.; Binkley, J.S.; Seeger, R.; Pople, J.A. Self-consistent molecular orbital methods. XX. A basis set for correlated wave functions. *J. Chem. Phys.* **1980**, *72*, 650–654. [CrossRef]
57. Kaloyanova, S.; Ivanova, I.; Tchorbanov, A.; Dimitrova, P.; Deligeorgiev, T. Synthesis of chloro-substituted analogs of Thiazole orange—Fluorophores for flow cytometric analyses. *J. Photochem. Photobiol. B Biol.* **2011**, *103*, 215–221. [CrossRef]
58. Vasilev, A.A.; Kandinska, M.I.; Stoyanov, S.S.; Yordanova, S.B.; Sucunza, D.; Vaquero, J.J.; Castaño, O.D.; Baluschev, S.; Angelova, S.E. Halogen-containing thiazole orange analogues—New fluorogenic DNA stains. *Beilstein J. Org. Chem.* **2017**, *13*, 2902–2914. [CrossRef] [PubMed]
59. Evenson, W.E.; Boden, L.M.; Muzikar, K.A.; O'Leary, D.J. 1H and 13C NMR Assignments for the Cyanine Dyes SYBR Safe and Thiazole Orange. *J. Org. Chem.* **2012**, *77*, 10967–10971. [CrossRef] [PubMed]
60. Runge, E.; Gross, E. Density-Functional Theory for Time-Dependent Systems. *Phys. Rev. Lett.* **1984**, *52*, 997–1000. [CrossRef]
61. Casida, M.E. Time-Dependent Density Functional Response Theory for Molecules. In *Recent Advances in Density Functional Methods*; Recent Advances in Computational Chemistry; World Scientific: Singapore, 1995; pp. 155–192. [CrossRef]

62. Jacquemin, D.; Mennucci, B.; Adamo, C. Excited-state calculations with TD-DFT: From benchmarks to simulations in complex environments. *Phys. Chem. Chem. Phys.* **2011**, *13*, 16987–16998. [CrossRef]
63. Toulouse, J.; Colonna, F.; Savin, A. Long-range–short-range separation of the electron-electron interaction in density-functional theory. *Phys. Rev. A* **2004**, *70*, 062505. [CrossRef]
64. Seminario, J.M. (Ed.) *Recent Developments and Applications of Modern Density Functional Theory*; Elsevier: Amsterdam, The Netherlands, 1996; pp. 327–354. ISBN 9780444824042.
65. Oberhofer, K.E.; Musheghyan, M.; Wegscheider, S.; Wörle, M.; Iglev, E.D.; Nikolova, R.D.; Kienberger, R.; Pekov, P.S.; Iglev, H. Individual control of singlet lifetime and triplet yield in halogen-substituted coumarin derivatives. *RSC Adv.* **2020**, *10*, 27096–27102. [CrossRef]
66. Hunt, P.A.; Robb, M.A. Systematic Control of Photochemistry: The Dynamics of Photoisomerization of a Model Cyanine Dye. *J. Am. Chem. Soc.* **2005**, *127*, 5720–5726. [CrossRef]
67. Yang, D.; Liu, Y.; Shi, D.; Sun, J. Theoretical study on the excited-state photoinduced electron transfer facilitated by hydrogen bonding strengthening in the C337–AN/MAN complexes. *Comput. Theor. Chem.* **2012**, *984*, 76–84. [CrossRef]
68. Pedone, A. Role of Solvent on Charge Transfer in 7-Aminocoumarin Dyes: New Hints from TD-CAM-B3LYP and State Specific PCM Calculations. *J. Chem. Theory Comput.* **2013**, *9*, 4087–4096. [CrossRef]
69. Karunakaran, V.; Lustres, J.L.P.; Zhao, L.; Ernsting, N.P.; Seitz, O. Large Dynamic Stokes Shift of DNA Intercalation Dye Thiazole Orange has Contribution from a High-Frequency Mode. *J. Am. Chem. Soc.* **2006**, *128*, 2954–2962. [CrossRef]
70. Sanchez-Galvez, A.; Hunt, P.; Robb, M.A.; Olivucci, M.; Vreven, A.T.; Schlegel§, H.B. Ultrafast Radiationless Deactivation of Organic Dyes: Evidence for a Two-State Two-Mode Pathway in Polymethine Cyanines. *J. Am. Chem. Soc.* **2000**, *122*, 2911–2924. [CrossRef]
71. Retting, W. Charge Separation in Excited States of Decoupled Systems—TICT Compounds and Implications Regarding the Development of New Laser Dyes and the Primary Process of Vision and Photosynthesis. *Angew. Chem. Int. Ed. Engl.* **1986**, *25*, 971–988. [CrossRef]
72. West, W.; Pearce, S. The Dimeric State of Cyanine Dyes. *J. Phys. Chem.* **1965**, *69*, 1894–1903. [CrossRef]
73. Biver, T.; Boggioni, A.; Secco, F.; Turriani, E.; Venturini, M.; Yarmoluk, S. Influence of cyanine dye structure on self-aggregation and interaction with nucleic acids: A kinetic approach to TO and BO binding. *Arch. Biochem. Biophys.* **2007**, *465*, 90–100. [CrossRef]
74. Ogul'Chansky, T.; Losytskyy, M.; Kovalska, V.; Yashchuk, V.; Yarmoluk, S. Interactions of cyanine dyes with nucleic acids. XXIV. Aggregation of monomethine cyanine dyes in presence of DNA and its manifestation in absorption and fluorescence spectra. *Spectrochim. Acta Part A Mol. Biomol. Spectrosc.* **2001**, *57*, 1525–1532. [CrossRef]
75. Nygren, J.; Svanvik, N.; Kubista, M. The interactions between the fluorescent dye thiazole orange and DNA. *Biopolymers* **1998**, *46*, 39–51. [CrossRef]
76. Mooi, S.M.; Heyne, B. Size Does Matter: How To Control Organization of Organic Dyes in Aqueous Environment Using Specific Ion Effects. *Langmuir* **2012**, *28*, 16524–16530. [CrossRef]
77. Mooi, S.M.; Keller, S.N.; Heyne, B. Forcing Aggregation of Cyanine Dyes with Salts: A Fine Line between Dimers and Higher Ordered Aggregates. *Langmuir* **2014**, *30*, 9654–9662. [CrossRef]

*Article*

# UPS, XPS, NEXAFS and Computational Investigation of Acrylamide Monomer

Luca Evangelisti [1,2,3], Sonia Melandri [2,3,4], Fabrizia Negri [4,5], Marcello Coreno [6], Kevin C. Prince [7,8] and Assimo Maris [2,4,*]

1. Department of Chemistry "G. Ciamician", University of Bologna, I-48123 Ravenna, Italy; luca.evangelisti6@unibo.it
2. Interdepartmental Centre for Industrial Aerospace Research (CIRI Aerospace), University of Bologna, I-47121 Forlì, Italy; sonia.melandri@unibo.it
3. Interdepartmental Centre for Industrial Agrifood Research (CIRI Agrifood), University of Bologna, I-47521 Cesena, Italy
4. Department of Chemistry "G. Ciamician", University of Bologna, I-40126 Bologna, Italy; fabrizia.negri@unibo.it
5. National Interuniversity Consortium of Materials Science and Technology (INSTM), UdR Bologna, I-40126 Bologna, Italy
6. CNR-ISM, Trieste LD2 Unit, I-34149 Trieste, Italy; marcello.coreno@elettra.eu
7. Elettra-Sincrotrone Trieste, Area Science Park, I-34149 Trieste, Italy; kevin.prince@elettra.eu
8. Department of Chemistry and Biotechnology, School of Science, Computing and Engineering Technology, Swinburne University of Technology, Hawthorn, VIC 3122, Australia
* Correspondence: assimo.maris@unibo.it

**Abstract:** Acrylamide is a small conjugated organic compound widely used in industrial processes and agriculture, generally in the form of a polymer. It can also be formed from food and tobacco as a result of Maillard reaction from reducing sugars and asparagine during heat treatment. Due to its toxicity and possible carcinogenicity, there is a risk in its release into the environment or human intake. In order to provide molecular and energetic information, we use synchrotron radiation to record the UV and X-ray photoelectron and photoabsorption spectra of acrylamide. The data are rationalized with the support of density functional theory and ab initio calculations, providing precise assignment of the observed features.

**Keywords:** acrylamide; synchrotron radiation; molecular modeling

**Citation:** Evangelisti, L.; Melandri, S.; Negri, F.; Coreno, M.; Prince, K.C.; Maris, A. UPS, XPS, NEXAFS and Computational Investigation of Acrylamide Monomer. *Photochem* **2022**, *2*, 463–478. https://doi.org/10.3390/photochem2030032

Academic Editors: Gulce Ogruc Ildiz and Licinia L.G. Justino

Received: 31 May 2022
Accepted: 20 June 2022
Published: 22 June 2022

**Publisher's Note:** MDPI stays neutral with regard to jurisdictional claims in published maps and institutional affiliations.

**Copyright:** © 2022 by the authors. Licensee MDPI, Basel, Switzerland. This article is an open access article distributed under the terms and conditions of the Creative Commons Attribution (CC BY) license (https://creativecommons.org/licenses/by/4.0/).

## 1. Introduction

Acrylamide, or prop-2-enamide in IUPAC notation, is a small organic compound widely used in industrial and laboratory processes. The monomer is incorporated into grout and soil-stabilizer products, is used in the synthesis of dyes, and serves as a precursor of a series of homo- and co-polymers whose properties strongly depend on the non-ionic, anionic, or cationic features.

Polyacrylamides are exploited in a wide range of applications [1–4]. They are used as flocculant or coagulant agents in drinking water and wastewater treatment and in mining processes, as binder and retention supports for fibres and pigments in paper production, for enhanced oil recovery, as drift control agents in commercial herbicide mixtures, as gel supports for chromatography and electrophoresis, in thickeners, in permanent-press fabrics, for contact lenses, in food-packaging adhesives, paper, and paperboard, to wash or peel fruits and vegetables, for water retention in diapers, as stabilizers and binder in cosmetic products, as media for hydroponic crops, as granules in arable lands to reduce the need of irrigation. Although polyacrylamides are quite stable, a small release of the monomer (present as residual of the polymerization process or formed by degradation) in the environment cannot be excluded [4,5]. This constitutes a potential hazard to health since acrylamide is a peripheral nerve toxin to man and to animals and causes birth defects and cancer in animals [6].

Moreover, acrylamide is present in the smoke of cigarettes [7], and it is formed from food components during heat treatment ($T \geq 393$ K, 120 °C) as a result of the Maillard reaction between the aminoacid asparagine ($COOH-CHNH_2-CH_2-CONH_2$) and reducing sugars [8]. Although it is not clear exactly what risk acrylamide poses to humans, in 2016, the Food and Drug Administration developed a Guidance for Industry (Docket Number: FDA-2013D-0715) that outlines strategies to help growers, manufacturers, and food service operators reduce acrylamide in the food supply. Similarly, in the European Union, the Commission Regulation (EU) 2017/2158 establishes mitigation measures and benchmark levels for the reduction of the presence of acrylamide in food.

Many spectroscopic studies are reported for acrylamide, but to the best of our knowledge, only a few of them concern the molecule in the isolated phase. The studies cover the rotational (20–60.5 GHz [9], 75–480 GHz [10]), far infrared (50–665 $cm^{-1}$ [11]) and photoelectron (7–27 eV [12]) spectral ranges. The scope of the present work is to investigate the behaviour of acrylamide by means of synchrotron-light techniques and quantum mechanical calculations. Here, we report new data on the inner shell photoionization and photoabsorption processes and also reinvestigate the valence band photoelectron spectrum of acrylamide. This study will enhance our understanding of the fundamental properties of this molecule which is in the spotlight because of its versatility and miscellaneous applications but also because of the global ecological risk posed by it.

## 2. Materials and Methods Details

*2.1. Experimental Methods*

Photoionization and photoabsorption spectra of acrylamide were recorded at the Gas Phase beamline [13] at the Elettra synchrotron light source (Trieste, Italy) using an electron energy analyser described previously [14]. Acrylamide purchased from Alfa Aesar (purity 98%) was introduced into the experimental chamber without further purification. The sample appears as white crystalline flakes at room temperature, with melting point $m.p.$ = 357 K and boiling point $b.p.$ = 398 K/25 mmHg. It was sublimated using a custom-built, resistively heated furnace, based on a stainless steel crucible, a Thermocoax® heating element, and a type K thermocouple. The temperature was increased up to 308 K (35 °C), reaching pressure of about $p$ = 0.256(2) mPa. The vapour pressure estimated at the same temperature by graphical interpolation of the data reported by Carpenter and Davis [15] is 2.25(10) Pa. During the whole experiment, the pressure in the ionization region remained constant and no evidence of contamination or thermal decomposition was found. The photoionization spectra were recorded using a VG-Scienta SES-200 photoelectron analyzer [16] mounted at the magic angle (54.7°). The use of this angle allows us to record spectra in which the intensities of all peaks is proportional to the angle-integrated intensity, that is, angular effects are eliminated. Thus, the intensity of the photoemission bands is proportional to the angle-integrated intensity, and angular effects are eliminated. The photoabsorption spectra were acquired in total ion yield mode, by means of channel electron multiplier placed near the ionization region.

*2.2. Computational Methods*

Geometry optimizations [17] and subsequent vibrational frequency calculations in the harmonic approximation to determine the nature of the stationary points were run both at the ab initio MP2 and density-functional theory (DFT) B3LYP and B2PLYP levels of calculation using the aug-cc-pVTZ basis set. The time-dependent (TD [18]) DFT approach was used in the case of electronic excited states. The ab initio geometries were used for the estimation of the vertical ionization energies by means of the Symmetry Adapted Cluster/Configuration Interaction method (SAC-CI [19]) with the cc-pVTZ basis set of the Electron Propagator Theory (EPT [20,21]) with the aug-cc-pVTZ basis, by applying the Outer Valence Green's Function (OVGF [22–24]) and the Partial Third Order (P3 [25]) self-energy approximations. The SAC-CI/cc-pVTZ approach was also used for the estimation of the core shell excitation energies. All the above quantum mechanical calculations were performed using the GAUSSIAN16™ (Gaussian is a registered trademark of Gaussian, Inc., 340 Quinnipiac St. Bldg. 40, Wallingford, CT, USA) software package (G16, Rev. A.03).

The vibronic structure associated with each ionization in the low energy region of the photoelectron spectrum was simulated including the Franck–Condon activities [26], namely by computing the Franck–Condon factors ($FC_k$) for each vibrational mode $k$ with frequency $\nu_k$. These were determined by the evaluation of the Huang–Rhys factors $S_k$ [27] as:

$$S_k = \frac{1}{2} \cdot B_k^2 \quad (1)$$

where $B_k$ is the dimensionless displacement parameter defined, assuming the harmonic approximation, and neglecting Duschinsky rotation [28], as:

$$B_k = \left(\frac{2\pi\nu_k}{\hbar}\right)^{1/2} \cdot (X_j - X_i) \cdot M^{1/2} \cdot Q_k(j) \quad (2)$$

where $X_i$ (or $X_j$) is the $3N$ dimensional vector of the equilibrium Cartesian coordinates of the $i$ (or $j$) electronic state (here $i$ is the neutral and $j$ is the cationic state), $M$ is the $3N \times 3N$ diagonal matrix of atomic masses, and $Q_k(j)$ is the $3N$ dimensional vector describing the $k$th normal coordinate of the $j$ (cationic) state in terms of mass weighted Cartesian coordinates. While the approach is less rigorous than that discussed in recent work [29–31], it is justified by the minor changes computed for the active vibrational modes upon ionization. For each normal mode $k$, the Franck–Condon factor $FC$ for a transition from a vibrational level $m$ (of the neutral molecule) to the vibrational level $n$ (of the cation) is [27]:

$$FC_k(m,n)^2 = e^{-S_k} \cdot S_k^{n-m} \cdot \frac{m!}{n!} \cdot [L_m^{n-m}(S_k)]^2 \quad (3)$$

where $L$ is a Laguerre polynomial. The intensity $I_k(m,n)$ of the $m$ to $n$ transition for the normal mode $k$ is given by the $FC$ factor, weighted for the population of the vibrational state $m$:

$$I_k(m,n) = FC_k(m,n)^2 \cdot \frac{1}{Z} \cdot \exp\left(-\frac{m\hbar\omega_k}{k_B T}\right) \quad (4)$$

with $k_B$ being the Boltzmann constant, $T$ is the temperature, and $Z$ is the partition function. The total intensity of the multimode vibrational transition, including all the active normal modes, is the simple product of the monodimensional intensities [27]. Photoelectron spectra were simulated at $T = 0$ K and a Gaussian broadening function with $FWHM = 80$ and $440$ cm$^{-1}$ was convoluted with each computed intensity.

## 3. Results and Discussion

### 3.1. Conformational Analysis

Acrylamide is the smallest conjugated amide. It is formed by the vynil and amide planar frames connected by a rotatable bond: $CH_2=CH-CONH_2$. Its conformation is usually described by the relative orientation of the carbon–carbon double bond and the carbonyl, through the CCCO torsional dihedral angle ($\tau$). We explored the conformational space of acrylamide by running a series of quantum mechanical calculations where $\tau$ was constrained to fixed values ($\tau = 0$–$360°$, $\Delta\tau = 10°$), whereas all the other parameters were freely optimized. The results achieved at three levels of calculation (B3LYP/aug-cc-pVTZ, B2PLYP/aug-cc-pVTZ and MP2/aug-cc-pVTZ) are qualitatively similar, as can be seen in Figure 1.

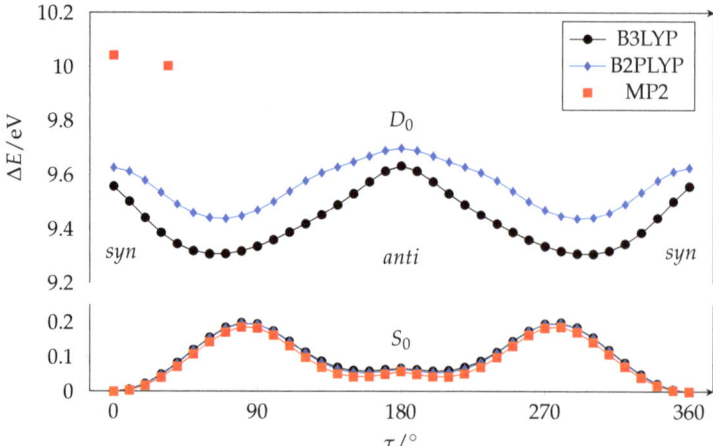

**Figure 1.** Fully relaxed PESs along the C-C torsion of acrylamide in its neutral ground state ($S_0$) and cation neutral ($D_0$) states calculated with aug-cc-pVTZ basis set. Only two stationary points are given for the $D_0$ state at the MP2/aug-cc-pVTZ level of calculation.

The potential energy surface (PES) of acrylamide in its neutral ground state ($S_0$) evidences the presence of two non-equivalent minima, see Figure 2. The global minimum is planar with the double bonds in *syn* orientation ($\tau = 0°$). The other local minimum has an *anti* arrangement, which is typical of amides in peptides and protein, and lies at 5.6/5.2/4.0 kJ mol$^{-1}$, while the *syn* to *anti* barrier is about 19/19/18 kJ mol$^{-1}$ (B3LYP/B2PLYP/MP2).

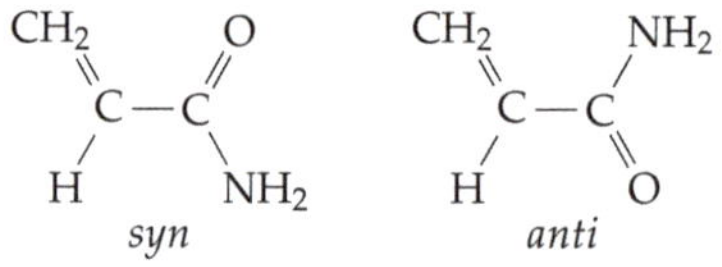

**Figure 2.** Conformers of acrylamide in its neutral ground state.

Taking into account the vibrational zero-point energy correction, this energy difference increases, becoming 6.2/6.2/5.6 kJ mol$^{-1}$ (B3LYP/B2PLYP/MP2). Because of the steric hindrance between the CH$_2$ and NH$_2$ groups, the *anti* geometry is slightly non planar as indicated by the presence of two equivalent minima at $\tau \simeq \pm 156°$ in the curve. Since the interconversion barrier between the equivalent conformations is quite low (0.75/1.08/1.61 kJ mol$^{-1}$ (B3LYP/B2PLYP/MP2) at $\tau = 180°$) due to tunnel effect, the ground vibrational state is expected to be observed in the form of a doublet. These results are in agreement with the rotational spectroscopy data of Marstokk et al. [9] and Kolesniková et al. [10]; indeed, the authors (i) assigned the transition lines of both the *syn* and *anti* conformers, (ii) estimated the conformers energy difference to be 6.5(6) kJ mol$^{-1}$ [9], and (iii) determined the doubling splitting energy of the *anti* conformer as 13.844586(2) cm$^{-1}$ [10]. In the crystal phase only, the *syn* species has been identified [32]. According to the Boltzmann distribution, and taking into account the effect of the degeneracy of *anti* species, the relative population of the two conformers at the experimental working temperature ($T$ = 308 K) is $N_{anti}/N_{syn}$ = 0.16(4).

We used the same approaches to calculate the PES of the acrylamide cation in its ground state ($D_0$), where the single occupied molecular orbital (MO) is a $\sigma$-type non bonding orbital centered on the oxygen ($\sigma_n$). The DFT results plotted in Figure 1 show that, for the cation, the *syn* and *anti* arrangements correspond to maxima while the minima correspond to two non-planar equivalent species with $\tau = 64/67°$ (B3LYP/B2PLYP). Unfortunately, we did not succeed in determining the whole path also at the MP2/aug-cc-pVTZ level, but we successfully optimized the geometries of the minimum and the *syn* arrangement (red squares in Figure 1). As in the case of DFT, the planar *syn* structure displays an imaginary frequency, albeit considerably smaller ($i120$ cm$^{-1}$) than that determined at the B3LYP level ($i436$ cm$^{-1}$), indicating a much shallower PES along the torsion axis. Thus, although the ab initio result predicts a non-planar minimum similar to DFT, such a minimum corresponds to a remarkably decreased torsion angle ($\tau = 34°$) accompanied by a drastically reduced stabilization compared to DFT. More precisely, DFT finds the *syn* form 24/18 kJ mol$^{-1}$ (B3LYP/B2PLYP) above the minimum, whereas the ab initio energy difference is only 3.8 kJ mol$^{-1}$. The DFT and ab initio landscapes are remarkably different, with the small energy difference predicted by MP2 calculations clearly suggesting a reduced relevance of twisting. This is further demonstrated by the almost unchanged bond lengths of the planar and twisted structures (Table 1), in contrast to the remarkable C=O bond length shortening occurring in the twisted DFT structures. Furthermore, the small energy difference predicted between the planar and twisted structures at the ab initio level is within the error of the method. As will be shown below, the simulation of vibronic structure strongly supports a planar or negligibly twisted $D_0$ structure.

Finally, we used TD-B3LYP/aug-cc-pTZVP to calculate the structure of the first excited state of the acrylamide cation ($D_1$), where the singly occupied *MO* is a $\pi$-type orbital composed of the non bonding out-of-plane $p$ orbitals of nitrogen and oxygen ($\pi_n$). The optimized geometry is found to be planar as in the case of the neutral species. The energy and structure parameters related to the calculated stationary points are summarized in Table 1.

**Table 1.** Energy and structure parameters of acrylamide calculated using the aug-cc-pVTZ basis set.

|  |  | $\tau/°$ | $E_e$/a.u. | $\Delta E_e$/eV | C=C/Å | C-C/Å | C-N/Å | C=O/Å | CCC/° | CCO/° | CCN/° |
|---|---|---|---|---|---|---|---|---|---|---|---|
| B3LYP | $S_0$ syn | 0 | −247.400743 | 0 | 1.327 | 1.493 | 1.366 | 1.219 | 121.1 | 114.5 | 123.4 |
|  | $S_0$ anti | 157 | −247.398607 | 0.06 | 1.328 | 1.491 | 1.367 | 1.220 | 126.2 | 117.2 | 120.6 |
|  | $D_0$ | 64 | −247.058750 | 9.31 | 1.350 | 1.487 | 1.307 | 1.246 | 120.5 | 126.3 | 106.1 |
|  | $D_0$ syn (TS) | 0 | −247.049500 | 9.56 | 1.335 | 1.465 | 1.311 | 1.280 | 123.7 | 122.2 | 118.1 |
| TD-B3LYP | $D_1$ syn | 0 | −247.045674 | 9.66 | 1.403 | 1.508 | 1.329 | 1.22 | 118.3 | 116.8 | 118.0 |
| B2PLYP | $S_0$ syn | 0 | −247.204462 | 0 | 1.330 | 1.491 | 1.365 | 1.221 | 120.7 | 114.5 | 123.3 |
|  | $S_0$ anti | 156 | −247.202496 | 0.05 | 1.332 | 1.489 | 1.368 | 1.222 | 125.7 | 116.9 | 120.8 |
|  | $D_0$ | 67 | −246.857625 | 9.44 | 1.356 | 1.486 | 1.308 | 1.240 | 119.7 | 127.4 | 103.5 |
|  | $D_0$ syn (TS) | 0 | −246.850666 | 9.63 | 1.337 | 1.459 | 1.309 | 1.287 | 122.8 | 118.4 | 122.6 |
| MP2 | $S_0$ syn | 0 | −246.852848 | 0 | 1.334 | 1.490 | 1.364 | 1.224 | 120.3 | 114.3 | 123.3 |
|  | $S_0$ anti | 154 | −246.851325 | 0.04 | 1.337 | 1.487 | 1.369 | 1.225 | 124.9 | 116.9 | 121.1 |
|  | $D_0$ | 34 | −246.485231 | 10.00 | 1.343 | 1.457 | 1.304 | 1.286 | 120.1 | 124.7 | 114.3 |
|  | $D_0$ syn (TS) | 0 | −246.483791 | 10.04 | 1.326 | 1.453 | 1.305 | 1.297 | 121.3 | 118.8 | 123.1 |

### 3.2. UPS

The photoionization spectrum of acrylamide has been recorded in the 6.5–26.5 eV spectral range using a 30.4 nm He(II) source by [12]. Here, we extend the investigation up to 37 eV, using 98 eV energy photons. The measured spectrum is shown in Figure 3.

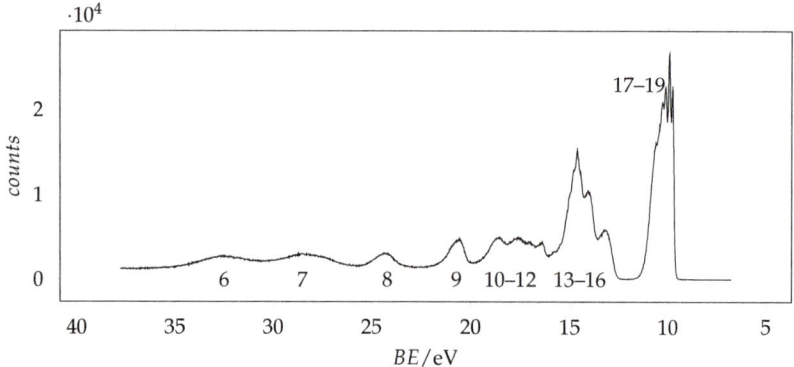

**Figure 3.** Full valence band photoemission spectrum of acrylamide recorded with 98 eV photon energy and 0.090 eV resolution and calibrated with the $3p_{3/2}$ Ar line at 15.76 [33]. The molecular orbitals are labelled according to Figure 4.

According to the Koopmans' theorem [34], a rough correlation can be found between the observed ionization energies and the calculated energy of the occupied *MOs* of the most stable conformer, whose shapes and energies are given in Figure 4 and in Table 2, respectively.

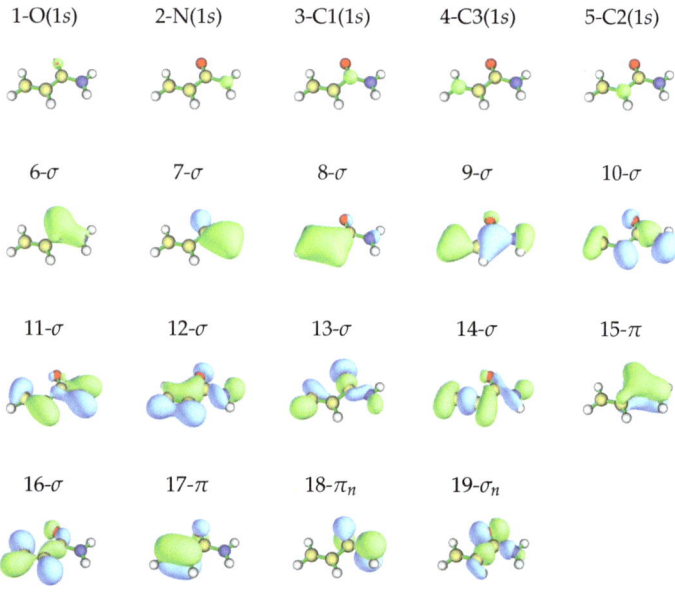

**Figure 4.** Occupied molecular orbitals of *syn*-acrylamide in the neutral ground state obtained at the MP2/aug-cc-pVTZ//HF/aug-cc-pVTZ level of calculation, isodensity surface at 0.05 a.u.

Table 2. Theoretical molecular orbital energy (*MOE*, eV), vertical binding energy (*BE*, eV) and pole strength (*PS*) values of *syn*-acrylamide compared to the experimental peaks' ionization energies.

| | | HF [a] MOE | B3LYP [b] MOE | B2PLYP [c] MOE | SAC-CI [d] BE | PS | P3 [e] BE | PS | OVGF [f] BE | PS | Exp. BE |
|---|---|---|---|---|---|---|---|---|---|---|---|
| 1 | O(1s) | −558.27 | −519.67 | −535.80 | 537.99 | 0.79 | - | - | - | - | 537.22 |
| 2 | N(1s) | −424.19 | −390.48 | −404.55 | 406.77 | 0.78 | - | - | - | - | 405.97 |
| 3 | C1(1s) | −309.01 | −279.87 | −292.06 | 294.88 | 0.77 | - | - | - | - | 294.16 |
| 4 | C3(1s) | −306.24 | −277.45 | −289.42 | 291.79 | 0.75 | - | - | - | - | 291.03 |
| 5 | C2(1s) | −306.07 | −277.36 | −289.38 | 291.60 | 0.76 | - | - | - | - | 291.03 |
| 6 | σ | −37.47 | −28.36 | −32.14 | 31.73 | 0.17 | - | - | - | - | 32.6 |
| 7 | σ | −33.05 | −24.90 | −28.30 | 27.26 | 0.14 | - | - | - | - | 28.3 |
| 8 | σ | −29.05 | −21.57 | −24.71 | 24.28 | 0.63 | - | - | - | - | 24.3 |
| 9 | σ | −23.91 | −17.81 | −20.37 | 20.55 | 0.83 | - | - | - | - | 20.6 |
| 10 | σ | −21.01 | −15.69 | −17.92 | 18.31 | 0.87 | 18.98 | 0.86 | 19.11 | 0.87 | 18.6 |
| 11 | σ | −19.61 | −14.65 | −16.74 | 17.37 | 0.89 | 17.95 | 0.87 | 18.04 | 0.88 | 17.6 |
| 12 | σ | −18.92 | −13.94 | −16.04 | 16.41 | 0.87 | 17.07 | 0.86 | 17.18 | 0.87 | 16.4/17.0 [g] |
| 13 | σ | −16.74 | −11.84 | −13.85 | 14.13 | 0.90 | 14.82 | 0.88 | 15.10 | 0.89 | 15.0 |
| 14 | σ | −16.27 | −11.76 | −13.69 | 14.18 | 0.90 | 14.86 | 0.89 | 14.77 | 0.90 | 14.63 |
| 15 | π | −15.69 | −11.53 | −13.28 | 13.87 | 0.89 | 14.33 | 0.86 | 14.53 | 0.87 | 14.07 |
| 16 | σ | −14.75 | −10.44 | −12.23 | 12.76 | 0.91 | 13.48 | 0.89 | 13.58 | 0.90 | 13.17 |
| 17 | π | −10.60 | −7.95 | −8.99 | 10.29 | 0.93 | 10.59 | 0.90 | 10.53 | 0.90 | 10.675 |
| 18 | $\pi_n$ | −11.35 | −7.69 | −9.25 | 9.71 | 0.91 | 10.33 | 0.88 | 10.44 | 0.89 | 10.296 |
| 19 | $\sigma_n$ | −11.69 | −7.28 | −9.09 | 9.19 | 0.90 | 9.86 | 0.88 | 10.43 | 0.89 | 9.756 |

[a] MP2/aug-cc-pVTZ//HF/aug-cc-pVTZ. [b] B3LYP/aug-cc-pVTZ. [c] B2PLYP/aug-cc-pVTZ. [d] MP2/aug-cc-pVTZ//SAC-CI/cc-pVTZ. [e] MP2/aug-cc-pVTZ//P3/aug-cc-pVTZ. [f] MP2/aug-cc-pVTZ//OVGF/aug-cc-pVTZ. [g] Shoulder.

We observe that, concerning the outer valence band, different methods provide different ordering of the *MOs*. However, using more sophisticated calculations, it is possible to obtain more reliable predictions. The vertical ionization energies obtained using the SAC-CI and EPT (P3 and OVGF) methods are compared to the *MOs* energy values in Table 2 and are used for the assignment of the observed peaks, whose energy values are given in the same table. Notably, all these methods find that the lower ionization energy is associated with the $\sigma_n$ orbital (n. 19), followed by the ionization of $\pi_n$ (n. 18) and $\pi_{C=C}$ (n. 17). The assignment of Åsbrink et al. [12] follows the same order, involving the features lying at 10.0, 10.3 and 10.7 eV. Since the first peak observed in our ionization spectrum has a lower energy (9.8 eV) and the spectral region between 9.5 and 11.5 eV appears quite crowded, we decided to record the spectrum at higher resolution. The resulting spectrum is shown in Figure 5 with the theoretical vertical ionization energies related to the three outer valence orbitals.

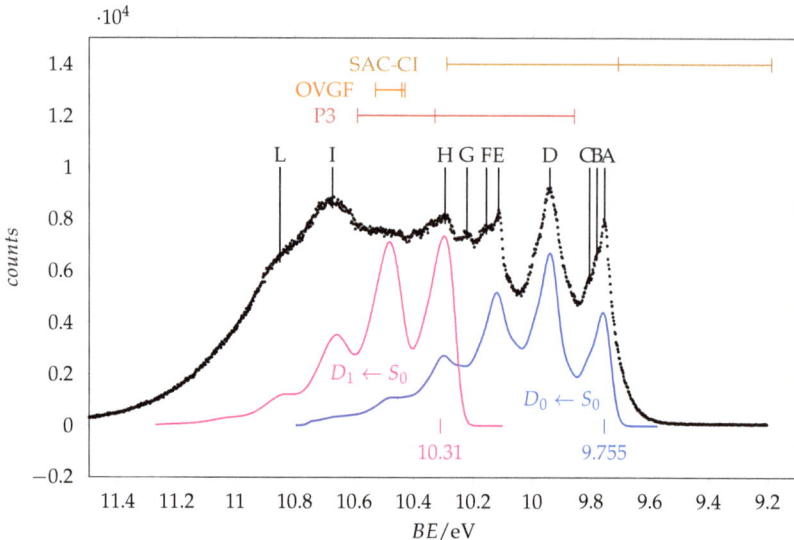

**Figure 5.** Outer valence band photoemission spectrum of acrylamide recorded with 25 eV photon energy and 0.011 eV resolution and calibrated according with the $3p_{3/2}$ Ar line at 15.76 [33]. Measured energy values: A = 9.756 eV, B = 9.782 eV, C = 9.808 eV, D = 9.943 eV, E = 10.116 eV, F = 10.158 eV, G = 10.223 eV, H = 10.296 eV, I = 10.675 eV, L = 10.851 eV. The theoretical vertical ionization energies for the three outer valence orbitals (17–19, from left to right) of *syn*-acrylamide are indicated in the top part. Blue and magenta traces are the $D_0 \leftarrow S_0$ and $D_1 \leftarrow S_0$ simulated vibrational resolved bands obtained at T = 0 K using a broadening Gaussian function (*FWHM* = 440 cm$^{-1}$/54.6 meV). Their origins have been placed at 9.755 and 10.31 eV, respectively.

The lower energy peak (labelled as A in Figure 5) lies at 9.756 eV. It has two shoulders (B and C) likely due to vibrational contributions: indeed, the energy differences are B-A = 27 meV and C-A = 52 meV, which are consistent with low frequency normal modes. Two strong peaks (D and E) are found at 9.943 and 10.116 eV. For the hypothesis of a vibrational progression starting from peak A, the involved quanta will be 187 meV (ca. 1508 cm$^{-1}$) and 173 meV (ca. 1395 cm$^{-1}$). A similar trend was also observed in the spectrum of formamide [35], and it was tentatively assigned to the asymmetric stretching of the N-C=O frame, an interpretation supported by recently reported vibronic structure simulations [36]. According to this interpretation, we observe that the shape of peak "D" is akin to that of "A".

To support this analysis, we modelled the vibronic structure of the $D_0 \leftarrow S_0$ and $D_1 \leftarrow S_0$ transitions using MP2/aug-cc-pVTZ and TD-B3LYP/aug-cc-pVTZ calculated geometries and vibrational normal coordinates to evaluate the Huang–Rhys factors governing vibronic progressions. As discussed above, the MP2 results for the $D_0$ state suggest a limited stabilization of the non-planar structure with a modest twisting and almost unchanged bond lengths. Thus, we simulated the vibronic structure taking the planar $D_0$ structure as a reference. The vibronically active vibrational frequencies and corresponding Huang–Rhys factors are listed in Table 3, while the calculated spectra are compared to the experimental ones in Figures 5 and 6.

**Table 3.** Calculated wavenumbers $\tilde{v}_k$ and Huang–Rhys factors $S_k$ for the vibrational normal modes of the $D_0 \leftarrow S_0$ (MP2/aug-cc-pVTZ) and $D_1 \leftarrow S_0$ (TD-B3LYP/aug-cc-pVTZ) transitions of *syn*-acrylamide.

| $D_0 \leftarrow S_0$ | | | $D_1 \leftarrow S_0$ | | |
|---|---|---|---|---|---|
| $\tilde{v}_k$/cm$^{-1}$ | $\tilde{v}_k$/meV | $S_k$ | $\tilde{v}_k$/cm$^{-1}$ | $\tilde{v}_k$/meV | $S_k$ |
| 260 | 32 | 0.272 | 183 | 23 | 0.018 |
| 456 | 57 | 0.326 | 235 | 29 | 0.004 |
| 550 | 68 | 0.083 | 305 | 38 | 0.854 |
| 873 | 108 | 0.106 | 467 | 58 | 0.003 |
| 1061 | 132 | 0.238 | 648 | 80 | 0.156 |
| 1078 | 134 | 0.038 | 835 | 104 | 0.088 |
| 1342 | 166 | 0.011 | 838 | 104 | 0.001 |
| 1444 | 179 | 1.011 | 1063 | 132 | 0.001 |
| 1481 | 184 | 0.028 | 1134 | 141 | 0.002 |
| 1584 | 196 | 0.247 | 1297 | 161 | 0.134 |
| 1689 | 209 | 0.046 | 1388 | 172 | 0.410 |
| 2200 | 273 | 0.003 | 1492 | 185 | 0.081 |

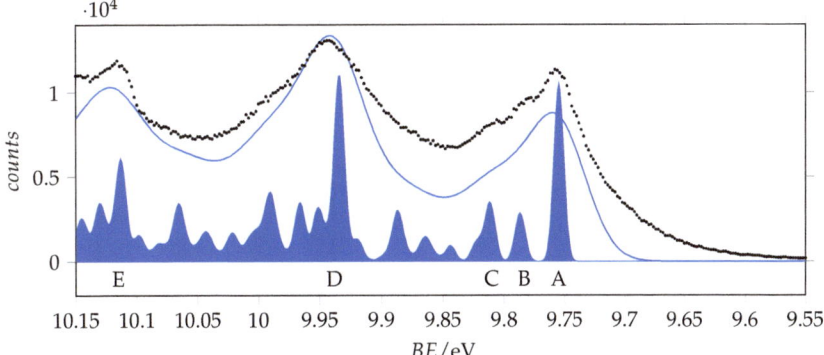

**Figure 6.** Outer valence band photoemission spectrum of acrylamide compared to the $D_0 \leftarrow S_0$ simulated vibrational resolved bands of *syn*-acrylamide obtained at $T = 0$ K using broadening Gaussian functions with $FWHM$ = 440 and 80 cm$^{-1}$ (54.6 and 9.9 meV). Labels A–E are the same as in Figure 5.

The simulated vibronic profiles of both the $D_0 \leftarrow S_0$ (blue trace in Figure 5) and $D_1 \leftarrow S_0$ (pink trace in Figure 5) transitions show a resolved structure with major vibronic peaks separated by about 200 meV, which is in agreement with the experimental observation. However, the shape and relative intensity of the vibronic bands are different for the two ionizations and comparison with the detected spectrum suggests that the two lower energy peaks must be assigned to the $D_0 \leftarrow S_0$ transition. By aligning the calculated and experimental spectra, we determine the adiabatic ionization energies to the $D_0$ state as $\Delta E_{0,0}$ = 9.755 eV. This value is 0.475 eV smaller than that of formamide (10.23(6) eV [36]) and 0.021 eV greater than that of acetamide (9.734(8) eV [36]). According to the Huang–Rhys factors (Figure 7), the main vibrational structure is related to the skeletal stretching motions associated with the largest geometry changes upon ionization, specifically $\nu_{C-C}$ = 179 meV (1444 cm$^{-1}$), $\nu_{C-N}$ = 196 meV (1584 cm$^{-1}$) and $\nu_{C=O}$ = 132 meV (1061 cm$^{-1}$). Notably, the C=O stretching displays a considerably reduced frequency as a result of the remarkable bond length elongation in the cation (compare 1.224 Å in $S_0$ with 1.29 Å in $D_0$, see Table 1). The "B" and "C" features of the lower energy peak can be associated with the $\delta_{CCC}$ = 32 meV (260 cm$^{-1}$) and $\delta_{CCN}$ = 57 meV (456 cm$^{-1}$) bending motions, respectively.

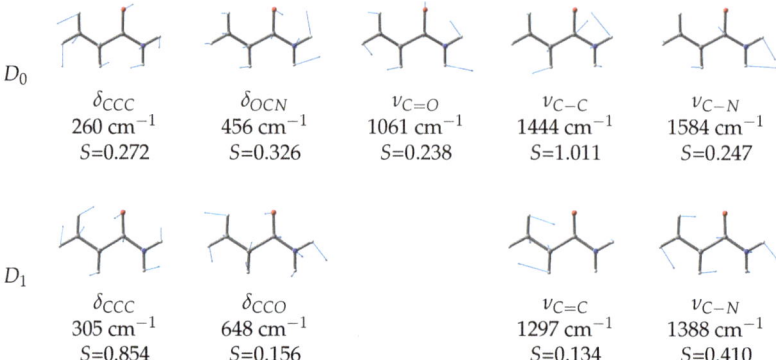

**Figure 7.** Calculated wavenumbers $\tilde{\nu}_k$ and Huang–Rhys factors $S_k$ for the vibrational normal modes of the $D_0 \leftarrow S_0$ (MP2/aug-cc-pVTZ) and $D_1 \leftarrow S_0$ (TD-B3LYP/aug-cc-pVTZ) transitions of *syn*-acrylamide.

The positioning of the $D_1 \leftarrow S_0$ transition is more challenging. In Figure 5, we propose a tentative assignment with an adiabatic ionization energy $\Delta E_{0,0} = 10.31$ eV, corresponding to peak H. This value is 0.033 eV smaller than that of formamide (10.64(2) eV [36]) and very close to that of acetamide (10.28(2) [36]). According to the Huang–Rhys factors (Figure 7), the most active normal modes are $\delta_{CCC} = 38$ meV (306 cm$^{-1}$), $\nu_{C-N} = 172$ meV (1388 cm$^{-1}$), $\delta_{CCO} = 80$ meV (648 cm$^{-1}$), and $\nu_{C=C} = 161$ meV (1297 cm$^{-1}$). Notably, different vibration modes are active in the ionization to $D_1$ compared to $D_0$, on account of the different geometry changes documented in Table 1 for the $D_1 \leftarrow S_0$ and $D_0 \leftarrow S_0$ transitions.

Finally, the peaks "I" and "L" can be reasonably assigned to the resolved ionization band of the C=C $\pi$-type orbital, their energy difference being 176 meV, a value very similar to that of the previously considered transitions.

### 3.3. XPS

The inner shell photoemission spectra of acrylamide are shown in Figures 8–10, with the values predicted at the SAC-CI/cc-pVTZ level of calculation for *syn*-acrylamide.

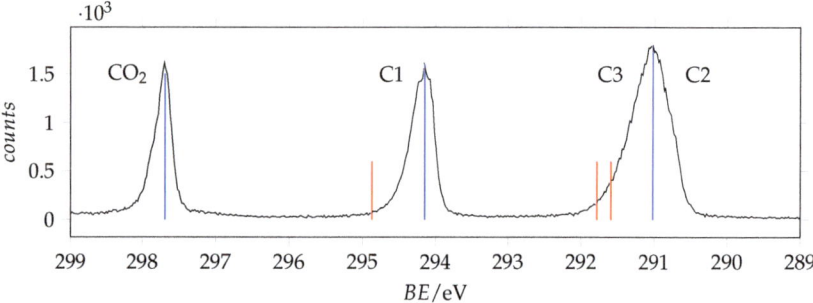

**Figure 8.** C(1s) photoemission spectrum of acrylamide and calibrant $CO_2$ recorded with photon energy 382 eV and resolution 0.22 eV. The calibrant line lies at 297.7 eV [37]. Blue and red lines are the measured and SAC-CI/cc-pVTZ predicted binding energy values.

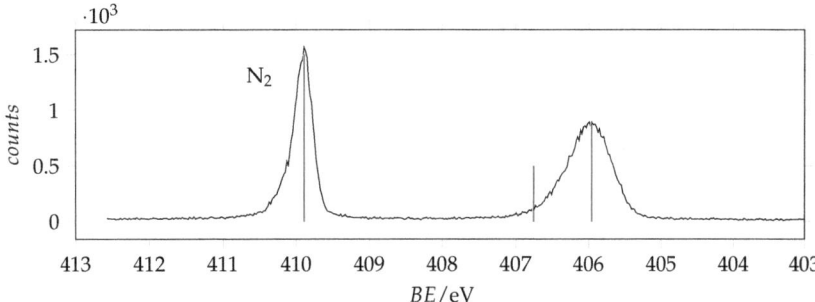

**Figure 9.** N(1s) photoemission spectrum of acrylamide and calibrant $N_2$ recorded with photon energy 495 eV and resolution 0.24 eV. The calibrant line lies at 409.9 eV [37]. Blue and red lines are the measured and SAC-CI/cc-pVTZ predicted binding energy values.

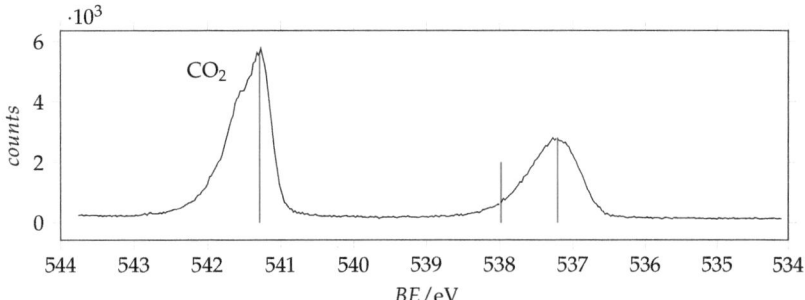

**Figure 10.** O(1s) photoemission spectrum of acrylamide and calibrant $CO_2$ recorded with photon energy 628 eV and resolution 0.38 eV. The calibrant line lies at 541.3 eV [37]. Blue and red lines are the measured and SAC-CI/cc-pVTZ predicted binding energy values.

The measured and theoretical ionization are listed in Table 2. The predictions are overestimated by about 0.7–0.8 eV with respect to the observations. As regards the amide frame, the observed binding energies are smaller (O(1s) −0.5 eV, N(1s) −0.5 eV, C1(1s) −0.3 eV) than those of formamide [38] suggesting an overall electron donating effect of the vinylic $\pi$ electron cloud. Differently, the observed acrylamide binding energies are greater (O(1s) +0.2 eV, N(1s) +0.2 eV, C1(1s) +0.4 eV) than those of N-methylacetamide [39]. In addition, in the case of 3-carbamoyl-2,2,5,5-tetramethyl-3-pyrrolin-1-oxyl ($C_9H_{15}N_2O_2$) that for our purposes can be described as an acrylamide frame embedded in cyclic nitroxyl radical through the C=C bond, the binding energies of acrylamide are blue shifted: O(1s) +0.32 eV, N(1s) +0.07 eV, C1(1s) +0.16 eV and C2/C3(1s) +0.33 [40]. Comparison with pyridin-2-one, where the amide frame is embedded in a conjugate ring system, shows that the O(1s) binding energy of acrylamide is +0.6 eV greater, whereas the N(1s) binding energies is −0.4 eV smaller [41].

### 3.4. NEXAFS

The observed K-edge photoabsorption spectra of acrylamide are given in Figures 11–13.

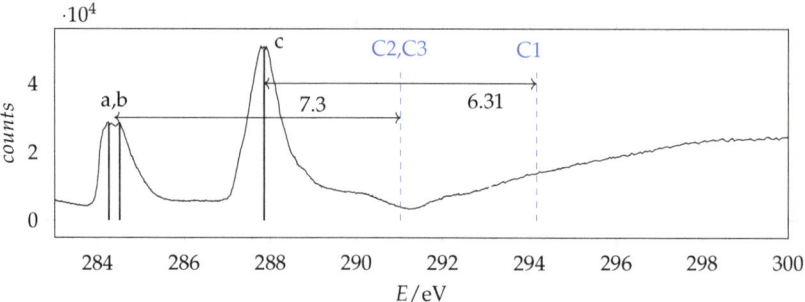

**Figure 11.** Photoabsorption spectrum of acrylamide in the region of C(1s) excitation recorded with 288 eV and 296 eV photon energy and 0.065 eV resolution and calibrated using the $1s \rightarrow \pi^*$ $CO_2$ line at 290.77 eV [42]. The dashed vertical lines indicate the location of the ionization thresholds. Continuous lines identify the measured energies.

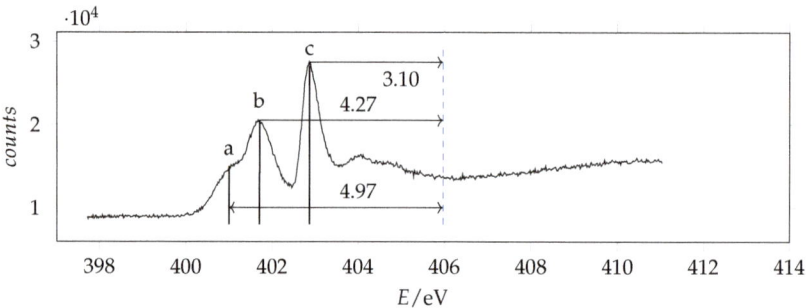

**Figure 12.** Photoabsorption spectrum of acrylamide in the region of N(1s) excitation recorded with 401 eV photon energy and 0.070 eV resolution and calibrated with respect to the $1s \rightarrow \pi^*$ $N_2$ line at 401.10 eV [43]. The dashed vertical line indicates the location of the ionization threshold. Continuous vertical lines identify the measured energies.

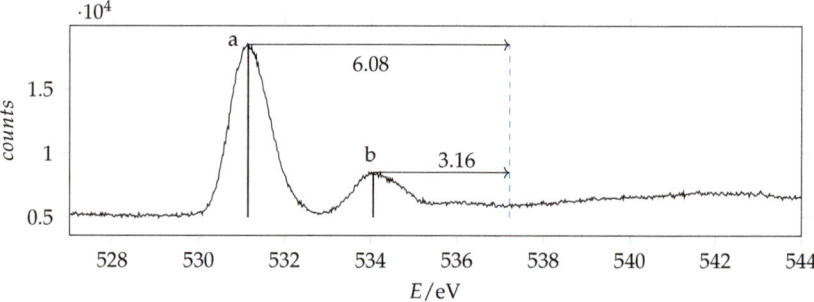

**Figure 13.** Photoabsorption spectrum of acrylamide in the region of O(1s) excitation recorded with 536 eV photon energy and 0.130 eV resolution and calibrated with respect to the $1s \rightarrow \pi^*$ $CO_2$ line at 535.4 eV [44]. The dashed vertical line indicates the location of the ionization threshold. Continuous vertical lines identify the measured energies.

The photoabsorption energy values of the main peaks are listed in Table 4 with the term values calculated as difference with the ionization energy:

$$T = BE - E \tag{5}$$

In the same table, we report the assigned features of related molecular systems formamide (NH$_2$–CHO [45]), acrolein (CH$_2$=CH–CHO [46]) and 3-carbamoyl-2,2,5,5-tetramethyl-3-pyrrolin-1-oxyl (C$_9$H$_{15}$N$_2$O$_2$ [47,48]), as well as the results of the theoretical simulation.

**Table 4.** Energies ($E$/eV), term values ($T$/eV) and assignment for the main features observed in K-shell spectra of acrylamide and analogous compounds. Calculated (SAC-CI/cc-pVTZ) energy values, term values, oscillator strengths ($f$) and shape of selected unoccupied MOs of *syn*-acrylamide are also given.

| | Acrylamide | | Formamide | Acrolein | C$_9$H$_{15}$N$_2$O$_2$ | |
|---|---|---|---|---|---|---|
| | $E_o/T_o$ | $E_c/T_c$ ($f_c$, 10$^{-2}$) | $E_o/T_o$ [45] | $E_o$ [46] | $E_o$ [48] | Assignment |
| C1 BE | 294.16 | 294.88 | 294.5 | - | 294.0 | |
| C2; C3 BE | 291.03 | 291.79; 291.60 | | | 290.7 | |
| C (a) | 287.85/6.31 | 288.40/6.48 (6.8) | 288.1/6.4 | 286.10 | | C1 1s $\rightarrow$ 1$\pi^*$ |
| C (b) | 284.52/7.27 | 285.18/5.85 (6.0) | | 284.19 | | C3 1s $\rightarrow$ 1$\pi^*$ |
| C (c) | 284.27/7.33 | 284.97/6.06 (3.0) | | 284.19 | | C2 1s $\rightarrow$ 1$\pi^*$ |
| N BE | 405.97 | 406.77 | 406.5 | | 405.9 | |
| N (a) | 401.00/4.97 | 402.83/3.94 (1.0) | | | | N 1s $\rightarrow$ 1$\sigma^*$ |
| N (b) | 401.70/4.27 | 402.59/4.18 (1.5) | 401.9/4.6 | | 401.95/3.95 | N 1s $\rightarrow$ 1$\pi^*$ |
| N (c) | 402.87/3.10 | 404.01/2.76 (3.1) | | | 403.15/2.75 | N 1s $\rightarrow$ 2$\sigma^*$ |
| O BE | 537.22 | 537.99 | 537.7 | - | 536.9 | |
| O (a) | 531.14/6.08 | 532.30/5.69 (2.9) | 531.5/6.2 | 530.59 | 531.3/5.6 | O 1s $\rightarrow$ 1$\pi^*$ |
| O (b) | 534.06/3.16 | 537.04/0.95 (0.6) | | 533.57 | 534.3/3.0 | O 1s $\rightarrow$ 2$\pi^*$ |

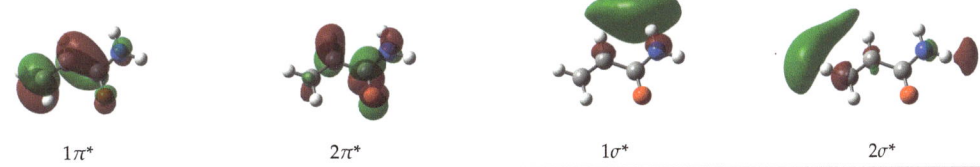

| 1$\pi^*$ | 2$\pi^*$ | 1$\sigma^*$ | 2$\sigma^*$ |

Most of the observed peaks can be assigned to transitions from the 1s atomic orbital to the first unoccupied $\pi^*$ orbital. The predictions are overestimated by about 0.6–1.2 eV with respect to the observations the shifts increasing in going from C to O. The 1s $\rightarrow$ $\pi^*$ absorption energies of acrylamide are smaller than formamide and larger than acrolein. As regards the other features, by comparison with acrolein, the observed peak O(b) can be assigned to the 1s $\rightarrow$ 2$\pi^*$ transition. We note that in this case the calculated value is considerably overestimated. Besides the 1s $\rightarrow$ $\pi^*$ peak, N(b), the nitrogen NEXAFS spectrum shows two other significant signals: a shoulder at lower energy, N(a), and an intense peak at higher energy, N(c). The same spectral profile has been observed in 3-carbamoyl-2,2,5,5-tetramethyl-3-pyrrolin-1-oxyl [47,48], although the shoulder is clearly overlapped with the vibrational resolved band of N$_2$, used as calibrant. Actually, in the spectrum of acrylamide, the N(a) signal does not show any additional vibrational structure. Thus, we infer that it pertains to acrylamide itself. Accordingly, calculations predict a lower energy, less intense 1s $\rightarrow$ 1$\sigma^*$ transition that can be related to N(a) and a higher energy more intense 1s $\rightarrow$ 2$\sigma^*$ transition that can be related to N(c). Despite comparison with other analogous molecular systems (see Table 4) support the assignment of peak N(b) to the 1s $\rightarrow$ 1$\pi^*$ transition, we observe that, based exclusively on the predicted energy values, the assignment of peaks N(a) and N(b) should be inverted. However, the predicted absorption intensities are in good agreement with the first proposed assignment.

## 4. Conclusions

We have performed an extensive core and valence level investigation of acrylamide and reported the theoretical and experimental photoelectron and photoabsorption spectra, as well as calculated along the C-C torsion and structural data of the ground and electronic excited states. Concerning the valence photoelectron spectra, most vibronic peaks have been successfully assigned on the basis of quantum-chemically computed geometry changes associated with ionization of the highest energy occupied $\sigma_n$ orbital and the following $\pi_n$-orbital. We find that ab initio and DFT results predict rather different equilibrium structures for the $D_0$ state, with the former level of theory indicating an almost negligible energy difference between planar and twisted structures, in contrast with DFT. Interestingly, the simulated vibronic structure based on the ab initio results agrees very closely with the observed spectra and is in line with recently reported vibronic structure simulations for the smaller formamide.

As regards the core photoelectron spectra, SAC-CI/cc-pVTZ calculations provide predictions slightly overestimated (about 2%) with respect to the experimental values. It is worth noting that the values of the inner shell orbitals calculated at the B2PLYP/aug-cc-pVTZ are of the same order of magnitude of the corresponding binding energies. SAC-CI/cc-pVTZ calculations were also useful to assign the core photoabsorption spectra. Thus, we find generally good agreement between theory and experiment, with few discrepancies that will be the subject of future investigations. More specifically, further investigation of the N(1s) absorption spectrum is warranted, and Resonant Auger–Meitner or angular resolved photoemission spectroscopy experiments could provide additional information for a refined assignment of the outer valence UPS spectrum.

**Author Contributions:** Conceptualization, A.M.; methodology, L.E., S.M., F.N., M.C., K.C.P. and A.M.; investigation, L.E., S.M., F.N., M.C., K.C.P. and A.M.; writing—original draft preparation, A.M.; writing—review and editing, L.E., S.M., F.N., M.C., K.C.P. and A.M.; funding acquisition, L.E., S.M. and A.M. All authors have read and agreed to the published version of the manuscript.

**Funding:** This work was supported by the Italian MIUR (Attività Base di Ricerca) and the University of Bologna (Ricerca Fondamentale Orientata). L.E. and A.M. acknowledge Elettra Sincrotrone Trieste for providing financial support as Italian Funded Users to attend shifts assigned to the Proposal No. 20195417 at the gas phase beamline.

**Institutional Review Board Statement:** Not applicable.

**Informed Consent Statement:** Not applicable.

**Data Availability Statement:** Not applicable.

**Acknowledgments:** We acknowledge the CINECA award under the ISCRA initiative, for the availability of high-performance computing resources and support.

**Conflicts of Interest:** The authors declare no conflict of interest.

## References

1. Smith, E.; Prues, S.; Oehme, F. Environmental Degradation of Polyacrylamides. 1. Effects of Artificial Environmental Conditions: Temperature, Light, and pH. *Ecotoxicol. Environ. Saf.* **1996**, *35*, 121–135. [CrossRef] [PubMed]
2. Huang, S.Y.; Lipp, D.; Farinato, R. Acrylamide polymers. In *Encyclopedia of Polymer Science and Technology*; American Cancer Society: Atlanta, GA, USA, 2001. [CrossRef]
3. NTP (National Toxicology Program). *Report on Carcinogens*, 14th ed.; U.S. Department of Health and Human Services, Public Health Service: Research Triangle Park, NC, USA, 2016.
4. Xiong, B.; Loss, R.; Shields, D.T.; Pawlik, R.H.; Zydney, A.; Kumar, M. Polyacrylamide degradation and its implications in environmental systems. *NPJ Clean Water* **2018**, *1*, 17. [CrossRef]
5. Tepe, Y.; Çebi, A. Acrylamide in Environmental Water: A Review on Sources, Exposure, and Public Health Risks. *Expos. Health* **2019**, *11*, 3–12. [CrossRef]
6. Exon, J.H. A Review of the Toxicology of Acrylamide. *J. Toxicol. Environ. Heal. Part B* **2006**, *9*, 397–412. [CrossRef] [PubMed]

7. Moldoveanu, S.; Gerardi, A. Acrylamide Analysis in Tobacco, Alternative Tobacco Products, and Cigarette Smoke. *J. Chromatogr. Sci.* **2011**, *49*, 234–242. [CrossRef]
8. Mottram, D.; Wedzicha, B.; Dodson, A. Acrylamide is formed in the Maillard reaction. *Nature* **2002**, *419*, 448–449. [CrossRef]
9. Marstokk, K.M.; Møllendal, H.; Samdal, S. Microwave spectrum, conformational equilibrium, 14N quadrupole coupling constants, dipole moment, vibrational frequencies and quantum chemical calculations for acrylamide. *J. Mol. Struct.* **2000**, *524*, 69–85. [CrossRef]
10. Kolesniková, L.; Belloche, A.; Koucký, J.; Alonso, E.R.; Garrod, R.T.; Luková, K.; Menten, K.M.; Müller, H.S.P.; Kania, P.; Urban, S. Laboratory rotational spectroscopy of acrylamide and a search for acrylamide and propionamide toward Sgr B2(N) with ALMA. *Astron. Astrophys.* **2022**, *659*, A111. [CrossRef]
11. Kydd, R.; Dunham, A. The infrared spectra and structure of acetamide and acrylamide. *J. Mol. Struct.* **1980**, *69*, 79–88. [CrossRef]
12. Åsbrink, L.; Svensson, A.; von Niessen, W.; Bier, G. 30.4-nm He(II) photoelectron spectra of organic molecules: Part V. Heterocompounds containing first-row elements (C, H, B, N, O, F). *J. Electron Spectrosc. Relat. Phenom.* **1981**, *24*, 293–314. [CrossRef]
13. Prince, K.C.; Blyth, R.R.; Delaunay, R.; Zitnik, M.; Krempasky, J.; Slezak, J.; Camilloni, R.; Avaldi, L.; Coreno, M.; Stefani, G.; et al. The gas-phase photoemission beamline at Elettra. *J. Synchrotron Radiat.* **1998**, *5*, 565–568. [CrossRef] [PubMed]
14. Lüder, J.; de Simone, M.; Totani, R.; Coreno, M.; Grazioli, C.; Sanyal, B.; Eriksson, O.; Brena, B.; Puglia, C. The electronic characterization of biphenylene-Experimental and theoretical insights from core and valence level spectroscopy. *J. Chem. Phys.* **2015**, *142*, 074305. [CrossRef] [PubMed]
15. Carpenter, E.; Davis, H. Acrylamide. Its preparation and properties. *J. Appl. Chem.* **1957**, *7*, 671–676. [CrossRef]
16. Mårtensson, N.; Baltzer, P.; Brühwiler, P.; Forsell, J.O.; Nilsson, A.; Stenborg, A.; Wannberg, B. A very high resolution electron spectrometer. *J. Electron Spectrosc. Relat. Phenom.* **1994**, *70*, 117–128. [CrossRef]
17. Li, X.; Frisch, M.J. Energy-Represented Direct Inversion in the Iterative Subspace within a Hybrid Geometry Optimization Method. *J. Chem. Theory Comput.* **2006**, *2*, 835–839. [CrossRef]
18. Runge, E.; Gross, E.K.U. Density-Functional Theory for Time-Dependent Systems. *Phys. Rev. Lett.* **1984**, *52*, 997–1000. [CrossRef]
19. Ehara, M.; Hasegawa, J.; Nakatsuji, H. SAC-CI method applied to molecular spectroscopy. In *Theory and Applications of Computational Chemistry. The First Forty Years.*; Elsevier: Amsterdam, The Netherlands, 2005; pp. 1099–1141. [CrossRef]
20. Linderberg, J.; Öhrn, Y. *Propagators in Quantum Chemistry*, 2nd ed.; John Wiley and Sons, Inc.: Hoboken, NJ, USA, 2004.
21. Danovich, D. Green's function methods for calculating ionization potentials, electron affinities, and excitation energies. *Wiley Interdiscip. Rev. Comput. Mol. Sci.* **2011**, *1*, 377–387. [CrossRef]
22. Cederbaum, L.S. One-body Green's function for atoms and molecules: theory and application. *J. Phys. B Atom. Mol. Phys.* **1975**, *8*, 290–303. [CrossRef]
23. von Niessen, W.; Schirmer, J.; Cederbaum, L. Computational methods for the one-particle green's function. *Comput. Phys. Rep.* **1984**, *1*, 57–125. [CrossRef]
24. Zakrzewski, V.; Ortiz, J.; Nichols, J.; Heryadi, D.; Yeager, D.; Golab, J. Comparison of perturbative and multiconfigurational electron propagator methods. *Int. J. Quantum Chem.* **1996**, *60*, 29–36. [CrossRef]
25. Ortiz, J.V. Partial third-order quasiparticle theory: Comparisons for closed-shell ionization energies and an application to the Borazine photoelectron spectrum. *J. Chem. Phys.* **1996**, *104*, 7599–7605. [CrossRef]
26. Negri, F.; Zgierski, M. Franck–Condon analysis of the S0->T1 absorption and phosphorescence spectra of biphenyl and bridged derivatives. *J. Chem. Phys.* **1992**, *97*, 7124–7136. [CrossRef]
27. Malagoli, M.; Coropceanu, V.; da Silva Filho, D.; Brédas, J. A multimode analysis of the gas-phase photoelectron spectra in oligoacenes. *J. Chem. Phys.* **2004**, *120*, 7490–7496. [CrossRef]
28. Duschinsky, F. The importance of the electron spectrum in multi atomic molecules. Concerning the Franck–Condon principle. *Acta Physicochim. URSS* **1937**, *7*, 551–566.
29. Santoro, F.; Lami, A.; Improta, R.; Bloino, J.; Barone, V. Effective method for the computation of optical spectra of large molecules at finite temperature including the Duschinsky and Herzberg–Teller effect: The Qx band of porphyrin as a case study. *J. Chem. Phys.* **2008**, *128*, 224311. [CrossRef]
30. Dierksen, M.; Grimme, S. An efficient approach for the calculation of Franck–Condon integrals of large molecules. *J. Chem. Phys.* **2005**, *122*, 244101. [CrossRef] [PubMed]
31. Barone, V.; Bloino, J.; Biczysko, M.; Santoro, F. Fully Integrated Approach to Compute Vibrationally Resolved Optical Spectra: From Small Molecules to Macrosystems. *J. Chem. Theory Comput.* **2009**, *5*, 540–554. [CrossRef]
32. Udovenko, A.; Kolzunova, L. Crystal structure of acrylamide. *J. Struct. Chem.* **2008**, *49*, 961–964. [CrossRef]
33. Velchev, I.; Hogervorst, W.; Ubachs, W. Precision VUV Spectroscopy of Ar I at 105 nm. *J. Phys. B* **1999**, *32*, L511–L516. [CrossRef]
34. Koopmans, T. Über die Zuordnung von Wellenfunktionen und Eigenwerten zu den einzelnen Elektronen eines Atoms. *Physica* **1934**, *1*, 104–113. [CrossRef]
35. Siegbarn, H.; Asplund, L.; Kelfve, P.; Hamrin, K.; Karlsson, L.; Siegbahn, K. ESCA applied to liquids. II. Valence and core electron spectra of formamide. *J. Electron Spectrosc. Relat. Phenom.* **1974**, *5*, 1059–1079. [CrossRef]
36. Bodi, A.; Hemberger, P. Low-Energy Photoelectron Spectrum and Dissociative Photoionization of the Smallest Amides: Formamide and Acetamide. *J. Phys. Chem. A* **2019**, *123*, 272–283. [CrossRef] [PubMed]

37. Thomas, T.D.; Shaw, R.W. Accurate core ionization potentials and photoelectron kinetic energies for light elements. *J. Electron Spectrosc. Relat. Phenom.* **1974**, *5*, 1081–1094. [CrossRef]
38. Jolly, W.; Bomben, K.; Eyermann, C. Core-electron binding energies for gaseous atoms and molecules. *At. Data Nucl. Data* **1984**, *31*, 433–493. [CrossRef]
39. Li, C.; Salén, P.; Yatsyna, V.; Schio, L.; Feifel, R.; Squibb, R.; Kamińska, M.; Larsson, M.; Richter, R.; Alagia, M.; et al. Experimental and theoretical XPS and NEXAFS studies of N-methylacetamide and N-methyltrifluoroacetamide. *Phys. Chem. Chem. Phys.* **2016**, *18*, 2210–2218. [CrossRef] [PubMed]
40. Kovač, B.; Ljubić, I.; Kivimäki, A.; Coreno, M.; Novak, I. Characterisation of the electronic structure of some stable nitroxyl radicals using variable energy photoelectron spectroscopy. *Phys. Chem. Chem. Phys.* **2014**, *16*, 10734–10742. [CrossRef] [PubMed]
41. Melandri, S.; Evangelisti, L.; Canola, S.; Sa'adeh, H.; Calabrese, C.; Coreno, M.; Grazioli, C.; Prince, K.; Negri, F.; Maris, A. Chlorination and tautomerism: A computational and UPS/XPS study of 2-hydroxypyridine-2-pyridone equilibrium. *Phys. Chem. Chem. Phys.* **2020**, *22*, 13440–13455. [CrossRef]
42. Tronc, M.; King, G.; Read, F. Carbon K-shell excitation in small molecules by high-resolution electron impact. *J. Phys. B At. Mol. Phys.* **1979**, *12*, 137–157. [CrossRef]
43. Sodhi, R.; Brion, C. Reference energies for inner shell electron energy-loss spectroscopy. *J. Electron Spectrosc. Relat. Phenom.* **1984**, *34*, 363–372. [CrossRef]
44. Wight, G.; Brion, C. K-Shell energy loss spectra of 2.5 keV electrons in $CO_2$ and $N_2O$. *J. Electron Spectrosc. Relat. Phenom.* **1974**, *3*, 191–205. [CrossRef]
45. Ishii, I.; Hitchcock, A. A quantitative experimental study of the core excited electronic states of formamide, formic acid, and formyl fluoride. *J. Chem. Phys.* **1987**, *87*, 830–839. [CrossRef]
46. Duflot, D.; Flament, J.P.; Walker, I.C.; Heinesch, J.; Hubin-Franskin, M.J. Core shell excitation of 2-propenal (acrolein) at the O 1s and C 1s edges: An experimental and ab initio study. *J. Chem. Phys.* **2003**, *118*, 1137–1145. [CrossRef]
47. Ljubić, I.; Kivimäki, A.; Coreno, M. An experimental NEXAFS and computational TDDFT and ΔDFT study of the gas-phase core excitation spectra of nitroxide free radical TEMPO and its analogues. *Phys. Chem. Chem. Phys.* **2016**, *18*, 10207–10217. [CrossRef]
48. Totani, R.; Ljubić, I.; Ciavardini, A.; Grazioli, C.; Galdenzi, F.; de Simone, M.; Coreno, M. Frontier orbital stability of nitroxyl organic radicals probed by means of inner shell resonantly enhanced valence band photoelectron spectroscopy. *Phys. Chem. Chem. Phys.* **2022**, *24*, 1993–2003. [CrossRef] [PubMed]

Article

# Decoding Breast Cancer Metabolism: Hunting BRCA Mutations by Raman Spectroscopy

Monika Kopec [1], Beata Romanowska-Pietrasiak [2] and Halina Abramczyk [1,*]

[1] Laboratory of Laser Molecular Spectroscopy, Institute of Applied Radiation Chemistry, Lodz University of Technology, Wroblewskiego 15, 93-590 Lodz, Poland

[2] Oncological Surgery Department, Medical Genetics Department, Copernicus Provincial Multidisciplinary Centre of Oncology and Traumatology in Lodz, Pabianicka 62, 93-513 Lodz, Poland

* Correspondence: halina.abramczyk@p.lodz.pl

**Abstract:** Presented study included human blood from healthy people and patients with BReast CAncer gene (BRCA) mutation. We used Raman spectroscopy for BRCA mutation detection and the bioanalytical characterization of pathologically changed samples. The aim of this study is to evaluate the Raman biomarkers to distinguish blood samples from healthy people and patients with BRCA mutation. We demonstrated that Raman spectroscopy is a powerful technique to distinguish between healthy blood and blood with BRCA mutation and to characterize the biochemical composition of samples. We applied partial least squares discriminant analysis (PLS-DA) to discriminate BRCA1/2 mutations and control samples without the mutations based on vibrational features. The sensitivity and specificity for calibration obtained directly from PLS-DA are equal to 94.2% and 97.6% and for cross-validation are equal to 93.3% and 97%. Our combination (Raman spectroscopy and PLS-DA) provides quick methods to reliably visualize the biochemical differences in human blood plasma. We proved that Raman spectroscopy combined with the chemometric method is a promising tool for hunting BRCA mutation in breast cancer.

**Keywords:** Raman spectroscopy; human blood plasma; BRCA mutation; PLS-DA

**Citation:** Kopec, M.; Romanowska-Pietrasiak, B.; Abramczyk, H. Decoding Breast Cancer Metabolism: Hunting BRCA Mutations by Raman Spectroscopy. *Photochem* **2022**, *2*, 752–764. https://doi.org/10.3390/photochem2030048

Academic Editors: Gulce Ogruc Ildiz and Licinia L.G. Justino

Received: 4 July 2022
Accepted: 23 August 2022
Published: 25 August 2022

**Publisher's Note:** MDPI stays neutral with regard to jurisdictional claims in published maps and institutional affiliations.

**Copyright:** © 2022 by the authors. Licensee MDPI, Basel, Switzerland. This article is an open access article distributed under the terms and conditions of the Creative Commons Attribution (CC BY) license (https://creativecommons.org/licenses/by/4.0/).

## 1. Introduction

The blood is an important fluid that circulates through the body and has a lot of functions that are fundamental for survival [1]. The blood transports oxygen to cells and tissues and delivers substances necessary for living cells, such as sugars or hormones, as well as removing waste products from cells [2]. The blood also regulates the right temperature and concentration for each component of the blood [3]. The important function of the blood is to protect the body from diseases and infections [2–4]. From monitoring specific markers in the blood, we can obtain essential molecular information, which reveals the real health status based on metabolism [5]. The blood contains biomarkers that characterize cancer pathology processes and altered metabolism and reprogramming [6–10]. Blood marker analysis including BRCA mutations may help to identify the early stages of pathological processes that can lead to the development of cancer [11]. Breast cancer is one of the most common cancer in the population and one of the most common cancer deaths [12]. One of the hallmarks of a higher risk of breast cancer is related to mutations in breast genes (BRCA). Normal BRCA genes produce proteins that repair DNA [13,14]. People with BRCA1/2 mutations have a higher level of risk of cancer disease, particularly breast cancer [15,16]. There are reports demonstrating that tumor grade correlates with the level of BRCA proteins [17].

The protein expression induced by BRCA mutations in breast cancer has been studied by numerous groups [6,18–25]. E. Gross et al. used chromatography to study BRCA mutations [26].

Recently, next-generation sequencing (NGS) technologies have been widely developed for BRCA mutations analysis [27,28]. E. Szczerba et al. studied BRCA mutations in tumor tissues by using bioinformatic NGS software programs to analyze BRCA. NGS procedures can improve clinical diagnosis in the near future [21,29–32].

The molecular diagnosis of genetic disorders is defined as searching for and revealing defects in deoxyribonucleic acid (DNA) and/or ribonucleic acid (RNA) samples. The methods for identification of the disease-causing mutations can be classified as methods for the detection of known and unknown mutations [33–35].

Peripheral blood is treated with ethylenediamine tetraacetic acid (EDTA) or sodium citrate to prevent clotting and the human genomic DNA can be isolated and purified [36,37].

In families carrying a pathogenic variant in BRCA1 and BRCA2 genes, usually, the gold-standard Sanger sequencing method is used to detect them. This method uses the selective incorporation of chain-terminating dideoxynucleotides by DNA polymerase during in vitro DNA replication [38,39].

The main technique dedicated to the detection of known mutations in BRCA1 and BRCA2 is the polymerase chain reaction (PCR) and its various options, for example, real-time PCR with high sensitivity and specificity [40–42].

In cases of large genomic rearrangements, such as the exons deletions or amplifications in hereditary breast cancer, another molecular diagnostic technique of multiplex ligation-dependent probe amplification (MLPA) is used for large genomic rearrangements [43–46].

The BRCA gene test is a standard method used for people who inherited this mutation based on family history. However, it is justified to find new methods for the detection of cancer mutations that will provide deeper insight into biochemical alterations. This may be provided by Raman spectroscopy and Raman imaging. By establishing changes in Raman spectroscopic profiles in the human blood, it will be able to determine molecular differences, leading to establishing biochemical alterations during cancer progression in humans. Moreover, Raman spectroscopy allows the label-free molecular sensing of a biological sample [47–51].

To the best of our knowledge, only a few papers report on Raman spectroscopy to study BRCA mutation. L.R. Allain et al. reported the use of Surface Enhanced Raman Spectroscopy (SERS) to monitor the BRCA1 breast cancer mutations on modified silver surfaces [52]. They proved that the SERS technique for gene expression studies is a significant step forward in enabling biological quantification [52]. Another group that used SERS to analyze BRCA mutation was that of M. Culha et al. [53]. The most important achievement of Pal's group was the preparation of a protocol for binding cresyl fast violet (CFV), a SERS-active dye (label) containing an aromatic amino group with a modified oligomer that has a carboxy-derivatized thymidine moiety using carbodiimide coupling [54]. The potential of using Raman spectroscopy for BRCA analyses was investigated also by Vo-Dinh et al. [55] and Coluccio et al. [56].

Raman imaging and Raman spectroscopy are tools that provide information about the biochemical composition of organelles in single cells and identify cancer biomarkers that can discriminate between the normal state and cancer pathology.

There are a vast popular machine learning tools that are used as feature selectors and classifiers. Among these, partial least squares discriminant analysis is the most common method used to identify vibrational biomarkers. PLS-DA is an algorithm for the predictive and descriptive modeling and discriminative variable selection.

PLS-DA analysis allows for the calculation of sensitivity and specificity. Sensitivity is described by the equation:

$$\text{sensitivity} = \frac{\text{TP}}{(\text{TP} + \text{FN})}$$

Specificity is described by the equation:

$$\text{specificity} = \frac{\text{TN}}{(\text{TN} + \text{FP})}$$

where TP is the true positive result, FN is the false negative result, TN is the true negative result, and FP is the false positive result.

The PLS-DA method also provides information about the Receiver Operating Characteristic (ROC). ROC is a measure of a classifier's predictive quality that compares and visualizes the tradeoff between the model's sensitivity and specificity [57,58].

The goal of this paper is to demonstrate that Raman spectroscopy combined with statistical methods is a quick and reliable tool for monitoring BRCA mutations. In this paper, we present that Raman spectroscopy results combined with statistical methods of artificial intelligence make it possible to discriminate between healthy people without BRCA mutations and patients with BRCA mutations by testing human blood plasma. In particular, we focus on the analysis of 104 Raman spectra of blood plasma from 15 patients without BRCA mutations (control) and 460 Raman spectra from 85 patients with BRCA mutations by using chemometric methods to differentiate between the samples. We demonstrate that Raman spectroscopy provides a new promise to provide a perfect tool for BRCA mutation diagnosis.

## 2. Materials and Methods

### 2.1. Sample Preparation

Blood samples were obtained from the Voivodeship Multi-Specialist Center for Oncology and Traumatology in Lodz. The spectroscopic analysis did not affect the scope of the course and type of hospital treatment undertaken. A written informed agreement was obtained from all patients. The Bioethical Committee at the Medical University of Lodz, Poland (RNN/17/20/KE) approved the measurement protocols.

Patients were diagnosed with or without BRCA mutations and treated at the Voivodeship Multi-Specialist Center for Oncology and Traumatology in Lodz. Real-time PCR kit for detection of 8 mutations, BRCA1 (185delAG, 4153delA, 5382insC, 3819delGTAAA, 3875delGTCT, T181G (Cys61Gly), 2080delA) and BRCA2 (6174delT) (Sacace Biotechnologies), was used to detect BRCA1 and BRCA2.

We examined 2 groups of patients participating in the experiment: (1) healthy people, i.e., without the BRCA 1/2 mutations and without oncological disease, and (2) patients with (a) mutation of the BRCA 1/2 genes who did not suffer from oncological diseases and (b) cancer patients with a mutation of the BRCA 1/2 patients after recovery for whom the treatment process was completed (with at least 5 years of survival free from the appearance of a new cancer pathology). The ratio of patients in groups 2a to 2b was estimated as 40%:60%. The demographics and clinical trials for cohort A were: Females aged $\geq 18$ (documented carrier of a germinale pathogenic mutation in BRCA1 or BRCA2 genes). Females aged $\geq 18$ (with a diagnosis of invasive breast cancer or/and invasive ovarian cancer or/and pancreatic cancer and documented carrier of a pathogenic mutation in BRCA1 or BRCA2 genes). The demographics and clinical trials for cohort B (control group) were: Females aged $\geq 18$ without any prior cancer diagnosis and without germinale pathogenic mutation in BRCA1 or BRCA2 genes (demonstrated normal genetic testing results). Exclusion criteria for (Cohort A and Cohort B) were: pregnant or breast-feeding subjects, major surgery within 4 weeks prior to the study entry, history of clinically significant disease or other immunosuppressive disease, HIV infection, receipt of any blood product within 2 weeks prior to the study entry, any illness or condition that in the opinion of the investigator could affect the safety of the subject or the evaluation of any study endpoint, and active drug use or dependence that, in the opinion of the investigator, could interfere with adherence to study requirements.

For all experiments, we used human fresh blood. Blood samples were collected in ethylenediamine tetraacetic acid (EDTA) vials and subsequently centrifuged at 3500 rpm for 5 min at 18 °C to obtain the plasma samples. A 10 µL drop of each blood plasma was placed on clean $CaF_2$ windows (Crystran). Raman spectra were recorded in the different localizations around the center of the dry drop.

### 2.2. Raman Spectroscopy

The Raman measurements were recorded using an alpha 300 RSA+ (WITec, Ulm, Germany) combined with a confocal microscope coupled via the fiber of 50 µm core diameter with a spectrometer UHTS (Ultra High Throughput Spectrometer) (WITec, Ulm, Germany) and a CCD Camera (Andor Newton DU970N-UVB-353) (Andor Technology, Belfast, Northern Ireland) operating in standard mode with 1600 × 200 pixels at −60 °C with full vertical binning. A standard calibration procedure was performed every day before measurement with the use of a silicon plate (520.7 cm$^{-1}$). Raman data were pre-processed using the WITec Project 4.1 Plus Software. Each Raman spectra were processed to remove cosmic rays (model: filter size: 2, dynamic factor: 10). The corrected Raman spectra were smoothed by a Savitzky and Golay procedure (model: order: 4; derivative: 0) and baseline subtraction (Raman peaks from the studied spectral range were fitted to second-degree polynomial). The Raman measurements were performed by using excitation of a laser beam (SHG of the Nd:YAG laser (532 nm)) and 40× dry objective (Nikon, objective type CFI Plan Fluor CELWD DIC-M, numerical aperture (NA) of 0.60 and a 3.6–2.8 mm working distance). Raman spectra for plasma blood samples were recorded using integration time 1s and 10 numbers of accumulations. All experiments were performed using a laser with a power 10 mW at the sample position.

The baseline subtraction (baseline mode: user-defined and baseline anchor points: 2nd derivative (zeroes)) and the normalization (model: divided by norm) were performed by using Origin software. The normalization was performed by dividing each spectrum by the spectrum norm, according to:

$$V' = \frac{V}{\|V\|}$$
$$\|V\| = \sqrt{v_1^2 + v_2^2 + \ldots v_n^2}$$
(1)

where:

$v_n$ is the nth V value [59].

The normalization procedure was performed for the spectral region of 400–1700 cm$^{-1}$.

*2.3. Statistical Analysis*

We analyzed samples with BRCA mutation from 85 patients and samples from control patients without BRCA mutations from 15 patients. A total of 460 Raman single spectra for BRCA mutations samples and 104 Raman single spectra for control patients were analyzed.

Statistical chemometric analysis was performed by using MATLAB and PLS_Toolbox Version 4.0 (Eigenvector Research, Wenatchee, WA, USA). Partial Least Squares Discriminant Analysis (PLS-DA) was used for building predictive classification models to validate the classification models and to calculate sensitivity and specificity. The PLS-DA used for the classification and cross-validation as well as permutation testing was used to validate the classification models. We used cross-validation: Venetian blinds w/10 splits, which means that each test set was performed by selecting every 10th object in the data set, starting at the data numbered 1. A ROC curve analysis was also performed. The PLS-DA analysis was performed using an imbalanced group approach (85 samples from patients with BRCA mutation and 15 samples from control patients). More details about chemometric methods were described in our previous papers [57].

## 3. Results and Discussion

In this section, we present the results for human blood plasma from patients without BRCA mutations (control) and patients with BRCA mutations. Figure 1 presents average normalized Raman spectra of human plasma with standard deviation (dark shadows) based on 104 single spectra (control, blue line) and 460 spectra of patients with BRCA mutations (red line). Additionally, the difference Raman spectrum (human blood plasma with BRCA mutation–control, marked with as a green line) is presented in Figure 1.

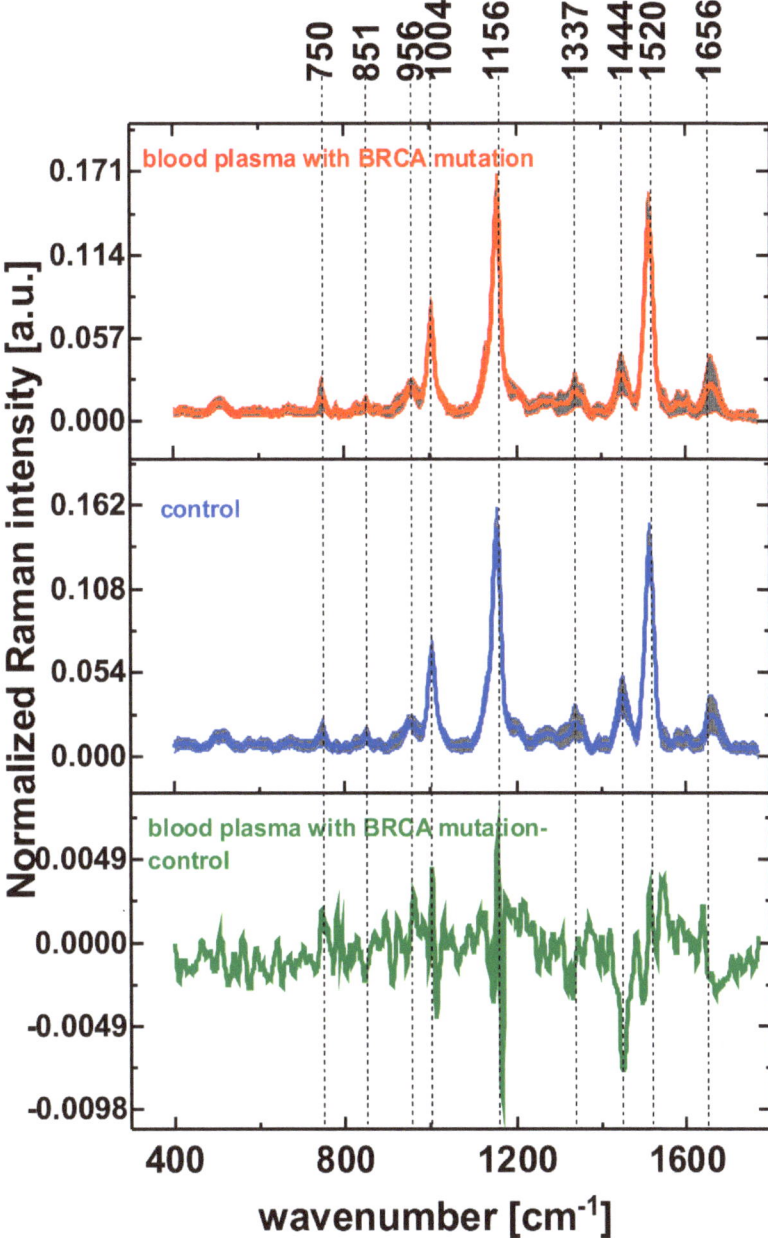

**Figure 1.** The normalized average Raman spectra typical for plasma from healthy people (control, blue line) with SD (blue shadow), plasma from patients with BRCA mutation (red line) with SD (red shadow) and the difference spectrum (blood plasma with BRCA, control) (green line). One can see from Figure 1 that the Raman spectra of the human blood plasma can be characterized by vibrational peaks at 750, 851, 956, 1004, 1156, 1337, 1444, 1520 and 1656 cm$^{-1}$. The tentative assignments of Raman peaks observed in human blood plasma are presented in Table 1.

Table 1. Tentative assignments of Raman peaks [60,61].

| Wavenumber [cm$^{-1}$] | Tentative Assignments |
| --- | --- |
| 750 | tryptophan, cytochrome c, hemoglobine |
| 851 | tyrosine |
| 956 | hydroxyproline/collagen backbone |
| 1004 | phenylalanine |
| 1156 | carotenoids |
| 1337 | cytochrome b |
| 1444 | fatty acids, triglycerides |
| 1520 | carotenoids |
| 1656 | amide I, lipids |

One can see from the difference spectrum at Figure 1 that the peak at 1004 cm$^{-1}$ assigned to phenylalanine and 750 cm$^{-1}$ corresponding to cytochrome c/hemoglobine are positive, which confirms the higher amount of cytochrome c in blood plasma with BRCA mutation. The positive difference is also observed at the Raman signal at 1656 cm$^{-1}$. This peak is assigned to amid I/lipids, which is stronger in the Raman spectra for blood plasma with BRCA mutation (positive correlation in difference spectrum).

Comparing the Raman signals at 1444 cm$^{-1}$ assigned to the C-H bending vibrations of fatty acids/triglycerides $CH_2$ or $CH_3$, one can see that this band is stronger for healthy blood plasma (negative correlation in difference spectrum). The same tendency is observed for the Raman signals at 1156 cm$^{-1}$ and 1520 cm$^{-1}$ assigned to carotenoids. This result supports our previous findings from our previous papers for cancer pathology [57,62].

To visualize chemical similarities and differences between human blood plasma for patients without BRCA mutations and blood plasma samples for patients with BRCA mutation, we used multivariate statistical methods for data interpretation. We performed statistical analysis for 564 single Raman spectra (460 Raman spectra of 85 human blood samples with BRCA mutation and 104 Raman spectra of 15 human blood samples without BRCA). A large number of samples and multidimensional Raman vectors (intensities vs. wavenumbers) were analyzed with dimension reduction by means of partial least squares discriminant analysis (PLS-DA).

Figure 2 shows the PLS-DA score plot for the Raman spectra of the human blood plasma samples with BRCA 1/2 mutations and the samples without BRCA mutations. The latent variable (LV) is the percent variance capture by model. The 95% confidence interval is presented by the PLS-DA ellipse (blue dashed line).

One can see from Figure 2 the evident separation between human blood plasma with BRCA and human blood plasma without BRCA mutations. The differences are clearly visible from grouping the results into two separate clusters: blue color samples without BRCA mutations and red for samples with the BRCA mutations. Raman spectra for human blood plasma without BRCA mutations (blue circle) are grouped in the left upper area of the plot, while the samples from human blood plasma with BRCA mutations (red circle) are grouped in the right upper area and in the lower area of the plot.

To extract the molecular information contained in the LV1, LV2, LV3 and LV4, we used the loading plots presented in Figure 3 that reveal the most important characteristic features in the Raman spectra.

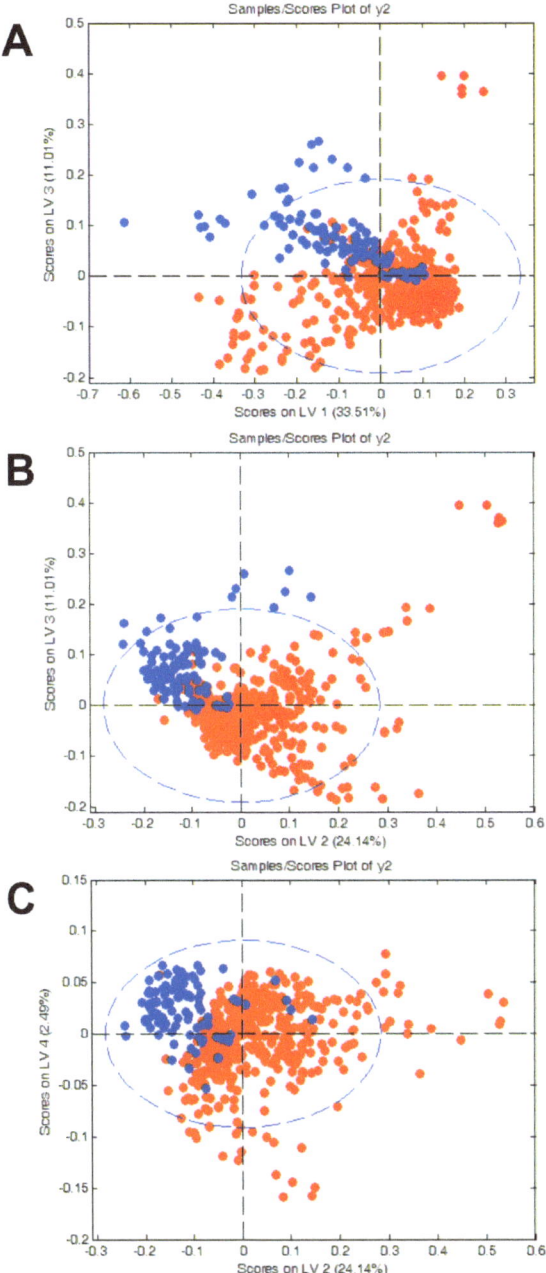

**Figure 2.** The scores plots (model: mean center) LV3 vs. LV1 (panel **A**), LV3 vs. LV2 (panel **B**) and LV4 vs. LV2 (panel **C**) obtained from the PLS–DA for the Raman spectra of blood plasma without BRCA mutations (blue circle) and for Raman spectra of blood plasma with BRCA mutations (red circle) and the 95% confidence interval in panel (**A**–**C**) (blue dashed line).

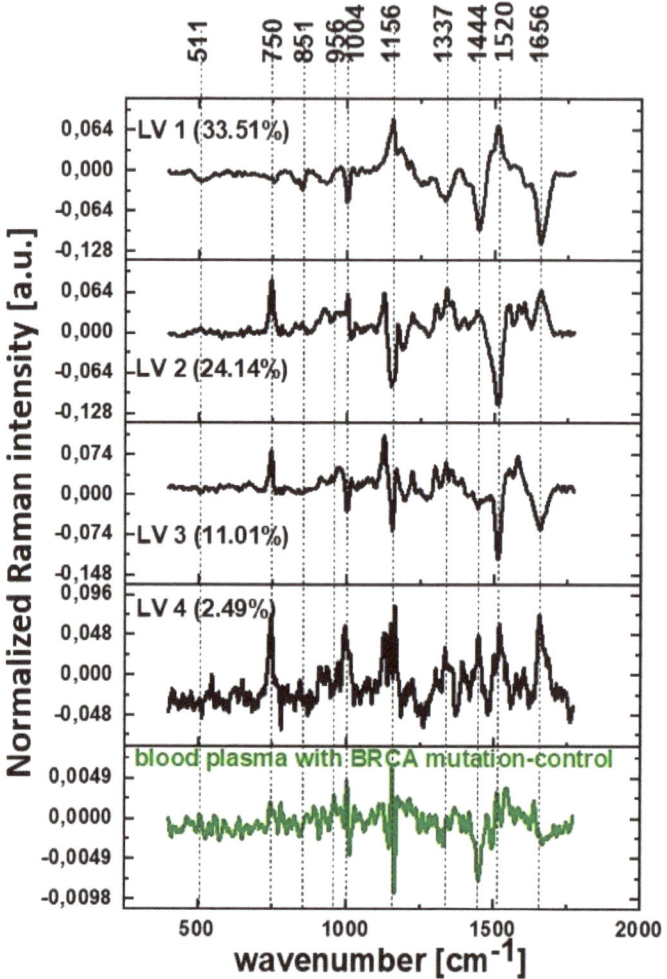

**Figure 3.** PLS–DA loading plot for LV1, LV2, LV3 and LV4 for the Raman spectra of the human blood plasma without BRCA mutations and Raman spectra of the human blood plasma with BRCA mutation and the difference spectrum (blood plasma with BRCA, control).

The loading plots of LV1, LV2, LV3 and LV4 versus wavenumbers obtained from PLS-DA methods for two classes of spectra typical for human blood plasma without BRCA mutations and for human blood plasma with BRCA mutation and comparison with the difference spectrum (blood plasma with BRCA–control) are shown in Figure 3. One can see that the loading plots show the changes around the characteristic Raman peaks of carotenoids, lipids and proteins.

One can see from Figure 3 that the first LV1 has a contribution of 33.51%, LV2 a contribution of 24.14%, LV3 a contribution of 11.01% to variance and LV4 a contribution of 2.49% to variance. LV1, LV2 and LV3 provide the dominant account for the maximum variance in the data.

The most characteristic minima in the loading plot of LV1 are at 511 cm$^{-1}$, 750 cm$^{-1}$, 851 cm$^{-1}$, 1004 cm$^{-1}$, 1337 cm$^{-1}$, 1444 cm$^{-1}$ and 1656 cm$^{-1}$ and the most characteristic maxima in the loading plot are at 1156 cm$^{-1}$ and 1520 cm$^{-1}$. The LV2 reaches its minima at 1156 cm$^{-1}$ and 1520 cm$^{-1}$ and maxima at 750 cm$^{-1}$, 1004 cm$^{-1}$, 1337 cm$^{-1}$ and 1656 cm$^{-1}$.

The LV3 reaches its minima at 1004 cm$^{-1}$, 1156 cm$^{-1}$, 1520 cm$^{-1}$ and 1656 cm$^{-1}$ and maximum at 750 cm$^{-1}$. The LV4 reaches its maxima at 750 cm$^{-1}$, 1004 cm$^{-1}$, 1156 cm$^{-1}$, 1337 cm$^{-1}$, 1444 cm$^{-1}$, 1520 cm$^{-1}$ and 1656 cm$^{-1}$.

Table 2 presents the results of the sensitivity and specificity obtained from the PLS-DA method.

**Table 2.** The values of sensitivity and specificity for the calibration and cross-validation procedure from PLS-DA analysis.

|  | Healthy Blood Plasma | Blood Plasma with BRCA Mutation |
|---|---|---|
| Sensitivity (calibration) | 0.942 | 0.976 |
| Specificity (calibration) | 0.976 | 0.942 |
| Sensitivity (cross validation) | 0.933 | 0.970 |
| Specificity (cross validation) | 0.970 | 0.933 |

In 2018, a survey of laboratories offering BRCA1/BRCA2 sequence analysis was reported, which revealed a large number of differences in technology, such as gene coverage, sensitivities, cost, single-site analyses and a variant of uncertain significance. The reported sensitivity was 97% and the median was 99.5% with a range from 85.2 to 100% [63].

We compared the sensitivity and specificity obtained by Raman method with other methods, such as immunohistochemistry, for which sensitivity to predict for BRCA1 mutation carriers was 80% and specificity was 100%, with a positive predictive value of 100% and a negative predictive value of 93% [64].

ROC curves (Receiver Operating Characteristic) for blood plasma samples without BRCA mutations (control) and for blood plasma samples with BRCA mutations are presented in Figure 4. Figure 4 confirms the high potential of Raman spectroscopy to differentiate between patients without BRCA mutations and with BRCA mutations.

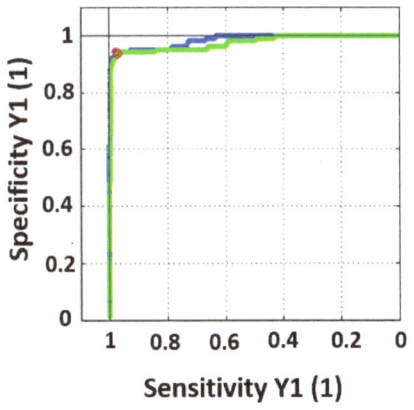
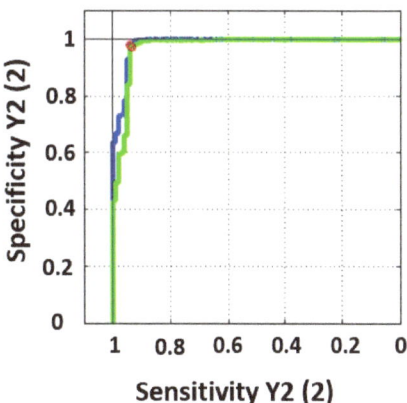

**Figure 4.** ROC curves for two classes of Raman spectra assigned to blood plasma without BRCA mutations (control) and blood plasma with BRCA mutation. The red dot means the value that maximized both the sensitivity and specificity values of the target class.

## 4. Conclusions

In this paper, we used Raman spectroscopy to monitor BRCA mutations in human blood plasma. Raman spectroscopy and PLS-DA are useful tools for the diagnosis of breast BRCA mutations. Raman spectroscopy and chemometric method were successfully applied to characterize and differentiate between human blood plasma without BRCA mutations and human blood plasma with BRCA mutation. The sensitivity and specificity for calibration obtained directly from PLS-DA are equal to 94.2% and 97.6%, respectively, and for cross-validation are equal to 93.3% and 97%, respectively, which were higher than the values of the conventional methods of sequence analysis in molecular biology used to date.

The results presented in this paper demonstrate that Raman biomarkers provide additional insight into the biology of human blood plasma. The differentiation by Raman spectroscopy between blood plasma without BRCA mutations and blood plasma with BRCA mutations is important because the high specificity and sensitivity can lead to better genomic breast diagnosis. Our results can help to implement Raman spectroscopy as a tool for blood analysis to investigate BRCA mutations.

**Author Contributions:** Conceptualization: H.A. and M.K.; Funding acquisition: H.A., Investigation: M.K. and B.R.-P., Methodology: H.A., M.K. and B.R.-P.; Writing—original draft: M.K., H.A. and B.R.-P.; Manuscript editing: M.K., H.A. and B.R.-P. All authors have read and agreed to the published version of the manuscript.

**Funding:** This work was supported by the National Science Centre of Poland (Narodowe Centrum Nauki, UMO-2019/33/B/ST4/01961).

**Institutional Review Board Statement:** The study was conducted according to the guidelines of the Declaration of Helsinki and approved by the Local Bioethical Committee at the Medical University of Lodz, Poland (RNN/17/20/KE).

**Informed Consent Statement:** Informed consent was obtained from all subjects involved in the study.

**Data Availability Statement:** The raw data underlying the results presented in the study are available from Lodz University of Technology Institutional Data Access for researchers who meet the criteria for access to confidential data. The data contain potentially sensitive information. Request for access to those data should be addressed to the Head of Laboratory of Laser Molecular Spectroscopy, Institute of Applied Radiation Chemistry, Lodz University of Technology. Data requests might be sent by email to the secretary of the Institute of Applied Radiation Chemistry: mitr@mitr.p.lodz.pl.

**Conflicts of Interest:** The authors declare no conflict of interest.

## References

1. Basu, D.; Kulkarni, R. Overview of blood components and their preparation. *Indian J. Anaesth.* **2014**, *58*, 529–537. [CrossRef] [PubMed]
2. Weiss, C.; Jelkmann, W. Functions of the Blood. In *Human Physiology*; Schmidt, R.F., Thews, G., Eds.; Springer: Berlin/Heidelberg, Germany, 1989; pp. 402–438. [CrossRef]
3. Kuhn, V.; Diederich, L.; Keller, T.C.S.; Kramer, C.M.; Lückstädt, W.; Panknin, C.; Suvorava, T.; Isakson, B.E.; Kelm, M.; Cortese-Krott, M.M. Red Blood Cell Function and Dysfunction: Redox Regulation, Nitric Oxide Metabolism, Anemia. *Antioxid. Redox Signal.* **2017**, *26*, 718–742. [CrossRef] [PubMed]
4. Seyoum, M.; Enawgaw, B.; Melku, M. Human blood platelets and viruses: Defense mechanism and role in the removal of viral pathogens. *Thromb. J.* **2018**, *16*, 16. [CrossRef] [PubMed]
5. Kirkman, M.S.; Mahmud, H.; Korytkowski, M.T. Intensive Blood Glucose Control and Vascular Outcomes in Patients with Type 2 Diabetes Mellitus. *Endocrinol. Metab. Clin. North Am.* **2018**, *47*, 81–96. [CrossRef] [PubMed]
6. Feng, Q.; Yu, M.; Kiviat, N.B. Molecular Biomarkers for Cancer Detection in Blood and Bodily Fluids. *Crit. Rev. Clin. Lab. Sci.* **2006**, *43*, 497–560. [CrossRef]
7. Mitchell, P.S.; Parkin, R.K.; Kroh, E.M.; Fritz, B.R.; Wyman, S.K.; Pogosova-Agadjanyan, E.L.; Peterson, A.; Noteboom, J.; O'Briant, K.C.; Allen, A.; et al. Circulating microRNAs as stable blood-based markers for cancer detection. *Proc. Natl. Acad. Sci. USA* **2008**, *105*, 10513–10518. [CrossRef]
8. El-Khoury, V.; Schritz, A.; Kim, S.-Y.; Lesur, A.; Sertamo, K.; Bernardin, F.; Petritis, K.; Pirrotte, P.; Selinsky, C.; Whiteaker, J.R.; et al. Identification of a Blood-Based Protein Biomarker Panel for Lung Cancer Detection. *Cancers* **2020**, *12*, 1629. [CrossRef]
9. Berghuis, A.M.S.; Koffijberg, H.; Prakash, J.; Terstappen, L.W.M.M.; Ijzerman, M.J. Detecting Blood-Based Biomarkers in Metastatic Breast Cancer: A Systematic Review of Their Current Status and Clinical Utility. *Int. J. Mol. Sci.* **2017**, *18*, 363. [CrossRef]
10. Kazarian, A.; Blyuss, O.; Metodieva, G.; Gentry-Maharaj, A.; Ryan, A.; Kiseleva, E.M.; Prytomanova, O.M.; Jacobs, I.J.; Widschwendter, M.; Menon, U.; et al. Testing breast cancer serum biomarkers for early detection and prognosis in pre-diagnosis samples. *Br. J. Cancer* **2017**, *116*, 501–508. [CrossRef]
11. Hampel, H.; O'Bryant, S.E.; Molinuevo, J.L.; Zetterberg, H.; Masters, C.L.; Lista, S.; Kiddle, S.J.; Batrla, R.; Blennow, K. Blood-based biomarkers for Alzheimer disease: Mapping the road to the clinic. *Nat. Rev. Neurol.* **2018**, *14*, 639–652. [CrossRef]

12. Harbeck, N.; Penault-Llorca, F.; Cortes, J.; Gnant, M.; Houssami, N.; Poortmans, P.; Ruddy, K.; Tsang, J.; Cardoso, F. Breast cancer. *Nat. Rev. Dis. Primers* **2019**, *5*, 66. [CrossRef] [PubMed]
13. Atchley, D.P.; Albarracin, C.T.; Lopez, A.; Valero, V.; Amos, C.I.; Gonzalez-Angulo, A.M.; Hortobagyi, G.N.; Arun, B.K. Clinical and Pathologic Characteristics of Patients with BRCA-Positive and BRCA-Negative Breast Cancer. *J. Clin. Oncol.* **2008**, *26*, 4282–4288. [CrossRef] [PubMed]
14. Force, U.P.S.T.; Owens, D.K.; Davidson, K.W.; Krist, A.H.; Barry, M.J.; Cabana, M.; Caughey, A.B.; Doubeni, C.A.; Epling, J.W.; Kubik, M.; et al. Risk Assessment, Genetic Counseling, and Genetic Testing for BRCA-Related Cancer: US Preventive Services Task Force Recommendation Statement. *JAMA* **2019**, *322*, 652–665. [CrossRef]
15. Narod, S.A. BRCA mutations in the management of breast cancer: The state of the art. *Nat. Rev. Clin. Oncol.* **2010**, *7*, 702–707. [CrossRef]
16. Alsop, K.; Fereday, S.; Meldrum, C.; DeFazio, A.; Emmanuel, C.; George, J.; Dobrovic, A.; Birrer, M.J.; Webb, P.M.; Stewart, C.; et al. BRCA Mutation Frequency and Patterns of Treatment Response in BRCA Mutation–Positive Women with Ovarian Cancer: A Report from the Australian Ovarian Cancer Study Group. *J. Clin. Oncol.* **2012**, *30*, 2654–2663. [CrossRef]
17. Hedau, S.; Batra, M.; Singh, U.R.; Bharti, A.C.; Ray, A.; Das, B.C. Expression of BRCA1 and BRCA2 proteins and their correlation with clinical staging in breast cancer. *J. Cancer Res. Ther.* **2015**, *11*, 158–163. [CrossRef]
18. Rakha, E.A.; El-Sheikh, S.E.; Kandil, M.A.; El-Sayed, M.E.; Green, A.R.; Ellis, I.O. Expression of BRCA1 protein in breast cancer and its prognostic significance. *Hum. Pathol.* **2008**, *39*, 857–865. [CrossRef]
19. Wang, Z.; Zhang, J.; Zhang, Y.; Deng, Q.; Liang, H. Expression and mutations of BRCA in breast cancer and ovarian cancer: Evidence from bioinformatics analyses. *Int. J. Mol. Med.* **2018**, *42*, 3542–3550. [CrossRef]
20. Al-Mulla, F.; Abdulrahman, M.; Varadharaj, G.; Akhter, N.; Anim, J.T. BRCA1 Gene Expression in Breast Cancer: A Correlative Study between Real-time RT-PCR and Immunohistochemistry. *J. Histochem. Cytochem.* **2005**, *53*, 621–629. [CrossRef]
21. Andres, J.L.; Fan, S.; Turkel, G.J.; Wang, J.-A.; Twu, N.-F.; Yuan, R.-Q.; Lamszus, K.; Goldberg, I.D.; Rosen, E.M. Regulation of BRCA1 and BRCA2 expression in human breast cancer cells by DNA-damaging agents. *Oncogene* **1998**, *16*, 2229–2241. [CrossRef]
22. Chen, H.; Wu, J.; Zhang, Z.; Tang, Y.; Li, X.; Liu, S.; Cao, S.; Li, X. Association Between BRCA Status and Triple-Negative Breast Cancer: A Meta-Analysis. *Front. Pharmacol.* **2018**, *9*, 909. [CrossRef] [PubMed]
23. Welcsh, P.L. BRCA1 and BRCA2 and the genetics of breast and ovarian cancer. *Hum. Mol. Genet.* **2001**, *10*, 705–713. [CrossRef] [PubMed]
24. Wagner, T.; Stoppa-Lyonnet, D.; Fleischmann, E.; Muhr, D.; Pagès, S.; Sandberg, T.; Caux, V.; Moeslinger, R.; Langbauer, G.; Borg, A.; et al. Denaturing High-Performance Liquid Chromatography Detects Reliably BRCA1 and BRCA2 Mutations. *Genomics* **1999**, *62*, 369–376. [CrossRef] [PubMed]
25. Roth, J.; Peer, C.J.; Mannargudi, B.; Swaisland, H.; Lee, J.-M.; Kohn, E.C.; Figg, W.D. A Sensitive and Robust Ultra HPLC Assay with Tandem Mass Spectrometric Detection for the Quantitation of the PARP Inhibitor Olaparib (AZD2281) in Human Plasma for Pharmacokinetic Application. *Chromatography* **2014**, *1*, 82–95. [CrossRef]
26. Gross, E.; Arnold, N.; Goette, J.; Schwarz-Boeger, U.; Kiechle, M. A comparison of BRCA1 mutation analysis by direct sequencing, SSCP and DHPLC. *Hum Genet.* **1999**, *105*, 72–78. [CrossRef]
27. D'Argenio, V.; Esposito, M.V.; Telese, A.; Precone, V.; Starnone, F.; Nunziato, M.; Cantiello, P.; Iorio, M.; Evangelista, E.; D'Aiuto, M.; et al. The molecular analysis of BRCA1 and BRCA2: Next-generation sequencing supersedes conventional approaches. *Clin. Chim. Acta* **2015**, *446*, 221–225. [CrossRef]
28. Wallace, A.J. New challenges for BRCA testing: A view from the diagnostic laboratory. *Eur. J. Hum. Genet.* **2016**, *24* (Suppl. S1), S10–S18. [CrossRef]
29. Szczerba, E.; Kamińska, K.; Mierzwa, T.; Misiek, M.; Kowalewski, J.; Lewandowska, M.A. BRCA1/2 Mutation Detection in the Tumor Tissue from Selected Polish Patients with Breast Cancer Using Next Generation Sequencing. *Genes* **2021**, *12*, 519. [CrossRef]
30. Behjati, S.; Tarpey, P.S. What is next generation sequencing? *Arch. Dis. Child.-Educ. Pract. Ed.* **2013**, *98*, 236–238. [CrossRef]
31. Jones, M.A.; Rhodenizer, D.; da Silva, C.; Huff, I.J.; Keong, L.; Bean, L.J.; Coffee, B.; Collins, C.; Tanner, A.K.; He, M.; et al. Molecular diagnostic testing for congenital disorders of glycosylation (CDG): Detection rate for single gene testing and next generation sequencing panel testing. *Mol. Genet. Metab.* **2013**, *110*, 78–85. [CrossRef]
32. Dahui, Q. Next-generation sequencing and its clinical application. *Cancer Biol. Med.* **2019**, *16*, 4–10. [CrossRef] [PubMed]
33. Konstantinos, K.V.; Panagiotis, P.; Antonios, V.T.; Agelos, P.; Argiris, N.V. PCR–SSCP: A Method for the Molecular Analysis of Genetic Diseases. *Mol. Biotechnol.* **2008**, *38*, 155–163. [CrossRef] [PubMed]
34. Evans, D.G.R.; Eccles, D.M.; Rahman, N.; Young, K.; Bulman, M.; Amir, E.; Shenton, A.; Howell, A.; Lalloo, F. A new scoring system for the chances of identifying a BRCA1/2 mutation outperforms existing models including BRCAPRO. *J. Med Genet.* **2004**, *41*, 474–480. [CrossRef]
35. Mahdieh, N.; Rabbani, B. An Overview of Mutation Detection Methods in Genetic Disorders. *Iran. J. Pediatr.* **2013**, *23*, 375–388. [PubMed]
36. Xue, Y.; Ankala, A.; Wilcox, W.R.; Hegde, M.R. Solving the molecular diagnostic testing conundrum for Mendelian disorders in the era of next-generation sequencing: Single-gene, gene panel, or exome/genome sequencing. *Genet. Med.* **2015**, *17*, 444–451. [CrossRef]

37. Shashi, V.; McConkie-Rosell, A.; Rosell, B.; Schoch, K.; Vellore, K.; McDonald, M.; Jiang, Y.-H.; Xie, P.; Need, A.; Goldstein, D.B. The utility of the traditional medical genetics diagnostic evaluation in the context of next-generation sequencing for undiagnosed genetic disorders. *Genet. Med.* **2014**, *16*, 176–182. [CrossRef]
38. Crossley, B.M.; Bai, J.; Glaser, A.; Maes, R.; Porter, E.; Killian, M.L.; Clement, T.; Toohey-Kurth, K. Guidelines for Sanger sequencing and molecular assay monitoring. *J. Veter-Diagn. Investig.* **2020**, *32*, 767–775. [CrossRef]
39. Sikkema-Raddatz, B.; Johansson, L.F.; de Boer, E.N.; Almomani, R.; Boven, L.G.; van den Berg, M.P.; van Spaendonck-Zwarts, K.Y.; van Tintelen, J.P.; Sijmons, R.H.; Jongbloed, J.D.H.; et al. Targeted Next-Generation Sequencing can Replace Sanger Sequencing in Clinical Diagnostics. *Hum. Mutat.* **2013**, *34*, 1035–1042. [CrossRef]
40. Kluska, A.; Balabas, A.; Paziewska, A.; Kulecka, M.; Nowakowska, D.; Mikula, M.; Ostrowski, J. New recurrent BRCA1/2 mutations in Polish patients with familial breast/ovarian cancer detected by next generation sequencing. *BMC Med Genom.* **2015**, *8*, 19. [CrossRef]
41. Tuffaha, H.W.; Mitchell, A.; Ward, R.L.; Connelly, L.; Butler, J.R.; Norris, S.; Scuffham, P.A. Cost-effectiveness analysis of germ-line BRCA testing in women with breast cancer and cascade testing in family members of mutation carriers. *Genet. Med.* **2018**, *20*, 985–994. [CrossRef]
42. Richards, S.; Aziz, N.; Bale, S.; Bick, D.; Das, S.; Gastier-Foster, J.; Grody, W.W.; Hegde, M.; Lyon, E.; Spector, E.; et al. Standards and guidelines for the interpretation of sequence variants: A joint consensus recommendation of the American College of Medical Genetics and Genomics and the Association for Molecular Pathology. *Genet. Med.* **2015**, *17*, 405–424. [CrossRef] [PubMed]
43. Hömig-Hölzel, C.; Savola, S. Multiplex Ligation-dependent Probe Amplification (MLPA) in Tumor Diagnostics and Prognostics. *Diagn. Mol. Pathol.* **2012**, *21*, 189–206. [CrossRef] [PubMed]
44. Bozsik, A.; Pócza, T.; Papp, J.; Vaszkó, T.; Butz, H.; Patócs, A.; Oláh, E. Complex Characterization of Germline Large Genomic Rearrangements of the *BRCA1* and *BRCA2* Genes in High-Risk Breast Cancer Patients—Novel Variants from a Large National Center. *Int. J. Mol. Sci.* **2020**, *21*, 4650. [CrossRef] [PubMed]
45. Engert, S.; Wappenschmidt, B.; Betz, B.; Kast, K.; Kutsche, M.; Hellebrand, H.; Goecke, T.O.; Kiechle, M.; Niederacher, D.; Schmutzler, R.K.; et al. MLPA screening in theBRCA1gene from 1506 German hereditary breast cancer cases: Novel deletions, frequent involvement of exon 17, and occurrence in single early-onset cases. *Hum. Mutat.* **2008**, *29*, 948–958. [CrossRef]
46. Durmaz, A.A.; Karaca, E.; Demkow, U.; Toruner, G.; Schoumans, J.; Cogulu, O. Evolution of Genetic Techniques: Past, Present, and Beyond. *BioMed Res. Int.* **2015**, *2015*, 461524. [CrossRef]
47. Abramczyk, H.; Brozek-Pluska, B.; Kopeć, M. Double face of cytochrome c in cancers by Raman imaging. *Sci. Rep.* **2022**, *12*, 2120. [CrossRef]
48. Abramczyk, H.; Brozek-Pluska, B.; Jarota, A.; Surmacki, J.; Imiela, A.; Kopec, M. A look into the use of Raman spectroscopy for brain and breast cancer diagnostics: Linear and non-linear optics in cancer research as a gateway to tumor cell identity. *Expert Rev. Mol. Diagn.* **2020**, *20*, 99–115. [CrossRef]
49. Nargis, H.F.; Nawaz, H.; Ditta, A.; Mahmood, T.; Majeed, M.I.; Rashid, N.; Muddassar, M.; Bhatti, H.N.; Saleem, M.; Jilani, K.; et al. Raman spectroscopy of blood plasma samples from breast cancer patients at different stages. *Spectrochim. Acta Part A Mol. Biomol. Spectrosc.* **2019**, *222*, 117210. [CrossRef]
50. Giamougiannis, P.; Silva, R.V.O.; Freitas, D.L.D.; Lima, K.M.G.; Anagnostopoulos, A.; Angelopoulos, G.; Naik, R.; Wood, N.J.; Martin-Hirsch, P.L.; Martin, F.L. Raman spectroscopy of blood and urine liquid biopsies for ovarian cancer diagnosis: Identification of chemotherapy effects. *J. Biophotonics* **2021**, *14*, e202100195. [CrossRef]
51. Giamougiannis, P.; Morais, C.L.M.; Grabowska, R.; Ashton, K.M.; Wood, N.J.; Martin-Hirsch, P.L.; Martin, F.L. A comparative analysis of different biofluids towards ovarian cancer diagnosis using Raman microspectroscopy. *Anal. Bioanal. Chem.* **2021**, *413*, 911–922. [CrossRef]
52. Allain, L.R.; Vo-Dinh, T. Surface-enhanced Raman scattering detection of the breast cancer susceptibility gene BRCA1 using a silver-coated microarray platform. *Anal. Chim. Acta* **2002**, *469*, 149–154. [CrossRef]
53. Culha, M.; Stokes, D.; Allain, L.R.; Vo-Dinh, T. Surface-Enhanced Raman Scattering Substrate Based on a Self-Assembled Monolayer for Use in Gene Diagnostics. *Anal. Chem.* **2003**, *75*, 6196–6201. [CrossRef] [PubMed]
54. Pal, A.; Isola, N.R.; Alarie, J.P.; Stokes, D.L.; Vo-Dinh, T. Synthesis and characterization of SERS gene probe for BRCA-1 (breast cancer). *Faraday Discuss.* **2006**, *132*, 293–301. [CrossRef] [PubMed]
55. Vo-Dinh, T.; Allain, L.R.; Stokes, D.L. Cancer gene detection using surface-enhanced Raman scattering (SERS). *J. Raman Spectrosc.* **2002**, *33*, 511–516. [CrossRef]
56. Coluccio, M.L.; Gentile, F.; Das, G.; Nicastri, A.; Perri, A.M.; Candeloro, P.; Perozziello, G.; Zaccaria, R.P.; Gongora, J.S.T.; Alrasheed, S.; et al. Detection of single amino acid mutation in human breast cancer by disordered plasmonic self-similar chain. *Sci. Adv.* **2015**, *1*, e1500487. [CrossRef]
57. Surmacki, J.; Brozek-Pluska, B.; Kordek, R.; Abramczyk, H. The lipid-reactive oxygen species phenotype of breast cancer. Raman spectroscopy and mapping, PCA and PLSDA for invasive ductal carcinoma and invasive lobular carcinoma. Molecular tumorigenic mechanisms beyond Warburg effect. *Analyst* **2015**, *140*, 2121–2133. [CrossRef]
58. Brozek-Pluska, B. Statistics assisted analysis of Raman spectra and imaging of human colon cell lines—Label free, spectroscopic diagnostics of colorectal cancer. *J. Mol. Struct.* **2020**, *1218*, 128524. [CrossRef]
59. Beton, K.; Wysocki, P.; Brozek-Pluska, B. Mevastatin in colon cancer by spectroscopic and microscopic methods—Raman imaging and AFM studies. *Spectrochim. Acta Part A Mol. Biomol. Spectrosc.* **2022**, *270*, 120726. [CrossRef]

60. Movasaghi, Z.; Rehman, S.; Rehman, I.U. Raman Spectroscopy of Biological Tissues. *Appl. Spectrosc. Rev.* **2007**, *42*, 493–541. [CrossRef]
61. Kopec, M.; Błaszczyk, M.; Radek, M.; Abramczyk, H. Raman imaging and statistical methods for analysis various type of human brain tumors and breast cancers. *Spectrochim. Acta Part A Mol. Biomol. Spectrosc.* **2021**, *262*, 120091. [CrossRef]
62. Abramczyk, H.; Brozek-Pluska, B. Raman Imaging in Biochemical and Biomedical Applications. Diagnosis and Treatment of Breast Cancer. *Chem. Rev.* **2013**, *113*, 5766–5781. [CrossRef] [PubMed]
63. Toland, A.E.; on behalf of the BIC Steering Committee; Forman, A.; Couch, F.J.; Culver, J.O.; Eccles, D.M.; Foulkes, W.D.; Hogervorst, F.B.L.; Houdayer, C.; Levy-Lahad, E.; et al. Clinical testing of BRCA1 and BRCA2: A worldwide snapshot of technological practices. *NPJ Genom. Med.* **2018**, *3*, 7. [CrossRef] [PubMed]
64. Vaz, F.H.; Machado, P.M.; Brandão, R.D.; Laranjeira, C.T.; Eugénio, J.S.; Fernandes, A.H.; André, S.P. Familial Breast/Ovarian Cancer and *BRCA1/2* Genetic Screening: The Role of Immunohistochemistry as an Additional Method in the Selection of Patients. *J. Histochem. Cytochem.* **2007**, *55*, 1105–1113. [CrossRef] [PubMed]

*Article*

# Photooxidation of 2,2′-(Ethyne-1,2-diyl)dianilines: An Enhanced Photocatalytic Properties of New Salophen-Based Zn(II) Complexes

Mahesh Subburu [1], Ramesh Gade [1], Prabhakar Chetti [2,*] and Someshwar Pola [1,*]

[1] Department of Chemistry, Osmania University, Hyderabad 500007, India; maheshsubburu@gmail.com (M.S.); rameshgade.chem@gmail.com (R.G.)
[2] Department of Chemistry, National Institute of Technology Kurukshetra, Kurukshetra 136119, India
* Correspondence: chetti@nitkkr.ac.in (P.C.); somesh.pola@osmania.ac.in (S.P.)

**Citation:** Subburu, M.; Gade, R.; Chetti, P.; Pola, S. Photooxidation of 2,2′-(Ethyne-1,2-diyl)dianilines: An Enhanced Photocatalytic Properties of New Salophen-Based Zn(II) Complexes. *Photochem* **2022**, *2*, 358–375. https://doi.org/10.3390/photochem2020025

Academic Editors: Gulce Ogruc Ildiz and Licinia L.G. Justino

Received: 30 March 2022
Accepted: 17 May 2022
Published: 23 May 2022

**Publisher's Note:** MDPI stays neutral with regard to jurisdictional claims in published maps and institutional affiliations.

**Copyright:** © 2022 by the authors. Licensee MDPI, Basel, Switzerland. This article is an open access article distributed under the terms and conditions of the Creative Commons Attribution (CC BY) license (https://creativecommons.org/licenses/by/4.0/).

**Abstract:** Under solvothermal conditions, the Zn(II) complexes formed from salophen-based ligands with N and O donor atoms are reported. These Zn(II) complexes were initially confirmed through elemental analysis and supported by mass spectral data. The purity of the ligands and Zn(II) complexes was confirmed by using NMR spectral studies. The functional group complexation was established by FT-IR analysis. Additional supportive information about the complexes is also reported through molar conductance and thermal studies. The bandgap energies of the ligands and Zn(II) complexes are estimated with UV-visible DRS studies. The rate of recombination of hole–electron pairs is directly related to photocatalytic activity, which is confirmed by using emission spectral analysis. The surface metaphors for ligands and complexes are obtained from FESEM analysis. These new sequences of Zn(II) complexes were used for the photooxidation of 2,2′-(ethyne-1,2-diyl)dianiline and its derivatives. Mechanistic studies on the fast degradation of dyes were supported in the presence of several scavengers. The rapid photooxidation process in the presence of [Zn(CPAMN)] has been demonstrated, and a highly efficient photocatalyst for the photooxidation of 2,2′-(ethyne-1,2-diyl) dianiline has been proposed. Furthermore, the experimental findings are supported by the DFT studies.

**Keywords:** Zn(II) complexes; DFT calculations; photocatalytic oxidation; rate of recombination; surface area; 2-(2-nitrophenyl)-3H-indol-3-one

## 1. Introduction

The coordination chemistry of transition metal complexes has been the subject of broad study in the past few decades. Moreover, metal–Schiff base complexes have continued to enjoy extensive magnitude owing to their structural diversity and potential applications in pharmacology and catalysis. Most of the studies have aimed to understand the role of M(II) cations in many metalloenzymes in terms of structure–function relationships [1]. The importance of transition metals in several biological systems [2] has motivated the study of the complexes of Zn(II) ions. Studies on lower/higher oxidation state complexes are of special importance because of their potential uses as oxidizing agents, catalysts [3–5] and electro-catalysts [6,7] for the oxidation of compounds such as alcohols, esters and water [8,9]. M (II) complexes with various Schiff base ligands play an important role in coordination chemistry, and a recognized Schiff base ligand is a salophen kind [10], with a bi-functional and tetradentate (-ONNO-) ligand. Some asymmetric salophen kinds of Schiff's base were described by R. Atkins [11] in 1985, who suggested a wide-ranging term for salophen kinds of tetradentate (-ONNO-) ligands. Because of the aromatic ring's substitution and outline hydroxyl group, salicylaldehyde and its analogues are suitable as building blocks for salophen-based ligands. As soon as the azomethine group is formed between the aldehyde and the primary amine, the alignment of the salophen kind of

Schiff's base [12,13] will form a new six-membered ring when it is complexed with different metal ions.

Schiff's base metal complexes with different d-block metals such as manganese [14], cobalt [15], copper [16], and zinc [17] are extensively used as catalysts in oxidation reactions. These metal complexes are active and selective catalysts in a diversity of organic transformations as well as being straightforwardly synthesized, inexpensive and stable [14]. Schiff's base ligands are typically 2-hydroxybenzaldehyde or salophen ligands. The high electron-donating capability of the Schiff bases is good for promoting the rate of electron transfer. As a result, these salophen-type ligands are being considered as potential candidates for increasing the overall catalytic performance of azomethine-functionalized complexes [18,19].

Jacobsen's catalyst is used in the well-known epoxidation reaction with a Schiff base catalyst [13,20]. At the same time, Japanese chemist, Katsuki et al., reported a very strong asymmetric epoxidation with a chiral catalyst that is correspondingly a salophen-based Mn(III) complex [21]. Several substituents at the 3,3′ and 5,5′ positions of Schiff base and an azomethine scaffold based on Jacobsen's catalyst were examined [22,23]. In the primary surveys of the (salophen) Mn(III)-catalyzed epoxidation, Zhang noticed that steric moieties such as phenyl and t-butyl groups at the 3,3′ location of the salophen system are crucial in the direction of attaining great enantioselectivity [13]. Metal–salophen-kind Schiff bases of uranyl complexes also play an important role as catalysts in Diels–Alder reactions [24]. Katsuki et al. reported the synthesis of asymmetric epoxidation of unfunctionalized olefins in the presence of chiral (salophen) manganese (III) complexes [23].

Zn(II) complexes include zinc carbonate, zinc acetate, zinc chloride, zinc nitrate, and zinc sulfate, along with zinc oxide, which may be of the greatest importance in affording its viable consequences as a semiconductor material [25]. Zn(II) Schiff base complexes [26] and Zn(II)-based frameworks [27] exhibit excellent photocatalytic activity.

Presently, one of the focused areas of research has been dedicated in recent years to the mixed ligand complexes of transition metals containing nitrogen and oxygen donors [28], with potential applications such as the luminescent and fluorescent detection of nitroaromatics [29,30]. In this current article, we designed and synthesized three Schiff base salophen ligands and their Zn(II) complexes and reported their photocatalytic and photooxidation studies. The significance of the present work is that all the photooxidation reactions were carried out in the presence of the Zn(II) complex and water as a solvent in visible light irradiation conditions. Moreover, the multiple photooxidation and cyclization reactions of new indol-3-one based compounds and also the amine group converted into a nitro group. Therefore, this photocatalytic reaction is a multiple photooxidation reaction.

## 2. Materials and Methods

$^1$H and $^{13}$C-NMR spectra were recorded on a Bruker AV400 MHz Spectrometer with chemical shifts referenced using the $^1$H resonance of residual CDCl$_3$ and $d_6$-DMSO. Mass spectra of the complexes were recorded by an HR-EI instrument (JEOL, Tokyo, Japan). The melting points of the complexes were verified by using a Cintex apparatus with range 50–450 °C. Thermograms of all the samples were obtained using a Shimadzu differential thermal analyzer (DTG-60H) with a heating rate of 10 °C min$^{-1}$ in the range from 50 to 1000 °C under a nitrogen-purging rate of 20 mL/minute. The surface morphology and cross-section images of the devices were taken by field-emission scanning electron microscopy (FE-SEM, ULTRA PLUS, a member of Carl Zeiss). FT-IR spectra were confirmed using Bruker spectrometer with 4 cm$^{-1}$ resolution. The phases of the new materials were established by powder X-ray diffractometer (Miniflex, Rigaku, Tokyo, Japan) (Cu K$\alpha$, $\lambda$ = 1.5406 Å angle range of 2$\theta$ = 10° to 45°). The electronic spectra were measured in CHCl$_3$/DMF (2:1) solutions on Shimadzu UV-3600 Plus UV-Vis Spectrophotometer. The photoluminescence (PL) spectra of the catalysts were documented on a Fluorescence Spectrophotometer F-7100 and with respect to their absorption maxima. All the photoreactions were performed by using a multi-tube photo reactor system with LED visible light, Lelesil Innovative Systems, India.

### 2.1. Synthesis of $N_2$, $O_2$—Donor-Based Schiff's Base Ligands

To a solution of 2-hydroxy-1-naphthaldehyde (50 mmol) in ethanol (50 mL) at 60 °C, we added 4-chlorobenzene-1,2-diamine/4-fluorobenzene-1,2-diamine/4-methoxybenzene-1,2-diamine (25 mmol) drop wise. The resulting reaction mixture was allowed to reflux for 4 h. After the reaction mass cooled down and the solvent was taken out, a standard aqueous workup was done to obtain the ligand (Scheme 1) as 1,1'-((1E,1'E)- ((4-chloro-1,2-phenylene)bis(azanylylidene))-bis(methanylylidene))bis(naphthalen-2-ol) (CPAMN), 1,1'-((1E,1'E)-((4-fluoro-1,2-phenylene)bis(azanylylidene))-bis(methanylylidene))bis(naphthalen-2-ol) (FPAMN) and 1,1'-((1E,1'E)-((4-methoxy-1,2-phenylene)bis(azanylylidene))bis(methanylylidene))bis-(naphthalen-2-ol) (MPAMN).

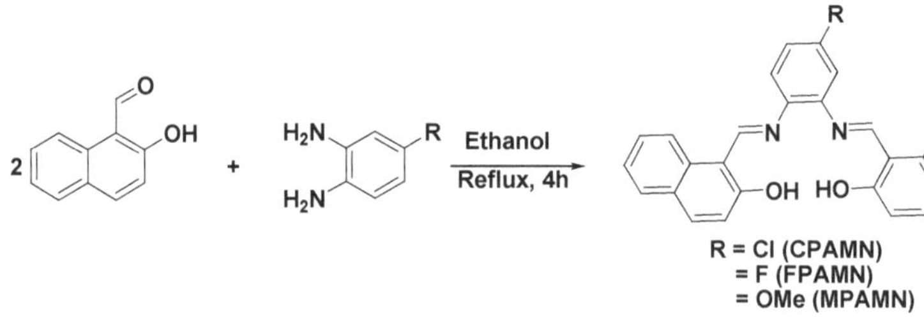

**Scheme 1.** Synthesis of new salophen-based Schiff's base ligands.

### 2.2. Synthesis of Zn(II) Complexes

The Schiff's base ligand (CPAMN/FPAMN/MPAMN) (1.00 mmol) and Zn(OAc)$_2$ (1.00 mmol) were dissolved in methanol (30 mL); then, they were heated at 60 °C under solvothermal conditions for 8 h. The solvent was evaporated under reduced pressure and recrystallized to yield microcrystalline materials with a yield of the materials 64–75% (Scheme 2).

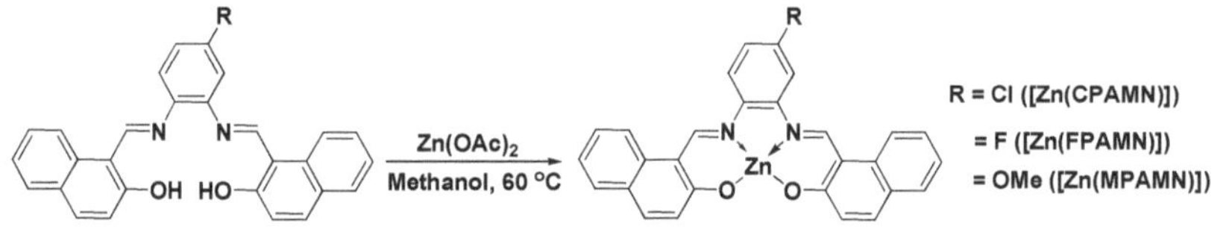

**Scheme 2.** Synthesis pathway for new Zn(II) complexes.

All the metal complexes were characterized using data from different instrumental techniques such as elemental analyses, mass spectra, $^1$H-NMR, $^{13}$C-NMR, IR studies, thermal studies, electronic spectra, and conductivity studies. The particle size and shape of complexes were examined through morphological studies by using FESEM.

## 3. Results and Discussion

The physicochemical and elemental data of both ligands (CPAMN, FPAMN and MPAMN) and Zn(II) complexes are given in Table S1. The experimental data of elements such as H, C, N, Cl, and Zn coincide with the calculated values and are equal to [Zn(CPAMN)], [Zn(FPAMN)], and [Zn(MPAMN)], respectively. The Maldi mass spectral data of the ligands and Zn(II) complexes were well-matched with the respective molecular weights of the molecules. The mass spectra of all ligands and its Zn(II) complexes (Figures S1–S6) were measured with the Maldi mass procedure (HRMS) and revealed that the base peak was given as [ML]$^+$. The $m/z$ values of 513.154, 497.075, and 508.843 for Zn(II) complexes correspond to molecular ions of [Zn(CPAMN)], [Zn(FPAMN)], and

[Zn(MPAMN)], respectively [31]. These formulae are supported well by the data acquired from mass spectral studies of all the ligands and complexes.

The purity of the ligands was checked by $^1$H-NMR spectra recorded in $d_6$-DMSO. The broad signals at δ 15.109 and 14.978 ppm are indicative of the presence of the protons of phenolic –OH groups of CPAMN ligand [32]. The singlet observed at δ 9.680 and 9.671 ppm can be assigned to the azomethine protons of the CPAMN ligand. The signals in the range of δ 8.584–7.020 ppm (Figure S7) may be attributed to the aromatic protons. Similarly, the $^1$H-NMR spectra of the FPAMN ligand also reveal phenolic –OH groups, azomethine, and aromatic protons at 15.109, 14.979, 9.692, 9.683, and 8.593–7.029 ppm, respectively (Figure S8) [28]. In the case of the MPAMN ligand, at 15.749 ppm, phenolic protons were observed (Figure S9). As shown in Figures S7–S9, the remaining azomethine and aromatic protons are arranged in nearly the same pattern in the CPAMN, FPAMN, and MPAMN ligands. Similarly, the diamagnetic nature of Zn(II) complexes was verified by using $^1$H-NMR spectra in $d_6$-DMSO at ambient condition (Figure S10). The disappearance of an Ar–OH protons resonance signal was observed as compared to ligands, which indicates that the ligand is complexed with $Zn^{2+}$ ions through phenolic hydroxy groups. The azomethine protons of all the three Zn complexes shifted to the deshielding by 0.12–0.20 ppm, and this indicates that the $Zn^{2+}$ ions complex with nitrogen atoms [32,33]. The remaining resonance signals of aromatic protons are moved very minor to upfield side in the $^1$H-NMR spectrum of each complex as related to those in the spectrum of each ligand.

The imine carbon resonance signal, shown at 169.827 ppm in the spectral data of the ligand FPAMN, is found to have altered to the downfield side by 3.931 ppm. Therefore, it indicates that imine carbons were shifted toward the lower frequency side and exposes that the $Zn^{2+}$ ions complexed with nitrogen atoms [34]. All the remaining resonance signals of carbon atoms present in aromatic rings have shown a very minor shift to the deshielding in the $^{13}$C-NMR spectrum of every complex as related to its ligand spectrum. The spectrum of the Zn(II) complex has shown a new signal at 188.668 ppm, which is characteristic of the coordinated phenolic carbon atom (Figure S11).

The bonding vibrational modes of all the ligands and Zn(II) complexes were recorded, and the comparison between IR spectra of ligands (CPAMN, FPAMN, and MPAMN) and their Zn(II) complexes is shown in Figure S12. The strong peaks noticed at 1645–1660 cm$^{-1}$ and 3295–3345 cm$^{-1}$ in the spectrum of ligands are recognized as the stretching vibrational modes of the imine (-C=N-) and phenolic (-OH) groups, respectively [35]. In all of the complexes, the phenolic group goes away compared to the ligand spectrum, and the imine peaks move to lower frequencies. This shows that the azomethine group is coordinated with zinc. Two new absorptions were observed in the IR spectra of the complexes when compared to their corresponding ligands; one is around 428, 436, and 419 cm$^{-1}$ (ν (Zn-N)) and the other is around 516, 522, and 518 cm$^{-1}$ (ν (Zn-O)) for [Zn(CPAMN)], [Zn(FPAMN)], and [Zn(MPAMN)], respectively. These absorptions in the IR spectra of complexes indicate that the Zn(II) ion is coordinated to the CPAMN/FPAMN/MPAMN through two imine and two phenolic -OH groups [35]. The IR spectral data of Zn(II) complexes along with details are presented in Table 1, and representative spectra are shown in Figure S12a–c.

**Table 1.** Infrared and electronic spectral data (cm$^{-1}$) of Zn(II) complexes.

| Complex | ν (Ar-OH) | ν (-C=N-) | ν (Zn-O) | ν (Zn-N) | $\lambda_{onset}$ (nm) | $E_g$ (eV) | $E_g$ (eV) (Solid State) | Surface Area (m$^2$/g) |
|---|---|---|---|---|---|---|---|---|
| CPAMN) | 3295 | 1645 | - | - | 503.66 | 2.46 | 2.75 | 65 |
| FPAMN | 3306 | 1652 | - | - | 512.39 | 2.42 | 2.72 | 59 |
| MPAMN | 3345 | 1660 | - | - | 514.52 | 2.41 | 2.87 | 57 |
| [Zn(CPAMN)] | - | 1628 | 516 | 428 | 596.15 | 2.08 | 2.18 | 208 |
| [Zn(FPAMN)] | - | 1634 | 522 | 436 | 553.57 | 2.24 | 2.15 | 184 |
| [Zn(MPAMN)] | - | 1642 | 518 | 419 | 568.80 | 2.18 | 2.24 | 169 |

The P-XRD peaks of the ligands CPAMN, FPAMN, and MPAMN exhibited 2θ values in between 5 and 3 with high-intensity peaks at 25.82°, 25.91°, and 24.79°, respectively. However, the [Zn(CPAMN), [Zn(FPAMN)], and [Zn(MPAMN)] complex peaks changed when they were linked to their respective ligands. The peaks revealed in the [Zn(CPAMN), [Zn(FPAMN), and [Zn(MPAMN)] complexes at 2θ = 9.26°, 9.28°, and 5.78° were new and intense after complexation with Zn(II) ions, whereas the peaks at 2θ = 25.82°, 25.91°, and 24.79° had a small shift and decreased in intensity (Figure 1). This indicates that Zn(II) complexes are formed with salophen ligands [36].

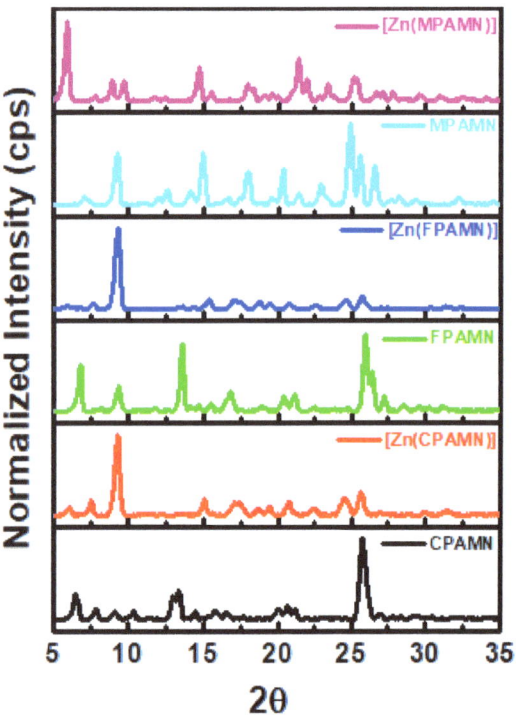

**Figure 1.** Powder XRD pattern of ligands and Zn(II) complexes.

In order to investigate the morphology features of samples, SEM characterization was performed. The results showed that there were sufficient Zn(II) complex nanofibers with a large aspect ratio, uniform diameter, and smooth surface. This was a foregone conclusion during the solvothermal synthesis of Zn(II) complex nanofibers [36]. Figure 2 shows the morphology of the molecules stirred in the ethanol solution. The results show that Zn(II) ions were able to be complexed with ligands, and the morphology of the ligands changed completely. Based on the data obtained from physicochemical and spectral studies of Zn(II) complexes, the tentative structures are shown in Scheme 2. The surface areas of the ligands CPAMN, FPAMN, and MPAMN were lower than those of Zn(II) complexes ([Zn(CPAMN], [Zn(FPAMN)], and [Zn(MPAMN)]). However, the ligand surface area is much smaller than all the complexes and is given in Table 1.

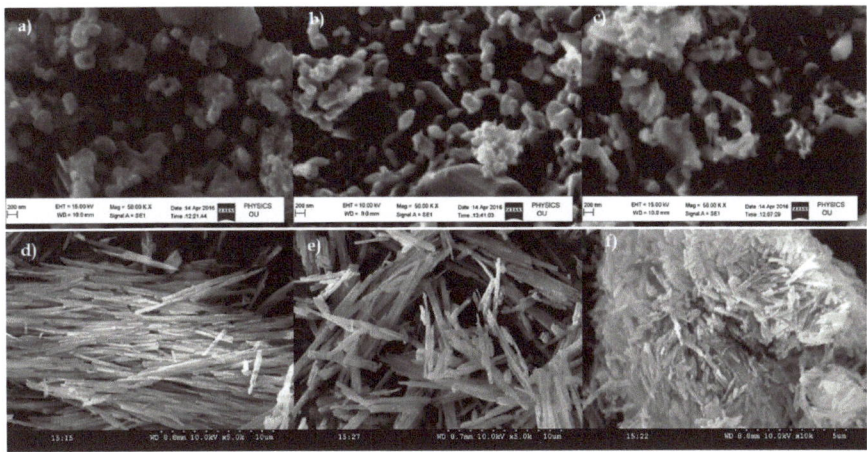

**Figure 2.** FESEM metaphors of (**a**) CPMAN, (**b**) FPMAN, (**c**) MPMAN ligands; (**d**) [Zn(CPMAN)], (**e**) [Zn(FPMAN)] and (**f**) [Zn(MPMAN)] complexes.

## 3.1. Molar Conductance and Thermal Analysis

The molar conductance of Zn(II) complexes was studied to establish the nature of the complexes, whether they are ionic or covalent in nature. The concentration of all the complexes was precise in dimethylformamide at $10^{-3}$ M. The outcomes are shown in Table S1, and the range of the molar conductance of all the complexes is 12.49 to 16.19 $ohm^{-1}$ $cm^2$ $mol^{-1}$. These data suggest that all the Zn(II) complexes were non-electrolytic in nature [37]. After 48 h of retest, the molar conductance values of each complex coincide with the initial data. Therefore, all the complexes are stable due to strong complexation with Zn(II) ions with salophen ligands.

The thermal stability of all the Zn(II) complexes was studied to understand the decomposition characteristics. The thermogravimetric data of all the Zn(II) complexes were recorded under an inert condition (N2 atmosphere) up to 800 °C with a heating rate of 10 °C $min^{-1}$ and are presented in Table S2. All the Zn(II) complexes of CPAMN, FPAMN, and MPAMN, undergo decomposition in a single stage, which indicates that the metal-to-ligand ratio is 1:1, as shown in Figure 3. The observed TG curves show that Zn(II) complexes lose their ligand part [38] at single-step weight losses of up to 84.12% (calculated 83.58%) between 250 and 700 °C. The TG curve shows a plateau from 250 to 700 °C and then no further decomposition up to 800 °C. The obtained values match with the theoretical values for the ligand part of each complex and the final part or residual mass specified as anhydrous ZnO as the final product. Therefore, the crucible consists of a small amount of undecomposed part of the complex left, which corresponds to ZnO [38].

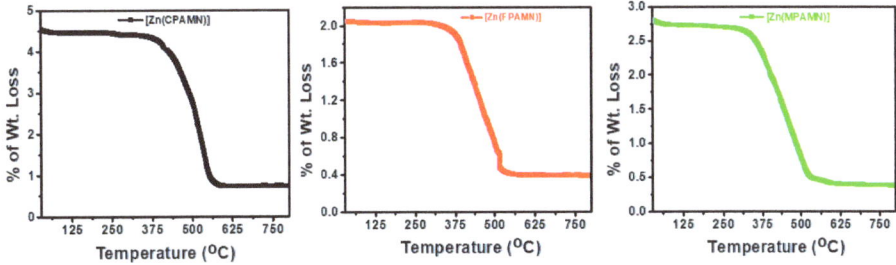

**Figure 3.** Thermograms of Zn(II) complexes.

## 3.2. Absorption and Emission Studies

The electronic spectra of Zn(II) complexes were recorded in DMF, and we also examined the solid-state UV-vis-DRS spectra, which are shown in Figure 4 and Figure S13. Each ligand consists of the three important transitions, such as $\pi \rightarrow \pi^*$ with two different aromatic rings, and another transition is $n \rightarrow \pi^*$ of the azomethine group. All of the bands in the ligand are shifted to a higher wavelength region after complexation with Zn(II) ions due to ligand to metal charge distribution via $T_{2g} \rightarrow T_{1u}$ [39]. The solution spectra of ligands CPAMN displays the bands at 318, 379, and 453 nm, respectively, whereas in the complexes [Zn(CPAMN)], they show at 331, 416, and 468 nm, respectively. Similarly, other ligands FPAMN, MPAMN, and their metal complexes also showed three bands.

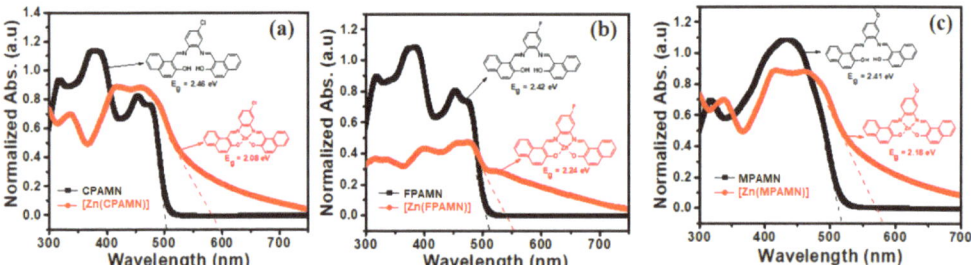

**Figure 4.** UV-visible spectra of (**a**) CPAMN and [Zn(CPAMN)], (**b**) FPAMN and [Zn(FPAMN)] and (**c**) MPAMN and [Zn(MPAMN)].

Figure S13 depicts the UV-visible diffuse reflectance spectra (UV-vis-DRS) of Zn(II) complexes, revealing that the absorption edge of the Zn(II) complexes is located in the visible light region [39]. It indicated that visible light should be selected as the driving force for the photocatalysis of these complexes. The bandgap (Eg) of the Zn(II) catalyst was used further for the improvement of optical performance. It can be calculated according to the following equation:

$$(\alpha h\nu) = A(E_g - h\nu)^2$$

where A is a constant, h is Planck's constant and $\nu$ is the frequency of the incident light [40]. The $E_g$ of the complexes [Zn(CPAMN)], [Zn(FPAMN)], and [Zn(MPAMN)] was 2.18, 2.15, and 2.24 eV, respectively, which was much smaller than that of standard ZnO (3.9 eV), indicating that the energy band structure of the complexes has been modified. Compared to the pure ligands CPAMN, FPAMN, and MPAMN, which were 2.75, 2.72, and 2.87 eV, respectively (Table 1), the reduced bandgap of the Zn(II) complexes made it easier to generate separated electrons and holes under the visible light radiation. Based on the above analysis, it can be concluded that the Zn(II) complexes could effectively utilize visible light to induce photocatalysis.

To get further insight into the observed electronic excitations, time-dependent density functional theory (TD-DFT) calculations have been performed for all the molecules under study [41]. The absorption properties of all the ligands have been calculated using the TD-B3LYP functional with a 6-31G(d,p) basis set. Absorption energies, oscillator strength, major transition, and percentage weight of ligand molecules are given in Table S3. The calculated absorption maximum for CPAMN, FPAMN, and MPAMN are at 437 nm, 442 nm, and 455 nm, respectively, and these are in good agreement with the experimentally observed absorption energies. All the ligands show a major transition from HOMO to LUMO.

The absorption properties of all the metal complexes have also been calculated using TD-B3LYP functional with a mixed basis set. The standard basis set of atomic functions 6-31G(d,p) was used for H, C, N, F, Cl and O atoms of metal complexes, and the LANL2DZ effective core potential basis set was used for zinc metal. The absorption values of three complexes are shown at 453 to 459 nm (shown in Table S3). The major transitions in all the metal complexes are also from HOMO to LUMO.

Frontier molecular orbital pictures of the molecules are shown in Table S4 for ligands and metal complexes, respectively. For all ligands, the distribution of electron density in HOMO is centered on chloro, fluoro, and methoxy substituted benzene. As for the LUMO of all the molecules, electron density is spread over the acceptor (naphthalene rings) through the π-spacer (bridge), as shown in Table S4. In the case of metal complexes, the electron density in HOMO and LUMO is spread over the backbone of the ligands (Table S4). The HOMO–LUMO energy gap (HLG), along with HOMO and LUMO energies, is calculated for all the ligand molecules and is given in Table S5. The HOMO energies of CPAMN and FPAMN are −5.25 eV and −5.16 eV, respectively, which indicate the destabilization of the HOMO level on replacing Cl with an F atom. Similarly, the LUMO energies of CPAMN and FPAMN are at −1.87 eV and −1.81 eV, respectively. This result reveals that in both cases, the energy gap is nearly the same, but as we replace the Cl atom and F atom with the -OMe group at the central benzene ring, there is an increase in HOMO as well as in LUMO energy levels (destabilization), as shown in Figure 5a. The destabilization of the HOMO level is higher than the destabilization of the LUMO level, as corresponds to chloro and fluoro-substituted molecules. The calculated HOMO, LUMO energies, and the HOMO–LUMO energy gap (HLG) for metal complexes is shown in Figure 5b. The HOMO and LUMO energy levels are destabilized by Cl → F → OMe substitution on the benzene ring, and the resultant HLG for all the three metal complexes is almost the same. It has also been calculated and found that there is not much difference in the energy gap of the three metal complexes shown in Figure 5b.

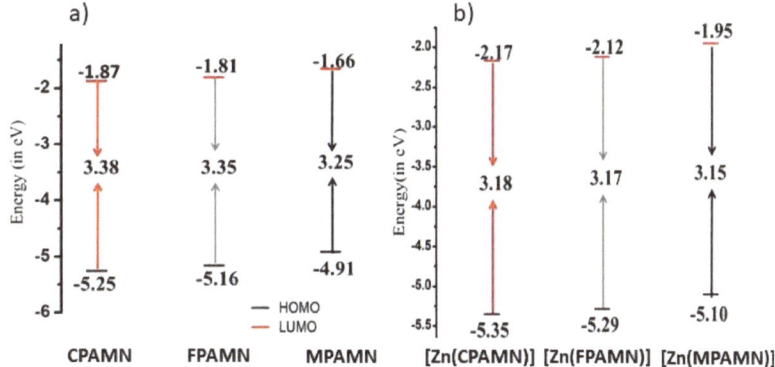

**Figure 5.** HOMO–LUMO energies and HOMO–LUMO gap of (**a**) ligands and (**b**) metal complexes.

The emission spectra of ligands and its Zn-complexes were recorded in solution. The compounds were excited at λ maxima of 480, 475, and 492 nm of ligands, whereas for Zn(II) complexes, they were at 505, 492, and 520 nm, respectively. The emission spectrum provides significant evidence about the rate of recombination of hole and electron pairs [41]. Based on this evidence, it is concluded that the ability to allocate carriers or trick charge to investigate the possible formation of electron–hole pairs from Zn(II) complexes [41]. As shown in Figure 6, the emission strength of the [Zn(CPAMN)] complex decreased completely when correlated with the [Zn(FPAMN)] and [Zn(MPAMN)]. The [Zn(CPAMN)] complex's lower emission strength indicates a low rate of recombination, implying that [Zn(CPAMN)] can benefit from hastening the separation of holes and electrons, resulting in a high photooxidation process.

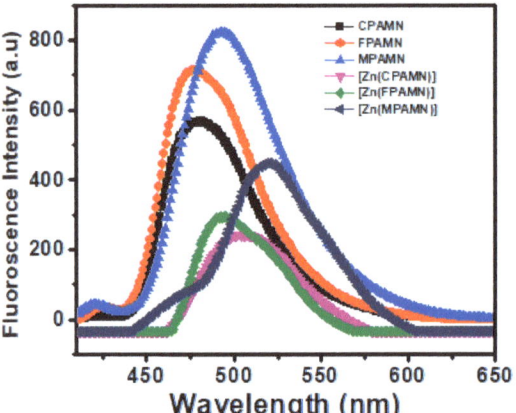

**Figure 6.** PL spectral pattern of ligands and Zn(II) complexes.

*3.3. Photocatalysis*

Under visible light source conditions, new 2-(2-nitrophenyl)-3H-indol-3-one compounds (2a–j) were synthesized in the presence of new Zn(II) complexes and an oxidant (PhI(OAc)2). The performance and stability of the catalyst play a vital role in the photocatalytic activities of the complexes [41]. The one-step synthetic pathway of 2-(2-nitrophenyl)-3H-indol-3-one and its derivatives from internal alkynes [42] is shown the in supporting information as Scheme S1. It is also noticed that the formation of 2-(2-nitrophenyl)-3H-indol-3-one (2a–j) as products at 500 watts with a tungsten lamp.

3.3.1. Optimization of Catalysis (Ligands and Zn(II) Complexes)

All the ligands and Zn(II) complexes are used for photooxidation and condensation in the occurrence of visible light. The conversion of 2-(2-nitrophenyl)-3H-indol-3-one was not observed in the presence of pure ligands, whereas in the presence of Zn(II) complexes, it was shown in TLC, and after 24 h, no starting material was present (Table 2).

**Table 2.** Optimization of catalysts.

| S.No. | Intensity of the Visible Light (Tungsten) | Time (h) | % Yield Compound 5a |
|---|---|---|---|
| 1 | CPAMN | 24 | – |
| 2 | FPAMN | 24 | – |
| 3 | MPAMN | 24 | – |
| 4 | [Zn(CPAMN)] | 24 | 84 |
| 5 | [Zn(FPAMN)] | 24 | 62 |
| 6 | [Zn(MPAMN)] | 24 | 40 |

3.3.2. Identifying the Active Species and Optimization of Oxidant for Photooxidation Process

The reactive species in any photocatalytic process are four types: superoxide radical ($O_2^{\bullet-}$), hydroxyl radical ($^{\bullet}OH$), hole ($h^+$), and electron ($e^-$) [43]. The photooxidation of internal alkynes was established in the presence of Zn(II) complexes and various types of oxidants, along with scavengers, are used in the visible light irradiation technique. The effect of several oxidants was tested for photooxidation reaction promotion: of all the oxidants tested, the strongest is PhI(OAc)$_2$ as compared to other oxidants, as shown in Table 3. Out of these four scavengers, a considerable decline in the reaction rate was observed in the presence of the O$_2^{\bullet-}$ scavenger (benzoquinone BQ). The existence of electron scavengers (K$_2$S$_2$O$_8$, KSO) affects the rate of photooxidation (is slower) of 2,2'-(ethyne-1,2-diyl)aniline, which indicates that the electrons also reactive species for the photooxidation procedure. As a result, these findings imply that the generated O$_2^{\bullet-}$ and electron (e$^-$) are the primary active species in photooxidation reactions. Hence, a photocatalytic mechanism [44] was proposed for the construction of compound 2a in the presence of Zn(II) complexes and the PhI(OAc)$_2$ system (Figure 7).

**Table 3.** Optimization of oxidant and identification active species in the presence of scavengers for conversion of 2-(2-nitrophenyl)-3H-indol-3-one.

| Oxidant | Scavenger | Time (min) | Conversion Rate |
|---|---|---|---|
| Tert-butyl peroxide | t-butyl alcohol | 24 | 10 |
| Pyridine N-oxide | t-butyl alcohol | 24 | 16 |
| 4-Methylpyridine N-oxide | t-butyl alcohol | 24 | 22 |
| (Diacetoxyiodo)benzene | t-butyl alcohol | 24 | 84 |
| Bis(tert-butylcarbonyloxy) iodobenzene | t-butyl alcohol | 24 | 69 |
| [Bis(trifluoroacetoxy)iodo]benzene | t-butyl alcohol | 24 | 74 |
| Tert-butyl peroxide | Benzoquinone | 24 | 5 |
| Pyridine N-oxide | Benzoquinone | 24 | 2 |
| 4-Methylpyridine N-oxide | Benzoquinone | 24 | trace |
| (Diacetoxyiodo)benzene | Benzoquinone | 24 | 15 |
| Bis(tert-butylcarbonyloxy) iodobenzene | Benzoquinone | 24 | 8 |
| [Bis(trifluoroacetoxy)iodo]benzene | Benzoquinone | 24 | 5 |

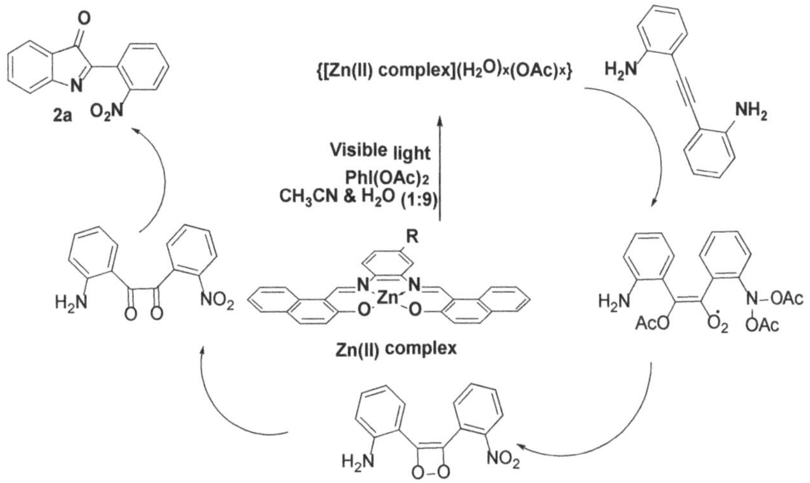

**Figure 7.** Proposal mechanism for formation of 2-(2-nitrophenyl)-3H-indol-3-one.

In this photocatalysis process, we focused mainly on three features, such as (i) variation in the strength of the energy source (effect of visible light), (ii) the effect of solvent and (iii) recyclability and stability of the catalytic system.

Effect of Visible Light Intensity

The various types of visible light sources [45] used with different intensities play a dynamic role in initializing the photooxidation and condensation in the occurrence of a photo-catalyst. The variation of the power is directly proportional to the visible light strength, which is related to the yield and time of the final compounds as presented in Table 3. In total, 500 watts of tungsten light was used for all the reactions, and the yield of by-products (diones) was reduced by two-fold. Finally, optimization with 0.02 mmol of a photocatalyst system is sufficient to yield the 2-(2-nitrophenyl)-3H-indol-3-one (Table 4) in twenty-four hours with an obtained yield of 84%.

**Table 4.** Optimization of the intensity of visible light.

| S.No. | Intensity of the Visible Light (Tungsten) | Time (h) | % Yield Compound 5 [a] |
|---|---|---|---|
| 1 | 300 Watts | 46 | 40 |
| 2 | 400 Watts | 38 | 62 |
| 3 | 500 Watts | 24 | 84 |

[a] After purification.

Effect of Solvent

The reaction has not progressed in the absence of acetonitrile solvent (Table 5), suggesting that acetonitrile is crucial for this oxidation process. We thus studied different protic and aprotic solvents, particularly acetonitrile as an additive in $H_2O$. As shown in Table 5, the reaction was much slower in methanol, ethanol, $CH_2Cl_2$, $CHCl_3$, DMSO, toluene, and DMF than in acetonitrile.

**Table 5.** Solvent effect for the synthesis of new 2-(2-nitrophenyl)-3H-indol-3-one.

| S.No. | Solvent and $H_2O$ | Time (h) | Yield (%) Compound 2a |
|---|---|---|---|
| 1 | Toluene | 24 | 32 |
| 2 | Dichloroethane | 24 | 58 |
| 3 | THF | 24 | 35 |
| 4 | 1,4-Dioxane | 24 | 18 |
| 5 | Acetone | 24 | 21 |
| 6 | Methanol | 24 | 30 |
| 7 | Ethanol | 24 | 42 |
| 8 | $CH_3CN$ | 24 | **84** |
| 9 | DMF | 24 | 13 |
| 10 | DMSO | 24 | 9 |
| 11 | Pure $H_2O$ | 24 | 65 |

One distinguishing feature that distinguishes acetonitrile from other solvents is its oxygen solubility. Oxygen is highly soluble in acetonitrile (8.1 mM) than in other usually used organic solvents [46], such as DMSO (2.1 mM) and DMF (2.1 mM). Dioxygen plays a crucial role in this conversion. For confirmation, a controlled experiment was performed in degassed acetonitrile; only 4% yield was observed in 24 h of reaction time, as compared to 84% of compound 2a yields in acetonitrile with water (1:1). The same was performed without organic solvent; only pure water as solvent had a 65% yield. This experiment shows that $O_2$ plays a role in the oxidation of the substrates in the presence of Zn(II) complexes and the PhI(OAc)$_2$ system. It also acts as a photocatalyst, which means it helps the process go faster. Therefore, in the acetonitrile and water solvent systems, these act as proper accelerators for the consequent solvent selection.

Recyclability and Stability of the Catalytic System

Subsequently, after the completion of reactions, the suspended catalyst residue at the bottom was separated by the centrifugation process. The collected catalyst samples are of their original color, such as dark brown, indicating that the photo-oxidative product washed out the entire product present on the catalyst surface with dichloromethane to obtain the dark brown-colored catalyst [Zn(CPAMN)]. After the third rotation of the reaction, all the $^1$H-NMR peaks are slightly broadened, as shown in Figure 8. Despite this, the [Zn(CPAMN)] photocatalyst performed 78% and slightly decreased the compound yield in the following cycle, which is due to the decreased purity of the complex [47,48]. After the photocatalysis process, the metal complexes recovered. Estimation of powder-XRD and FESEM analysis were carried out and are shown in Figures 9 and 10, respectively. As compared to the powder-XRD pattern and FESEM metaphors of pure Zn(II) complex with after photocatalysis, there was a very slight change in 2θ values, whereas there was no change in the morphology of pure and after photocatalysis.

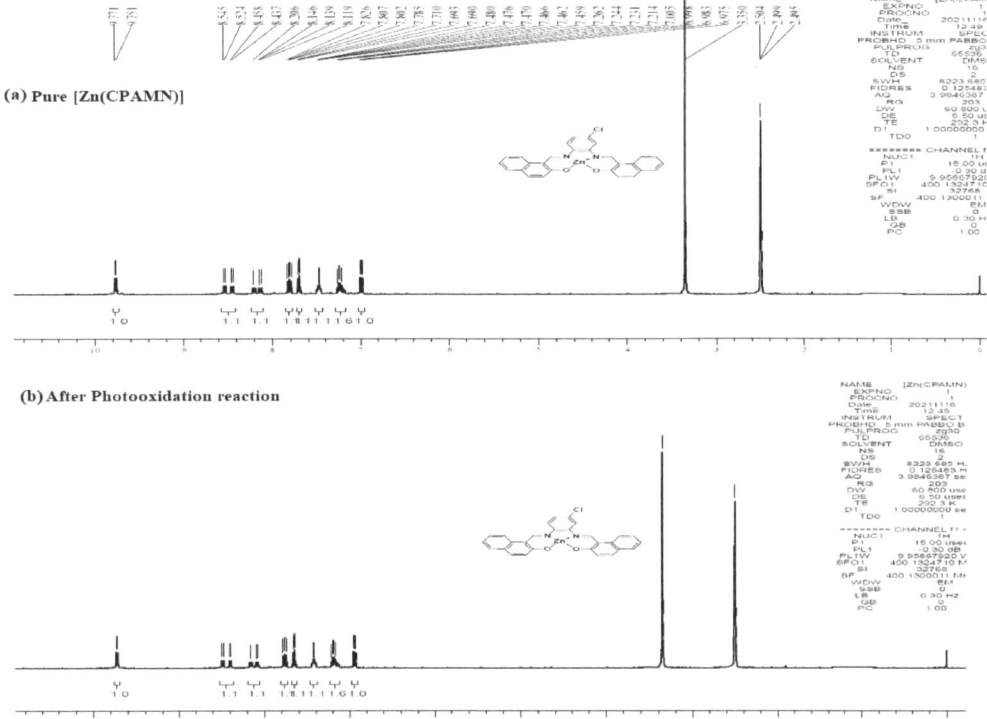

**Figure 8.** $^1$H-NMR spectrum of (**a**) pure and (**b**) after photocatalysis [Zn(CPAMN)] complex.

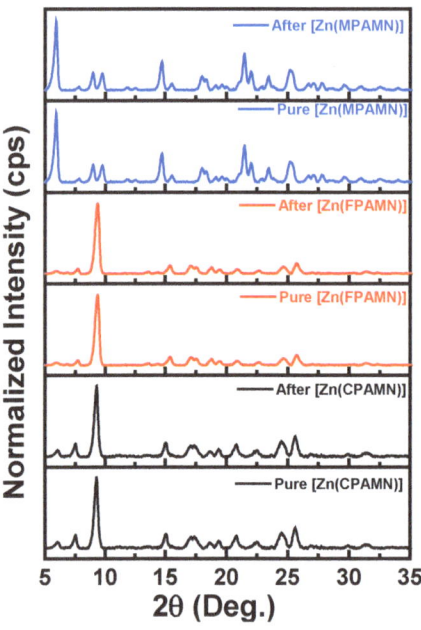

**Figure 9.** Powder XRD pattern of pure and after photocatalysis Zn(II) complexes.

**Figure 10.** FESEM images of pure (**a**) [Zn(CPAMN)]; (**b**) [Zn(FPAMN)]; (**c**) [Zn(MPAMN)]; and after photocatalysis (**d**) [Zn(CPAMN)]; (**e**) [Zn(FPAMN)] and (**f**) [Zn(MPAMN)] complexes.

### 3.3.3. The Efficiency of the Photocatalyst System

As shown in Figure 11, the order of efficiency of the photocatalyst system with all substrates shows that [Zn(CPAMN)] and PhI(OAc)$_2$ are more active than both [Zn(FPAMN)] and [Zn(MPAMN)] complexes. In the Scheme S1 (supporting information), synthesis of the compound 2a is a unique pathway such as the photooxidation method, and the formation of the indolone ring containing by-product is a minor yield. All the reactants and final target molecules are shown in Table 6, and the data of the final molecules are shown in Figures S14–S33. Finally, the new [Zn(CPAMN)] complex performance was compared with some reported Zn-complexes as given in Table 7; the reported new Zn(CPAMN)] shows the better photocatalytic activity.

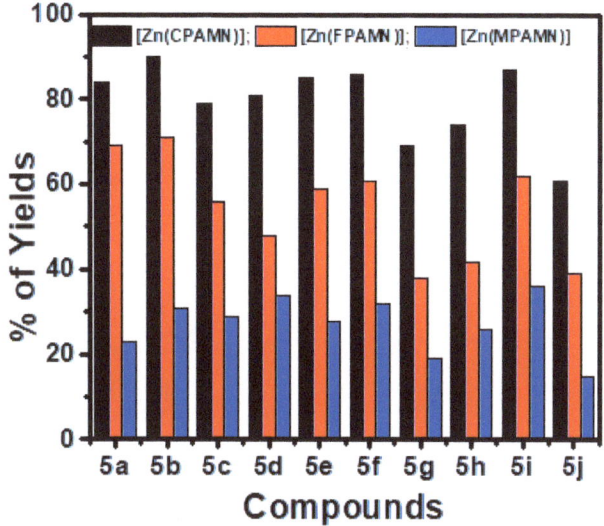

**Figure 11.** Optimization of photocatalytic performance of Zn(II) complexes.

**Table 6.** Synthesis of new 2-(2-nitrophenyl)-3H-indol-3-one and its derivatives.

| Diamine Compound | Product | Complexes | Yield % |
|---|---|---|---|
| 1a | 2a | [Zn(CPAMN)] | 84 |
| | | [Zn(FPAMN)] | 59 |
| | | [Zn(MPAMN)] | 48 |
| 1b | 2b | Zn(CPAMN)] | 90 |
| | | [Zn(FPAMN)] | 61 |
| | | [Zn(MPAMN)] | 39 |

Table 6. *Cont.*

| Diamine Compound | Product | Complexes | Yield % |
|---|---|---|---|
| 1c | 2c | Zn(CPAMN)] | 79 |
|  |  | [Zn(FPAMN)] | 55 |
|  |  | [Zn(MPAMN)] | 42 |
| 1d | 2d | Zn(CPAMN)] | 81 |
|  |  | [Zn(FPAMN)] | 49 |
|  |  | [Zn(MPAMN)] | 31 |
| 1e | 2e | Zn(CPAMN)] | 85 |
|  |  | [Zn(FPAMN)] | 51 |
|  |  | [Zn(MPAMN)] | 43 |
| 1f | 2f | Zn(CPAMN)] | 86 |
|  |  | [Zn(FPAMN)] | 49 |
|  |  | [Zn(MPAMN)] | 37 |
| 1g | 2g | Zn(CPAMN)] | 69 |
|  |  | [Zn(FPAMN)] | 50 |
|  |  | [Zn(MPAMN)] | 41 |
| 1h | 2h | Zn(CPAMN)] | 74 |
|  |  | [Zn(FPAMN)] | 38 |
|  |  | [Zn(MPAMN)] | 21 |
| 1i | 2i | Zn(CPAMN)] | 87 |
|  |  | [Zn(FPAMN)] | 56 |
|  |  | [Zn(MPAMN)] | 32 |
| 1j | 2j | Zn(CPAMN)] | 64 |
|  |  | [Zn(FPAMN)] | 35 |
|  |  | [Zn(MPAMN)] | 19 |

Table 7. Optimization of photocatalyst (PC) for conversion of 2,2′-(ethyne-1,2-diyl)dianiline into 2-(2-nitrophenyl)-3H-indol-3-one with various commercially available and new Zn(II) complexes.

| S.No. | Photocatalyst | Time (h) | Yield (%) [a] |
|---|---|---|---|
| PC-1 | Dichloro(N,N,N′,N′-tetramethylethylenediamine)zinc (28308-00-1) | 24 | – |
| PC-2 | Zinc bis[bis(trimethylsilyl)amide] (14760-26-0) | 24 | 5 |
| PC-3 | Zinc di[bis(trifluoromethylsulfonyl)imide] (168106-25-0) | 24 | 8 |
| PC-4 | Ziram (CAS No. 137-30-4) | 24 | 10 |
| PC-5 | Zinc phthalocyanine (14320-04-8) | 24 | – |
| PC-6 | [Zn(CPAMN)] | 24 | 84 |
| PC-7 | [Zn(FPAMN)] | 24 | 59 |
| PC-8 | [Zn(MPAMN)] | 24 | 48 |

[a] After purification.

## 4. Conclusions

In this study, we concluded the synthesis and account of the new three Schiff base salophen ligands and their Zn(II) complexes. The newly obtained Zn(II) complexes are examined by analytical, thermal, and spectroscopic studies. We studied the photooxidation of 2,2′-(ethyne-1,2-diyl)dianiline and its derivatives in the presence of Zn(II) complexes and converted them into new 2-(2-nitrophenyl)-3H-indol-3-one and its analogues under visible light irradiation. When compared to ligands, the bandgap energies of all Zn(II) complexes change only marginally, but their surface area increases three-fold. The outstanding photooxidation catalytic performance of [Zn(CPAMN)] is due to the low rate of recombination of hole–electron pairs, larger surface area, and low bandgap. Hence, the [Zn(CPAMN)] complex is suggested as an effective photocatalyst for the oxidation process.

**Supplementary Materials:** The following supporting information can be downloaded at: https://www.mdpi.com/article/10.3390/photochem2020025/s1, Figures S1–S11: H, $^{13}$C NMR and mass spectral data of ligands and its metal complexes; Figure S12: IR spectral images of ligands and Zn(II) complexes; Figure S13: UV-vis-DRS spectra of ligands and Zn(II) complexes; Figures S14–S33: NMR and mass spectral data of 2,2′-(ethyne-1,2-diyl)dianilines; Table S1: Analytical and physicochemical data of Zn(II) complexes; Table S2: Thermal data of Zn(II) complexes; Table S3: Electronic excitations ($\lambda_{CAL}$ in nm) oscillator strength ($f$) major transitions (MT) and % weight (%Ci) of ligands and complexes at TD-B3LYP/6-31G(d,p) method; Table S4: Molecular orbitals pictures of HOMO LUMO for ligands; Table S5: Calculated HOMO, LUMO energies and HLG (in eV) of ligands and complexes. Scheme S1: Synthesis process for 2-(2-nitrophenyl)-3H-indol-3-one and its derivatives.

**Author Contributions:** Conceptualization, S.P. and P.C.; methodology, M.S.; software, P.C.; validation, S.P. and P.C.; formal analysis, M.S. and R.G.; investigation, M.S.; resources, S.P.; data curation, S.P.; writing—original draft preparation, M.S.; writing—review and editing, S.P. and P.C.; visualization, P.C.; supervision, S.P.; project administration, S.P. and P.C.; funding acquisition, S.P. and P.C. All authors have read and agreed to the published version of the manuscript.

**Funding:** This research was funded by SERB (EMR//2014/000452), UGC-UPE-FAR and DST-PURSE, New Delhi, India and CSIR [02(0339)/18/EMR-II], New Delhi, India.

**Institutional Review Board Statement:** Not applicable.

**Informed Consent Statement:** Not applicable.

**Data Availability Statement:** All the data related to this article are provided in supporting information.

**Acknowledgments:** The authors special thank DST—FIST schemes and UGC, New Delhi. Mahesh Subburu thanks the University Grants Commission (UGC), New Delhi for the award of Junior Research Fellowship. S.P. and P.C. are also thanking to for SERB (EMR//2014/000452), UGC-UPE-FAR & DST-PURSE, New Delhi, India and CSIR [02(0339)/18/EMR-II], New Delhi, India respectively for financial support.

**Conflicts of Interest:** The authors declare no conflict of interest.

## References

1. Rauscher, F.J., III.; Morris, J.F.; Tournay, O.E.; Cook, D.M.; Curran, T. Binding of the Wilms' tumor locus zinc finger protein to the EGR-1 consensus sequence. *Science* **1990**, *250*, 1259–1262. [CrossRef] [PubMed]
2. Wieghardt, K. The active sites in manganese-containing metalloproteins and inorganic model complexes. *Angew. Chem. Int. Ed. Engl.* **1989**, *28*, 1153–1172. [CrossRef]
3. Deeney, F.A.; Harding, C.J.; Morgan, G.G.; McKee, V.; Nelson, J.; Teat, S.J.; Clegg, W. Response to steric constraint in azacryptate and related complexes of iron-(II) and-(III). *J. Chem. Soc. Dalton Trans.* **1998**, *11*, 1837–1844. [CrossRef]
4. Srinivasan, K.; Michaud, P.; Kochi, J.K. Epoxidation of olefins with cationic (salophen) manganese (III) complexes. The modulation of catalytic activity by substituents. *J. Am. Chem. Soc.* **1986**, *108*, 2309–2320. [CrossRef] [PubMed]
5. Ray, M.; Mukherjee, R.; Richardson, J.F.; Buchanan, R.M. Spin-state regulation of iron (III) centres by axial ligands with tetradentate bis (picolinamide) in-plane ligands. *J. Chem. Soc. Dalton Trans.* **1993**, *16*, 2451–2457. [CrossRef]
6. Goodson, P.A.; Oki, A.R.; Glerup, J.; Hodgson, D.J. Design, synthesis, and characterization of bis (.mu.-oxo) dimanganese (III, III) complexes. Steric and electronic control of redox potentials. *J. Am. Chem. Soc.* **1990**, *112*, 6248–6254. [CrossRef]
7. Devereux, M.; McCann, M.; Leon, V.; Geraghty, M.; McKee, V.; Wikaira, J. Synthesis and fungitoxic activity of manganese (II) complexes of fumaric acid: X-ray crystal structures of [Mn(fum)(bipy)($H_2O$)] and [Mn(Phen)$_2$($H_2O$)$_2$](fum)·4$H_2O$ (fumH$_2$=fumaric acid; bipy=2,2'-bipyridine; phen=1,10-phenanthroline). *Polyhedron* **2000**, *19*, 1205–1211. [CrossRef]
8. Gultneh, Y.; Yisgedu, T.B.; Tesema, Y.T.; Butcher, R.J. Dioxo-bridged dinuclear manganese (III) and-(IV) complexes of pyridyl donor tripod ligands: Combined effects of steric substitution and chelate ring size variations on structural, spectroscopic, and electrochemical properties. *Inorg. Chem.* **2003**, *42*, 1857–1867. [CrossRef]
9. Biswas, S.; Mitra, K.; Chattopadhyay, S.K.; Adhikary, B.; Lucas, C.R. Mononuclear manganese (II) and manganese (III) complexes of N2O donors involving amine and phenolate ligands: Absorption spectra, electrochemistry and crystal structure of [Mn(L$_3$)$_2$](ClO$_4$). *Transit. Met. Chem.* **2005**, *30*, 393–398. [CrossRef]
10. Ebrahimipour, S.Y.; Maryam, M.; Masoud, T.M.; Jim, S.; Joel, T.M.; Iran, S. Synthesis and structure elucidation of novel salophen-based dioxo-uranium (VI) complexes: In-vitro and in-silico studies of their DNA/BSA-binding properties and anticancer activity. *Eur. J. Med. Chem.* **2017**, *140*, 172–186. [CrossRef]
11. Atkins, R.; Brewer, G.; Kokot, E.; Mockler, G.M.; Sinn, E. Copper (II) and nickel (II) complexes of unsymmetrical tetradentate Schiff base ligands. *Inorg. Chem.* **1985**, *24*, 127–134. [CrossRef]
12. Kushwah, N.P.; Pal, M.K.; Wadawale, A.P.; Jain, V.K. Diorgano-gallium and-indium complexes with salophen ligands: Synthesis, characterization, crystal structure and C–C coupling reactions. *J. Organomet. Chem.* **2009**, *694*, 2375–2379. [CrossRef]
13. Zhang, W.; Loebach, J.L.; Wilson, S.R.; Jacobsen, E.N. Enantioselective epoxidation of unfunctionalized olefins catalyzed by salophen manganese complexes. *J. Am. Chem. Soc.* **1990**, *112*, 2801–2803. [CrossRef]
14. Rong, M.; Wang, J.; Shen, Y.; Han, J. Catalytic oxidation of alcohols by a novel manganese Schiff base ligand derived from salicylaldehyd and l-Phenylalanine in ionic liquids. *Catal. Commun.* **2012**, *20*, 51–53. [CrossRef]
15. Kervinen, K.; Korpi, H.; Leskelä, M.; Repo, T. Oxidation of veratryl alcohol by molecular oxygen in aqueous solution catalyzed by cobalt salophen-type complexes: The effect of reaction conditions. *J. Mol. Catal. A Chem.* **2003**, *203*, 9–19. [CrossRef]
16. Soroceanu, A.; Cazacu, M.; Shova, S.; Turta, C.; Kožíšek, J.; Gall, M.; Breza, M.; Rapta, P.; Mac Leod, T.C.; Pombeiro, A.J. Copper (II) complexes with Schiff bases containing a disiloxane unit: Synthesis, structure, bonding features and catalytic activity for aerobic oxidation of benzyl alcohol. *Eur. J. Inorg. Chem.* **2013**, *2013*, 1458–1474. [CrossRef]
17. Huang, S.; Zou, L.-Y.; Ren, A.-M.; Guo, J.-F.; Liu, X.-T.; Feng, J.-K.; Yang, B.-Z. Computational design of two-photon fluorescent probes for a zinc ion based on a salophen ligand. *Inorg. Chem.* **2013**, *52*, 5702–5713. [CrossRef]
18. Tarade, K.; Shinde, S.; Sakate, S.; Rode, C. Pyridine immobilised on magnetic silica as an efficient solid base catalyst for Knoevenagel condensation of furfural with acetyl acetone. *Catal. Commun.* **2019**, *124*, 81–85. [CrossRef]
19. Hu, J.; Li, K.; Li, W.; Ma, F.; Guo, Y. Selective oxidation of styrene to benzaldehyde catalyzed by Schiff base-modified ordered mesoporous silica materials impregnated with the transition metal-monosubstituted Keggin-type polyoxometalates. *Appl. Catal. A Gen.* **2009**, *364*, 211–220. [CrossRef]
20. Jacobsen, E.N.; Zhang, W.; Muci, A.R.; Ecker, J.R.; Deng, L. Highly enantioselective epoxidation catalysts derived from 1,2-diaminocyclohexane. *J. Am. Chem. Soc.* **1991**, *113*, 7063–7064. [CrossRef]
21. Irie, R.; Noda, K.; Ito, Y.; Matsumoto, N.; Katsuki, T. Catalytic asymmetric epoxidation of unfunctionalized olefins. *Tetrahedron Lett.* **1990**, *31*, 7345–7348. [CrossRef]
22. McGarrigle, E.M.; Gilheany, D.G. Chromium– and manganese–salophen promoted epoxidation of alkenes. *Chem. Rev.* **2005**, *105*, 1563–1602. [CrossRef] [PubMed]
23. Irie, R.; Noda, K.; Ito, Y.; Matsumoto, N.; Katsuki, T. Catalytic asymmetric epoxidation of unfunctionalized olefins using chiral (salophen) manganese (III) complexes. *Tetrahedron Asymmetry* **1991**, *2*, 481–494. [CrossRef]
24. Dalla Cort, A.; Mandolini, L.; Schiaffino, L. Exclusive transition state stabilization in the supramolecular catalysis of Diels–Alder reaction by a uranyl salophen complex. *Chem. Commun.* **2005**, *30*, 3867–3869. [CrossRef]
25. Chen, C.; Liu, P.; Lu, C. Synthesis and characterization of nano-sized ZnO powders by direct precipitation method. *Chem. Eng. J.* **2008**, *144*, 509–513. [CrossRef]
26. Patel, U.; Parmar, B.; Dadhania, A.; Suresh, E. Zn(II)/Cd(II)-Based Metal–Organic Frameworks as Bifunctional Materials for Dye Scavenging and Catalysis of Fructose/Glucose to 5-Hydroxymethylfurfural. *Inorg. Chem.* **2021**, *60*, 9181–9191. [CrossRef]

27. Chimupala, Y.; Kaeosamut, N.; Yimklan, S. Octahedral to Tetrahedral Conversion upon a Ligand-Substitution-Induced Single-Crystal to Single-Crystal Transformation in a Rectangular Zn(II) Metal–Organic Framework and Its Photocatalysis. *Cryst. Growth Des.* **2021**, *21*, 5373–5382. [CrossRef]
28. Bhattacharjee, C.H.; Das, G.; Mondal, P.; Rao, N.V.S. Novel photoluminescent hemi-disclike liquid crystalline Zn(II) complexes of [$N_2O_2$] donor 4-alkoxy substituted salicyldimine Schiff base with aromatic spacer. *Polyhedron* **2010**, *29*, 3089–3096. [CrossRef]
29. Song, X.; Yu, H.; Yan, X.; Zhang, Y.; Miao, Y.; Ye, K.; Wang, Y. A luminescent benzothiadazole-bridging bis(salicylaldiminato)zinc(II) complex with mechanochromic and organogelation properties. *Dalton Trans.* **2018**, *47*, 6146–6155. [CrossRef]
30. Germain, M.E.; Vargo, T.R.; Khalifah, P.G.; Knapp, M.J. Fluorescent detection of nitroaromatics and 2,3-dimethyl-2,3-dinitrobutane (DMNB) by a Zinc complex: (salophen)Zn. *Inorg. Chem.* **2007**, *46*, 4422–4429. [CrossRef]
31. Dalla Cort, A.; Mandolini, L.; Pasquini, C.; Rissanen, K.; Russo, L.; Schiaffino, L. Zinc–salophen complexes as selective receptors for tertiary amines. *New J. Chem.* **2007**, *31*, 1633–1638. [CrossRef]
32. Swamy, S.J.; Pola, S. Spectroscopic studies on Co(II), Ni(II), Cu(II) and Zn(II) complexes with a $N_4$-macrocylic ligands. *Spectrochim. Acta Part A Mol. Biomol. Spectrosc.* **2008**, *70*, 929–933. [CrossRef] [PubMed]
33. Tiwari, A.; Mishra, A.; Mishra, S.; Mamba, B.; Maji, B.; Bhattacharya, S. Synthesis and DNA binding studies of Ni (II), Co (II), Cu (II) and Zn(II) metal complexes of N1, N5-bis [pyridine-2-methylene]-thiocarbohydrazone Schiff-base ligand. *Spectrochim. Acta Part A Mol. Biomol. Spectrosc.* **2011**, *79*, 1050–1056. [CrossRef] [PubMed]
34. Asadi, M.; Asadi, Z.; Torabi, S.; Lotfi, N. Synthesis, characterization and thermodynamics of complex formation of some new Schiff base ligands with some transition metal ions and the adduct formation of zinc Schiff base complexes with some organotin chlorides. *Spectrochim. Acta Part A Mol. Biomol. Spectrosc.* **2012**, *94*, 372–377. [CrossRef] [PubMed]
35. Nawar, N.; Hosny, N.M. Synthesis, spectral and antimicrobial activity studies of o-aminoacetophenone o-hydroxybenzoylhydrazone complexes. *Transit. Met. Chem.* **2000**, *25*, 1–8. [CrossRef]
36. Dharnipathi, V.R.; Subburu, M.; Gade, R.; Basude, M.; Chetti, P.; Simhachalam, N.B.; Nagababu, P.; Bhongiri, Y.; Pola, S. A new Zn(II) complex-composite material: Piezoenhanced photomineralization of organic pollutants and wastewater from the lubricant industry. *Environ. Sci. Water Res. Technol.* **2021**, *7*, 1737–1747.
37. Geary, W.J. The use of conductivity measurements in organic solvents for the characterisation of coordination compounds. *Coord. Chem. Rev.* **1971**, *7*, 81–122. [CrossRef]
38. Ahemed, J.; Pasha, J.; Kore, R.; Gade, R.; Bhongiri, Y.; Chetti, P.; Pola, S. Synthesis of new Zn(II) complexes for photo decomposition of organic dye pollutants, industrial wastewater and photo-oxidation of methyl arenes under visible-light. *J. Photochem. Photobiol. A Chem.* **2021**, *419*, 113455. [CrossRef]
39. Manoj Kr., P.; Singh, Y.D.; Singh, N.B.; Sarkar, U. Emissive bis-salicylaldiminato schiff base ligands and their zinc(II)complexes: Synthesis, photophysical properties, mesomorphism and DFT studies. *J. Mol. Struct.* **2015**, *1081*, 316–328.
40. Subburu, M.; Gade, R.; Guguloth, V.; Chetti, P.; Ravulapelly, K.R.; Pola, S. Effective photodegradation of organic pollutantsin the presence of mono and bi-metallic complexes under visible-light irradiation. *J. Photochem. Photobiol. A Chem.* **2021**, *406*, 112996. [CrossRef]
41. Frisch, M.J.; Trucks, G.W.; Schlegel, H.B.; Scuseria, G.E.; Robb, M.A.; Cheeseman, J.R.; Scalmani, G.; Barone, V.; Mennucci, B.; Petersson, G.A.; et al. *Gaussian 09, Revision E.01*; Gaussian, Inc.: Wallingford, CT, USA, 2009.
42. Yang, X.-J.; Chen, B.; Li, X.-B.; Zheng, L.-Q.; Wu, L.-Z.; Tung, C.-H. Photocatalytic organic transformation by layered double hydroxides: Highly efficient and selective oxidation of primary aromatic amines to their imines under ambient aerobic conditions. *Chem. Commun.* **2014**, *50*, 6664–6667. [CrossRef] [PubMed]
43. Luís, M.T.F.; Amin, I.; Maria, L.S.C. Photochemical Transformations of Tetrazole Derivatives: Applications in Organic Synthesis. *Molecules* **2010**, *15*, 3757–3774.
44. Shi, L.; Liang, L.; Ma, J.; Wang, F.; Sun, J. Enhanced photocatalytic activity over the $Ag_2O$–$gC_3N_4$ composite under visible light. *Catal. Sci. Technol.* **2014**, *4*, 758–765. [CrossRef]
45. Zhu, X.; Li, P.; Shi, Q.; Wang, L. Thiyl radical catalyzed oxidation of diarylalkynes to α-diketones by molecular oxygen under visible-light irradiation. *Green Chem.* **2016**, *18*, 6373–6379. [CrossRef]
46. Guguloth, V.; Ahemed, J.; Subburu, M.; Guguloth, V.C.; Chetti, P.; Pola, S. A very fast photodegradation of dyes in the presence of new Schiff's base $N_4$-macrocyclic Ag-doped Pd (II) complexes under visible-light irradiation. *J. Photochem. Photobiol. A Chem.* **2019**, *382*, 111975. [CrossRef]
47. Li, Y.; Lee, T.B.; Wang, T.; Gamble, A.V.; Gorden, A.E. Allylic C–H Activations Using Cu(II) 2-Quinoxalinol Salophen and tert-Butyl Hydroperoxide. *J. Org. Chem.* **2012**, *77*, 4628–4633. [CrossRef]
48. Pola, S.; Subburu, M.; Guja, R.; Muga, V.; Tao, Y.-T. New photocatalyst for allylic aliphatic C–H bond activation and degradation of organic pollutants: Schiff base Ti(IV) complexes. *RSC Adv.* **2015**, *5*, 58504–58513. [CrossRef]

MDPI AG  
Grosspeteranlage 5  
4052 Basel  
Switzerland  
Tel.: +41 61 683 77 34

*Photochem* Editorial Office  
E-mail: photochem@mdpi.com  
www.mdpi.com/journal/photochem

Disclaimer/Publisher's Note: The title and front matter of this reprint are at the discretion of the Guest Editors. The publisher is not responsible for their content or any associated concerns. The statements, opinions and data contained in all individual articles are solely those of the individual Editors and contributors and not of MDPI. MDPI disclaims responsibility for any injury to people or property resulting from any ideas, methods, instructions or products referred to in the content.

www.ingramcontent.com/pod-product-compliance
Lightning Source LLC
LaVergne TN
LVHW072349090526
838202LV00019B/2508